S0-CEE-349

WORLD HEALTH ORGANIZATION

INTERNATIONAL AGENCY FOR
RESEARCH ON CANCER

UNITED NATIONS

ENVIRONMENT PROGRAMME

ENVIRONMENTAL CARCINOGENS SELECTED METHODS OF ANALYSIS

VOLUME 7 – Some Volatile
Halogenated Hydrocarbons

EDITORS
L. FISHBEIN & I.K. O'NEILL

IARC PUBLICATIONS No. 68

INTERNATIONAL AGENCY FOR RESEARCH ON CANCER
LYON
1985

The International Agency for Research on Cancer (IARC) was established in 1965 by the World Health Assembly as an independently financed organization within the framework of the World Health Organization. The headquarters of the Agency are at Lyon, France.

The Agency conducts a programme of research concentrating particularly on the epidemiology of cancer and the study of potential carcinogens in the human environment. Its field studies are supplemented by biological and chemical research carried out in the Agency's laboratories in Lyon and, through collaborative research agreements, in national research institutions in many countries. The Agency also conducts a programme for the education and training of personnel for cancer research.

The publications of the Agency are intended to contribute to the dissemination of authoritative information on different aspects of cancer research.

Distributed for the International Agency for Research on Cancer
by Oxford University Press, Walton Street, Oxford OX2 6DP, UK

London New York Toronto Delhi Bombay Calcutta Madras Karachi
Kuala Lumpur Singapore Hong Kong Tokyo Nairobi Dar es Salaam
Cape Town Melbourne Auckland

Oxford is a trade mark of Oxford University Press

Distributed in the United States by Oxford University Press, New York

ISBN 92 832 1168 5

The authors alone are responsible for the views expressed in the signed articles in this publication.

PRINTED IN SWITZERLAND

IARC MANUAL SERIES:

ENVIRONMENTAL CARCINOGENS - SELECTED METHODS OF ANALYSIS

Volume 1 (1978)	*Analysis of volatile Nitrosamines in Food* (*IARC Scientific Publications No. 18*), International Agency for Research on Cancer, Lyon
Volume 2 (1978)	*Methods for the Measurement of Vinyl Chloride in Poly (Vinyl Chloride), Air, Water and Foodstruffs* (*IARC Scientific Publications No. 22*), International Agency for Research on Cancer, Lyon
Volume 3 (1979)	*Analysis of Polycyclic Aromatic Hydrocarbons in Environmental Samples* (*IARC Scientific Publications No. 29*), International Agency for Research on Cancer, Lyon
Volume 4 (1982)	*Some Aromatic Amines and Azo Dyes in the General and Industrial Environments* (*IARC Scientific Publications No. 40*), International Agency for Research on Cancer, Lyon
Volume 5 (1982)	*Some Mycotoxins* (*IARC Scientific Publications No. 44*), International Agency for Research on Cancer, Lyon
Volume 6 (1982)	N-*Nitroso Compounds* (*IARC Scientific Publications No. 45*), International Agency for Research on Cancer, Lyon

CONTENTS

II. METHODS OF SAMPLING AND ANALYSIS

The Determination of Halogenated Alkanes and Alkenes in Air

Determination of Halogenated Alkanes and Alkenes in Water

Determination of organic-bound halogen in water samples

Determination of volatile organic halogen compounds in water samples by head-space gas chromatography

Determination of Residues of Halogenated Fumigants in Foodstuffs

Biological Monitoring

Breath analysis

FOREWORD

The widespread, large-scale use of volatile halogenated alkanes and alkenes, as solvents, intermediates, food fumigants, etc. - as well as their ubiquitous presence in the general environments - emphasise the need for standard detection techniques both in exposure control and in epidemiological studies. The provision of clear, unambiguous techniques and an overall approach are the aims of this series. This is of particular importance since there is experimental evidence of carcinogenicity for several of them.

Food, water and air exposure routes are covered in this volume, with emphasis on appropriate sampling techniques so as to make possible exposure assessment. Included also are very recent advances in biological monitoring via breath analysis, as well as establishd blood and urine measurements.

Thanks are due to the many workers in government and research laboratories who freely contributed chapters or re-wrote and clarified their techniques for inclusion in these pages. It is with particular pleasure that I record the support given by the United Nations Environment Programme for this series.

L. Tomatis, M.D.
Director
International Agency for Research on Cancer

PREFACE

The major objectives of the volumes on Environmental Carcinogens - Selected Methods of Analysis (Manuals Series) remain as initially formulated a decade ago, that is to address more definitively the needs for sampling and analysis of environmental carcinogens, both known and suspected, in the context of epidemiological and analytical studies. Hence this series augments the Monograph Series[1], in terms of compounds selected, and provides the epidemiologist with a view of past and present sampling and analytical procedures, which is necessary for a more definitive retrospective and prospective analysis of exposure and populations at risk.

The present volume deals with some lower-molecular-weight halogenated alkanes and alkenes, a number of which have been assessed in a previous IARC Monograph (Volume 20) and will be further evaluated at a forthcoming IARC Monograph meeting in 1986. These agents, which are environmentally ubiquitous, are produced in very large quantities and are employed in a great variety of applications as chemical intermediates, solvents, degreasing agents, aerosol propellants, refrigerants, food fumigants, flame retardants, anesthetics, etc. A number of these agents have also been employed in a wide range of commonly-available household items, such as degreasing agents, paint and varnish removers and adhesives. In addition, some of these agents found in air and water can arise as a consequence of water chlorination, as well as from emissions. Hence there is a broad potential for exposure of relatively large segments of the population, including the general public, with exposures arising at the workplace and in outdoor and indoor environments.

In order to furnish as comprehensive a treatment as possible, reviews concerning the use and occurrence, carcinogenicity, genetic toxicity, metabolism and epidemiology of these agents are included in this volume. The importance of sampling and analytical techniques for various environments and biological materials is stressed in 3 review chapters. A total of 30 analytical methods, written in the ISO format, are presented for the determination of specific agents in air, water and food. The increasing recognition of the utility of biological monitoring in providing data relating to recent or accumulated exposures is reflected in the 8 procedures described for the sampling and analysis of breath, blood, tissues and urine.

---

[1] Critical reviews of data on carcinogenicity of chemicals and evaluations of carcinogenicity, published by the IARC.

The Editorial Board is aware of the need to recognize advances in the state-of-the-art and to incorporate procedures reflecting enhanced sampling proficiency, accuracy, sensitivity, specificity, reproducibility, ease of manipulation and, as much as possible, economy, particularly since the readers of the Manual Series include analytical chemists in developing countries with facilities less advanced than those often found in more affluent societies. In future volumes, efforts to include procedures (e.g., spectrophotometric) requiring less formidable techniques and expensive instrumentation will therefore be extended, provided they do not unduly sacrifice sensitivity and selectivity.

Manuals in preparation include Metals (Vol. 8), Passive Smoking (Vol. 9) and Benzene-Toluene-Xylene (Vol. 10). The increasing recognition of the importance and consequences of concurrent or sequential short-term to lifetime exposures to mixtures of environmental carcinogens is further demonstrated by the preparation of forthcoming manuals which will cover process industries (e.g., styrene-acrylonitrile-butadiene), chlorinated dibenzodioxins and chlorinated dibenzofurans in indoor environments. Additional Manuals planned include volumes dealing with asbestos and mineral fibers, nitrates-nitrites and formaldehyde.

Lawrence Fishbein, Ph.D.
Chairman
Editorial Board
(from 1983)

INTRODUCTION:

VOLATILE HALOGENATED ALKANES AND ALKENES - IARC CARCINOGENIC RISK EVALUATIONS AND SELECTED ANALYTICAL METHODS

The 24 substances covered by this volume are widely used or present in the environment and thus the selection of analytical methods for their determination is rather important. The substances chosen by the Review Board for coverage in this volume are listed in Table 1, together with Chemical Abstracts Services registry numbers and frequently used names. The substances were chosen on the basis of known carcinogenicity or potential carcinogenic risk and widespread human exposure. Although for some substances, exposures are well-defined and mostly confined to occupational circumstances, others are found at low levels in ambient and indoor air (EPA, 1984) or in water. Recently developed, trace-level methods of environmental and biological monitoring are included, so that the extent and nature of these exposures can be better appreciated.

Sixteen of the substances have been considered by IARC working groups, and their evaluations are listed in Table 2. With regard to the classification of evidence for carcinogenicity, the terms "sufficient", "limited" and "inadequate" have precise definitions in the context of their use in reports of IARC working groups, and IARC Monograph volumes should be consulted. For 1,2-dichloroethane and 1,2-dibromo-3-chloropropane, it was considered that in the absence of adequate data for humans, it is reasonable, for practical purposes, to regard these substances as presenting a carcinogenic risk to humans. For methyl chloroform, hexachloroethane and allyl chloride, the available data did not permit an evaluation of their carcinogenicity. No monograph was published for methyl bromide, because adequate carcinogenicity studies were not available at the time of the meeting. Several studies are presently in progress.

The highest exposures to the listed substances usually occur in the workplace; for twenty-two there are airborne exposure limits set by at least one country in the most recent ILO compilation of exposure limits (ILO, 1980) and also by the American Conference of Governmental Industrial Hygienists (ACGIH, 1983). Most of the analytical methods for air in the present volume are adapted from publications of the U.S. National Institute for Occupational Safety and Health (NIOSH, 1977, 1984), and are tabulated with exposure limits in Chapter 6.

Methods for the detection of these compounds in other environmental or biological samples receive most attention in this volume and are indexed in Table 3. Those for water, food and exhaled air are aimed at monitoring exposure in the general environment, whereas the small number of biological monitoring methods for blood and urine reflects the relatively small proportion of substances for which there are reliable methods using these fluids.

I.K. O'Neill

REFERENCES

ACGIH (1983) Threshold limit values for chemical substances and physical agents in the work environment with intended changes for 1983-84, American Conference of Governmental Industrial Hygienists, Cincinnati, OH

EPA (1984) Total exposure assessment methodology (TEAM) study: first session - northern New Jersey. Pellizzari, E.D., Hartwell, T.D., Sparacino, C.M., Sheldon, L.S., Whitmore, R., Leininger, C., Zeln, H. & Wallace, L., U.S. Environmental Protection Agency Office of Research and Development, Washington, D.C.

IARC Monographs on the Evaluation of the Carcinogenic risk of Chemicals to Humans:

Volume 11 (1976) Cadmium, nickel, some epoxides, miscellaneous industrial chemicals and general considerations on volatile anaesthetics

Volume 15 (1977) Some fumigants, the herbicides 2,4-D and 2,4,5-T, chlorinated dibenzodioxins and miscellaneous industrial chemicals

Volume 19 (1979) Some monomers, plastics and synthetic elastomers and acrolein

Volume 20 (1979) Some halogenated hydrocarbons

Volume 36 (1985) Some allyl and allylic compounds, aldehydes, epoxides and peroxides

Supplement 1 (1979) Chemicals and industrial processes associated with cancer in humans - IARC Monographs, Volumes 1 to 20

Supplement 4 (1982) Chemicals, industrial processes and industries associated with cancer in humans. IARC Monogrpahs, volumes 1 to 29

ILO (1980) Occupational exposure limits for airborne toxic substances; a tabular compilation of values from selected countries; International Labour Office, Geneva

NIOSH (1977) NIOSH Manual of Analytical Methods, 2nd ed., Vol. 1 to 7, NIOSH, Cincinnati, OH

NIOSH (1984) NIOSH Manual of Analytical Methods, 3rd ed., Vol. 1, NIOSH, Cincinnati, OH

Table 1. Volatile halogenated alkanes and alkenes covered by this volume

Formula	CAS No.	Commonly-used names or abbreviations
1. CH_3Cl	74-87-3	Methyl chloride, chloromethane
2. CH_3Br	74-83-9	Methyl bromide
3. CH_2Cl_2	75-09-2	Methylene chloride; dichloromethane
4. $CHCl_3$	67-66-3	Chloroform; trichloromethane
5. $CHBrCl_2$	75-27-4	Bromodichloromethane; dichlorobromomethane
6. $CHBr_2Cl$	124-48-1	Dibromochloromethane; chlorodibromomethane
7. $CHBr_3$	75-25-2	Bromoform; tribromomethane; methenyltribromide
8. $CFCl_3$	75-69-4	F11; fluorocarbon 11; trichlorofluoromethane
9. $CHClF_2$	75-45-6	F22; fluorocarbon 22; chlorodifluoromethane
10. CCl_4	56-23-5	Carbon tetrachloride; tetrachloromethane
11. CH_2ClCH_2Cl	107-06-2	1,2-Dichloroethane; ethylene dichloride
12. CH_2BrCH_2Br	106-93-4	EDB; ethylene dibromide; 1,2-dibromoethane
13. CCl_3CH_3	71-55-6	1,1,1-Trichloroethane; methylchloroform
14. CCl_3CCl_3	67-72-1	Hexachloroethane
15. $CF_3CHBrCl$	151-67-7	Halothane; 1,1,1-trifluoro-2-bromo-2-chloroethane
16. $CCl_2{=}CH_2$	75-35-4	Vinylidene chloride; 1,1-dichloroethylene
17. $CHCl{=}CHCl$	540-59-0	1,2-Dichloroethylene; 1,2-dichloroethene
18. $CCl_2{=}CHCl$	79-01-6	Trichloroethylene; TCE; trichloroethene
19. $CCl_2{=}CCl_2$	127-18-4	Tetrachloroethylene; perchloroethylene; tetrachloroethene
20. $CH_2{=}CH{-}CH_2Cl$	107-05-1	Allyl chloride; 3-chloropropene
21. $CH_2BrCHBrCH_2Cl$	96-12-8	DBCP; 1,2-dibromo-3-chloropropane
22. $CH_2\ CH{-}CH_2Cl$ (bridged by O)	106-89-8	Epichlorohydrin, chloromethyloxirane
23. $ClCH_2OCH_2Cl$	432-88-1	BCME; bis(chloromethyl)ether; dichloromethyl ether
24. $ClCH_2OCH_3$	107-30-2	CMME; chloromethyl methyl ether; chloromethoxymethane

Table 2. Summary of IARC Monograph Evaluations

Formula or designation	IARC summary evaluation of carcinogenic risk to humans in Supplement 4[a]	Evidence for carcinogenicity in humans[a]	Evidence for carcinogenicity in animals[a]	IARC Monograph reference[b]
1. CH_3Cl	-	-	-	-
2. CH_3Br	-	-	-	-
3. CH_2Cl_2	3	Inadequate	Inadequate	20, 449-465
4. $CHCl_3$	2B	Inadequate	Sufficient	20, 401-417
5. $CHBrCl_2$	-	-	-	-
6. $CHBr_2Cl$	-	-	-	-
7. Bromoform	-	-	-	-
8. F-11	-	-	-	-
9. F-22	-	-	-	-
10. CCl_4	2B	Inadequate	Sufficient	20, 371-339
11. 1,2-Dichloroethane	*	-	Sufficient	20, 429-448
12. EDB	2B	Inadequate	Sufficient	15, 195, 209
13. Methylchloroform		-	Inadequate	20, 515-531
14. Hexachloroethane	-	-	Inadequate	20, 467-476
15. Halothane	3	Inadequate	Inadequate	Suppl. 4, 41-46
16. Vinylidene chloride	3	Inadequate	Limited	19, 439-459
17. 1,2-Dichloroethylene	-	-	-	-
18. Trichloroethylene	3	Inadequate	Limited	20, 545-572
19. Tetrachloroethylene	3	Inadequate	Limited	20, 491-514
20. Allyl chloride	-	-	Inadequate	36, in press
21. DBCP	*	-	Sufficient	20, 83-96
22. Epichlorhydrin	2B	Inadequate	Sufficient	11, 131-139
23. BCME	1	Sufficient	Sufficient	Suppl. 1, 26-27
24. CMME	1	Sufficient	Sufficient	Suppl. 1, 26-27

[a] Group 1 - The chemical is carcinogenic to humans. This category was used only when there was sufficient evidence from epidemiological studies to support a causal association between the exposure and cancer.

Group 2 - The chemical is probably carcinogenic to humans. This category includes exposures for which, at one extreme, the evidence of human carcinogenicity is almost "sufficient" as well as exposures for which, at the other extreme, it is inadequate. To reflect this range, the category was divided into higher (Group A) and lower (Group B) degrees of evidence. Usually, category 2A was reserved for exposures for which there was at least limited evidence of carcinogenicity to humans. The data from studies in experimental animals played an important role in assigning studies to category 2, and particularly those in group B; thus the combination of sufficient evidence in animals and inadequate data in humans usually resulted in a classification of 2B.

Group 3 - The chemical cannot be classified with respect to its carcinogenicity to humans.

* = In the absence of adequate data in humans, it is reasonable, for practical purposes, to regard the compound as if it presented a carcinogenic risk to humans.

[b] Volume and page numbers

Table 3. Analytical method number in this volume

Formula or designation	Air[a]	Water[b]	Food fumigant residues	Exhaled breath	Other biological monitoring
1. CH_3Cl	-	15,16	-	-	
2. CH_3Br	-	15,16	17,20,21	-	
3. CH_2Cl_2	-	15,16	-	-	
4. $CHCl_3$	12	15,16	17,18,19	23/24	25
5. $CHBrCl_2$	-	16		23/24	25
6. $CHBr_2Cl$	-	15,16	-	23/24	25
7. Bromoform	12	15,16	-	23/24	25
8. F-11	-	15,16	-	-	-
9. F-22	-	16	-	-	-
10. CCl_4	12	15,16	17,18,19	23/24	25
11. 1,2-dichloroethane	12	15,16	17	23/24	25
12. EDB	12	15,16	17,18,19,22	23/24	25
13. Methyl chloroform	-	15,16	17	23/24	25,26
14. Hexachloroethane	-	15,16	-	23/24	25
15. Halothane	-	15,16	-	23/24	25,28
16. Vinylidene chloride	-	15,16	-	-	-
17. 1,2-Dichloroethylene	-	15,16	-	-	-
18. Trichloroethylene	12	15,16	17,18,19	23/24	25,29,30
19. Tetrachlorethylene	12	16	17	23/24	25,27
20. Allyl chloride	-	15,16	-	23/24	25
21. DBCP	-	15,16	-	23/24	25
22. Epichlorohydrin	-	-	-	23/24	25
23. BCME	-	-	-	23/24	25
24. CMME	-	-	-	23/24	25

[a] For halogenated alkanes and alkenes in workplace air, Methods 1 through to 11 are tabulated in Chapter 6, together with information on exposure limits.

[b] For organic-bound halogen in water, Methods 13 and 14 give screening methods.

SCOPE OF MANUAL AND CRITERIA FOR THE SELECTION OF ANALYTICAL METHODS

Scope

1. The Manual consists of a number of individual volumes, each concerned with a specific group of compounds, the purpose of which is to present selected (preferably validated) methods in systematic format, based on ISO Guide 78 (ISO/R.78.1969[E]), to analysts and others interested in the field.

2. Each volume will normally comprise a general introduction to the field of analysis concerned, representing up to one-third of its length, and a description of the selected analytical and sampling methods, comprising not less than two-thirds of its length. The overall balance of each volume reflects the needs for analytical methods and the importance of the introductory material in relation to IARC and WHO requirements.

3. The chemicals (or groups of chemicals) considered have been evaluated in an IARC Monograph on the Evaluation of the Carcinogenic Risk of Chemicals to Humans and thus indicated to have, or to be likely to have, some carcinogenic effect in experimental animals and/or man, with evidence that a risk of human exposure exists. Chemicals for which carcinogenicity evaluations are still in progress and for which evidence of occurrence is needed, may also be included.

4. The methods of analysis and sampling selected are related primarily to the environmental substrates in which potential carcinogenic risks have been established or from which the major human exposures are known to occur.

Criteria for selection

1. Preference is given to methods of analysis and sampling for which the reliability (i.e., accuracy, precision and inter- and intra-laboratory variations) has been statistically established in collaborative or co-operative analytical studies.

2. Preference is also given to methods that have already been recommended or adopted by relevant international organizations or that have been adopted by a national organization and subsequently entered into wide use.

3. When an international organization has made separate provision for reference, routine screening or field test methods, these provisions are to be adopted.

4. If other methods have been shown to be equivalent to these methods, they may be accepted as alternatives. When inclusion of analytical procedures for additional substances is necessary in order to complete the description of a group of substances covered by the volume, a short review of available methods is provided, if no particular method is deemed suitable for selection.

5. When appropriate methods for analysis are uniformly applicable to various substrates, these are selected in preference to those which apply only to individual substrates.

6. When no method that has been subjected to full international collaborative study is available, methods are selected from those in the published literature to guide those who need to make a choice from the large field of published methods.

7. In selecting methods, particular consideration is given to the requirements of epidemiologists, hygienists and others concerned with the evaluation of carcinogenic and other toxic effects. Particular consideration is given to biological test methods that establish individual past exposure to environmental carcinogens.

8. The need has frequently been expressed by governments of developing countries for simple, specific, low-cost methods for use in the field; and, while it is recognized that such methods are often desirable, it is unlikely, in view of the very low levels and complexity of the contaminants concerned, that these can be developed easily.

***

MEMBERS ATTENDING THE EDITORIAL BOARD MEETING

MEMBERS ATTENDING THE REVIEW BOARD

L. FISHBEIN (*Chairman*)	National Center for Toxicological Research, Jefferson, Arkansas, USA
P.A. GREVE	Rijks Instituut voor Volkgezondheid, Bilthoven, Netherlands
I.K. O'NEILL	Unit of Environmental Carcinogens and Host Factors, Division of Environmental Carcinogenesis, International Agency for Research on Cancer, Lyon, France
R.A. RASMUSSEN	Oregon Graduate Center, Beaverton, Oregon, USA
M. SONNEBORN	Institut für Wasser-Boden-und Lufthygiene, Bundesgeshundhermsamt, Berlin, FRG
D.T. WILLIAMS	Environmental Health Centre, Tunney's Pasture, Ottawa, Ontario, Canada

ACKNOWLEDGEMENTS

Dr Harold Egan, Chairman Editorial Board, 1975-1983

Dr Egan, who was at the time the U.K. Government Chemist, assumed chairmanship of the Editorial Board at its inception in 1975. Under his chairmanship the Editorial Board guided volumes 1 to 6 into print and also saw the commencement of the present volume. The establishment of this series owes much to his organizational skills in recruiting eminent scientists to freely contribute or to participate in the Boards, and also to his expert chairmanship and knowledge of international analytical progress. Therefore, we were saddened to learn of his death, working for the international community even up to his last day, in June 1984. His contribution to this series will be greatly missed.

Contributors to Volume 7

We should like to thank the following persons whose considerable efforts have greatly expedited publication of this volume: B. Dodet (compiling editor), E. Heseltine and her staff (technical editing), Dr A. MacKenzie Peers (copy and revising editor). The detailed advice of members of the Review Board, during and subsequent to the Review Board meeting in Seattle, USA, has been most helpful. For secretarial assistance, we are grateful to M. Wrisez (IARC) and B. Honeycutt (NCTR).

The Editors

I. INTRODUCTION AND GENERAL ASPECTS

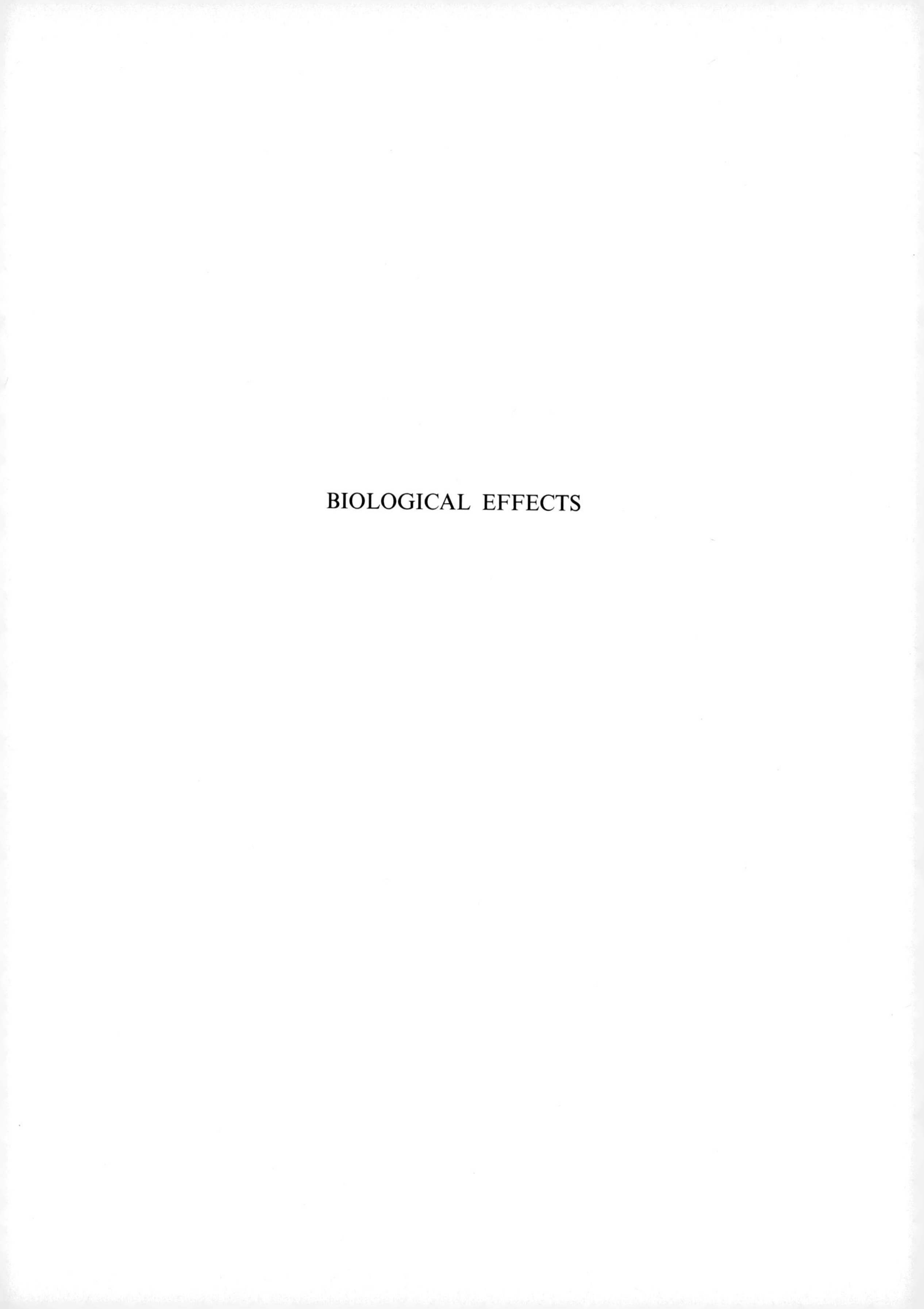

BIOLOGICAL EFFECTS

CHAPTER 1

HALOGENATED ALKANES AND ALKENES AND CANCER: EPIDEMIOLOGICAL ASPECTS

O. Axelson

Department of Occupational Medicine
University Hospital
S. 581 85 Linköping, Sweden

INTRODUCTION

Halogenated hydrocarbons are widely-used compounds with heterogenous chemistry and a variety of biological effects. They are found in agriculture and industry; for example, as pesticides, solvents, degreasing agents, cutting fluids, propellants and refrigerants. Some of them are also used as raw-materials for plastics and textiles, and several of the anaesthetic gases for surgery belong to this group of compounds. The halogenated hydrocarbons are alien to nature, although some fungi and algae can introduce chlorine or bromine into organic molecules, and thyroid hormone contains iodine. Through industrial and other pollution, organochlorine molecules can now be found almost anywhere and appear as residues in food and drinking water.

Many occupations and consumer products give rise to exposure to various halogenated hydrocarbons. This has become of increasing concern, as several of these compounds have been found to possess mutagenic properties in bacterial test systems or to cause cancer in animal experiments. Epidemiological data are limited or even nonexistent for most of these substances, however, and their carcinogenicity to humans is presently uncertain. This review will attempt to summarize epidemiological and other data concerning human cancers and the following substances: methyl chloride, methylene chloride, chloroform, carbon tetrachloride, methyl bromide, bromoform, trichlorofluoromethane (F 11), dichlorofluoromethane (F 22), methylchloroform, ethylene chloride, ethylene dibromide, halothane, hexachloroethane, trichloroethylene, perchloroethylene, allyl chloride, dibromochloropropane, epichlorohydrin, chloromethylmethyl ether and bischloromethyl ether.

Although several of these halogenated compounds have been used for decades, it was not until the late 1970s that information appeared on genotoxic effects; e.g., the carcinogenicity of trichloroethylene (NCI, 1976) and perchloroethylene (NCI, 1977) in animal experiments, or the mutagenic properties of methyl chloroform (Simmon _et al._, 1977) and methylene chloride (Jongen _et al._ 1978) in bacterial test systems. Epidemiological information concerning these compounds is still rather scanty, however, and is non-existent for others such as chloroform and carbon tetrachloride, in spite of their well-known general toxicity and wide-spread use.

HALOGENATED METHANES

Absorption of the chlorinated methanes (methyl chloride, methylene chloride, chloroform and carbon tetrachloride) usually takes place by inhalation, but may also occur by absorption through the skin. They all exert neurotoxic effects. These appear to be the main hazard with the first two compounds cited (Collier, 1936; Weiss, 1967; Scharnweber *et al*., 1974), whereas toxic effects on the liver and kidneys might be thought of as more characteristic of chloroform, and of carbon tetrachloride in particular (cf. IARC, 1979a).

Methylene chloride

With regard to the chlorinated methanes, epidemiological studies of occupationally or otherwise exposed groups seem to be available only for methylene chloride. Thus, Friedlander and coworkers (Friedlander *et al.*, 1978; Hearne & Friedlander, 1981) have reported both a proportional mortality ratio (PMR) analysis and a cohort follow-up of two somewhat differently selected, but overlapping, populations. Methylene chloride had been used as a solvent and the time-weighted average exposure levels were estimated to have been in the range of 30 to 125 ppm, with 350 ppm given as the highest concentration registered. It may be recalled that the biotransformation of methylene chloride leads to carbon monoxide and that even some formaldehyde is obtained, in addition to various non-toxic end products (Rodkey & Collison, 1977; Ahmed & Anders, 1978).

The PMR analysis cited above was based on death certificates from 1956 to 1976, and revealed 71 malignancies *versus* 73.4 expected. There were 9 unspecified gastrointestinal tumours *versus* 5.0 expected, but otherwise there was close agreement between observed and expected numbers of tumours.

The cohort analysis concerned 751 individuals, 252 of them forming a subcohort having been exposed to methylene chloride for 20 years or more prior to 1964. A total of 78 deaths had occurred, but no excess of cancer was found either in the total cohort or in the subcohort, regardless of whether the cohorts were compared with the general population of New York State (New York City excluded) or with other employees in the company. An up-date of the cohort through 1980 has also been reported (Hearne & Friedlander, 1981) and encompassed 110 deaths, but again there was no excess of cancers, nor of deaths from cardiovascular disease.

A further epidemiological evaluation of methylene chloride appeared in 1983 (Ott *et al.*, 1983) and dealt particularly with possible ischemic events due to the formation of carbon monoxide in the metabolism of methylene chloride, but cancer mortality was also reported. The study concerned workers (producing cellulose-acetate fibers) who had been exposed to methylene chloride, methanol and acetone. A cohort approach was applied encompassing 1 271 individuals, and their mortality was compared not only to expected deaths calculated from US national rates (for white males, non-white males and white females), but also to that of an unexposed cohort of 948 cellulose-acetate fiber workers. Because of the similar work situations for these two populations (the second group had also been exposed to acetone), the comparisons made should be quite valid.

Five neoplasms were reported among males and two among females in both the exposed and the non-exposed cohort. These numbers were lower than expected from the national rates; i.e., 6.3 and 10.0, respectively, for men and 5.2 and 2.3 for women. The mortality from diseases of the circulatory system was somewhat high when compared with that of the reference cohort, however, but did not permit any definite conclusions.

Since there was 18% loss with regard to follow-up for the exposed and 12% for the comparison cohort, the validity of this study is somewhat compromised, especially as the numbers are relatively small. Considering the two studies together, however, it nevertheless seems relatively unlikely that methylene chloride is a strong human carcinogen, but no definite conclusions can be drawn at this time.

Some other halomethanes

With regard to other chlorinated methanes, there are some case reports describing the appearance of liver tumours following exposure to carbon tetrachloride. Initially, this chemical caused fibrosis of the liver, later followed by cancer development (Johnstone, 1948; Simler *et al.*, 1964; Tracey & Sherlock, 1968). No epidemiologically relevant information concerning cancer seems to exist for methyl chloride and chloroform, nor for methyl bromide and bromoform.

Fluorocarbons

Two other commonly-used halogenated methanes, namely trichlorofluoromethane (or F 11) and dichlorofluoromethane (F 22), both belong to a group of compounds referred to as fluorocarbons or freons (trade name). Only one, rather small, epidemiological study with relevance for cancer is available for the fluorocarbons. Szmidt *et al.* (1981) conducted a mortality study of 539 refrigerator workers exposed to various fluorocarbons, including F 12 and F 22. A total of 18 deaths were recorded *versus* 26.7 expected, and 6 were from cancer, against 5.7 expected. Two of the cancers were lung cancers, whereas one case was expected. Since this cohort was relatively young and limited in size and had a rather short follow-up time, no definite conclusions can be drawn concerning the cancer hazard from fluorocarbons, the 95% confidence interval for the total cancer risk ratio being 0.4-2.3. Since those working with chemicals of this type are usually exposed to a number of different fluorocarbons, it will probably always be difficult, now and in the future, to obtain worker groups sufficiently large so that the effects of a specific fluorocarbon can be evaluated. Even users of medical spray products tend to be exposed to a mixture of fluorocarbons.

Halomethanes in drinking water

Trihalomethanes in drinking water are a possible cancer hazard for the general population. A major problem in this context is the low validity of the pertinent studies, as they have to be based on relatively diffuse measurements of exposure, both for these impurities in drinking water and with regard to various background variables. There might also be severe and uncontrolable interference, due, for example, to various industrial exposures and associated sociological risk factors occurring in areas using chlorinated surface water.

Bladder cancer, and to some extent brain tumours, have been associated particularly with bromine-substituted methanes. Kidney cancers among males have been found to correlate with trichloromethanes, and so have non-Hodgkin lymphomas (Cantor et al., 1978). In that study, however, stomach cancer in females was negatively correlated with halomethane levels in the water, whereas another investigation (in Ohio) has shown an increased mortality from both stomach and bladder cancer in areas with surface water supply (Kuzma et al., 1977). Other studies on water chlorination and cancer have indicated a significant association with colon cancer (Young et al., 1981) and rectal cancer (Gottlieb & Carr, 1982), but there were less clear relations for the other sites in these two studies. Alltogether, the available information concerning cancer and halomethanes in water is rather inconsistent and confusing and does not permit definite conclusions at this time.

HALOGENATED ETHANES AND ETHENES

As is the case for the halogenated methanes, epidemiological information concerning ethanes and ethenes is rather limited, despite the fact that some of them are widely used as solvents. A few studies, however, have been reported for trichloroethylene, perchloroethylene (tetrachloroethylene) and ethylene dibromide (dibromoethane). In addition, some observations of interest have been made with regard to halothane. For some other, more or less commonly-occurring, compounds such as methyl chloroform (1,1,1-trichloroethane), hexachloroethane and ethylene chloride (dichloroethane), there seems to be no epidemiological information available that is relevant to cancer.

Trichloroethylene

Trichloroethylene is widely used for degreasing in the metal manufacturing industry, but occurs as well in the production of rubber and plastics and solvent-based glues. It has also been used in anesthesia for surgery and obstetrics. The biotransformation of trichloroethylene through an epoxide (Bonse et al., 1975) and its mutagenicity in bacterial test systems (Greim et al., 1975), together with its induction of hepatocellular carcinoma in mice (particularly in males; NCI, 1976), have prompted epidemiological evaluations of exposed groups. So far, two non-positive studies have been presented (Axelson et al., 1978; Tola et al., 1980), both of a cohort type. An increased frequency of sister chromatid exchanges in a small group of chronically-exposed workers (Gu et al., 1981) has also been reported.

The work of Axelson et al. (1978) originally concerned 518 men with exposure prior to 1970 and with a follow-up from 1955 to 1975. This cohort was based on a register of urinary determinations of trichloroacetic acid, a trichloroethylene metabolite that is used for the control of exposed workers. The study has recently been extended and updated through 1979 and now includes 1 424 men, 65% of whom were exposed in 1970 through 1975. There was a deficit in total cancer mortality, with 22 versus 36.9 expected, but a significant excess in the incidence of urinary tract (11 versus 4.85) and hematolymphatic tumours (5 versus 1.20). Specifically, for 2 years or more of exposure and 10 years of latency, there were 3 urinary bladder cancers versus 0.83 expected, 4 cancers of the prostate versus 2.35 expected and 2 lymphomas versus 0.27 expected (Axelson et al., 1984).

The second study (Tola et al., 1980) was based on the files of a biochemical laboratory performing urinary determinations of trichloroacetic acid, but some other exposed individuals were also included (a total of 2 205 individuals were identified in the register and 89 individuals with earlier trichloroethylene poisoning were added). Out of this group of 2 294 persons, 90.8% (2 084) could be traced and another 33 subjects with otherwise known exposure were also enrolled in the cohort, which finally encompassed 1 148 males and 969 females. Only about 9% of the samples of trichloroacetic acid in the urine exceeded 100 mg/L, so that average exposure levels were quite low, just as in the study by Axelson et al. (1978, 1984).

The total number of deaths was 58 against 84.3 expected, and among these were 11 cancer deaths versus 14.3 expected (based on comparison with the national mortality statistics). A further, detailed analysis of the material did not reveal any excess of cancer for those with early exposure, i.e., prior to 1970.

These two studies are apparently rather small with regard to the number of cases observed and do not provide any definite basis for an evaluation of the cancer hazard for man from exposure to trichloroethylene. It would seem that low-grade exposure to trichloroethylene is not a serious cancer hazard, but the appearance of some excess of tumours in the updated study deserves attention and a follow-up of these cohorts.

Three studies dealing with trichloroethylene exposure and the incidence of liver cancer have also reported no positive effects (Malek et al., 1979; Novotna et al., 1979; Paddle, 1983). The reports on laundry and dry-cleaning workers (reviewed under tetrachloroethylene) may also have some relevance to trichloroethylene, as mixed exposures occur.

Perchloroethylene

Perchloroethylene (tetrachloroethylene) has been used for dry-cleaning since the 1950s, and also, to some extent, for degreasing and for lipid extractions. Perchloroethylene, like trichloroethylene, was found to cause hepatocellular carcinoma in mice when given by gavage (NCI, 1977), but again, it is unclear whether or not this compound would cause cancer in humans as well.

Only one epidemiological study of any relevance to carcinogenesis seems to exist (Blair et al., 1979), but there are also two case reports where exposure to perchloroethylene is mentioned as of possible etiological importance for neoplastic disorders. One of these refers to the familial appearance of chronic lymphatic leukemia among four of five siblings, whose father succumbed to the same disease. All had worked in the dry-cleaning business since the 1940s (Blattner et al., 1976). The other report deals with the familial occurrence of polycytemia vera (Ratnoff & Gress, 1980), and the exposure to perchloroethylene is discussed as a potential, although less likely, etiological factor for one of the cases.

None of the case reports contributes substantially to the evaluation of the potential human cancer hazard from perchloroethylene, but it is of interest that the epidemiological study of laundry and dry-cleaning workers (Blair et al., 1979) also indicated a formal but non-significant excess of leukemias. This study was of the proportional mortality type and was based on 330 death certificates obtained from two local trade union registers in the USA.

There were 87 deaths from cancer, compared with an expected number of 67.9, calculated from the national mortality statistics by the proportional method, allowing for age, race, gender and calendar year. Thus the PMR [proportional mortality ratio = (observed/expected) × 100] for all cancer deaths was 128 ($p < 0.05$), and was particularly high for lung cancers (17 vs 10.0; PMR = 170), skin (3 vs 0.7; PMR = 429), and cancers of cervix uteri (10 vs 4.8; PMR = 208), all these differences being significant at the 5% level or less. Liver cancer also came out somewhat high, 4 cases against 1.7 expected, and so did the leukemias, as already mentioned, i.e., with 5 cases versus 2.2 expected. The formal significance level of 5% was not reached for these last two disorders, however. Since the cancer rate was high, it is natural that circulatory diseases came out correspondingly and significantly low in a proportional mortality study of this type, i.e., there were only 100 circulatory deaths against 125.9 expected.

The evidence presented above concerning perchloroethylene and human cancer do not permit any definite conclusions, but there is certainly some justification for concern. However, other factors might be involved as well, e.g., of a social character, particularly in the case of the excess of cervical cancers, since it is difficult to imagine, at present, any plausible biological mechanism to account for the relation between this disorder and the exposure.

Ethylene dibromide

Ethylene dibromide is used as a fumigant for soil, grain, vegetables and fruit, as a lead scavenger in tetraalkyl lead gasoline, and also as a solvent for fats, waxes, etc. Due to its wide-spread use, large populations are exposed to low levels of this chemical; e.g., along well-travelled highways and around fumigation centres (IARC, 1979a). Consequently, the appearance of a high frequency of squamous-cell carcinomas of the stomach among rats given ethylene dibromide by intragastric tube (Olson et al., 1973), has created some concern with regard to humans. So far, however, there seems to be only one epidemiological study of interest, reporting mortality data for 161 employees of two production units of ethylene dibromide (Ott et al., 1980). However, these workers were exposed to a variety of other chemicals as well, more than a dozen compounds, including ethylene and bromine being mentioned in the report. Other chemicals which might have contributed to the overall exposure pattern were carbon tetrachloride, chloroform, vinyl bromide, benzene, nickel acetate and other organic and inorganic compounds. The average air concentrations of ethylene bromide were estimated to have been 30 ppm, sometimes reaching 75 ppm. Bromide in blood had been determined to some extent and found to range from 10-170 mg/L. However, these values reflected exposure to other bromine compounds, drugs included, as well as to ethylene dibromide.

The follow-up of this group of men showed 36 deaths versus 32.5 expected; these included 7 tumour deaths vs 5.8 expected. When an exposure time of six years or more and a 15 year or more induction-latency time was required, 4 tumours remained versus 1.4 expected, but all were of different types (the 95% confidence interval for the rate ratio is 0.8-7.3). The study is unfortunately quite inconclusive due to small numbers, but tends to show an increased number of neoplasms. The authors noticed, however, that fewer neoplasms had been observed than might have been expected from a model conceived by Ramsey et al. (1978) by extrapolation of data from animal experiments (Olson et al., 1973).

Further epidemiological evaluations of ethylene dibromide are apparently needed and it might be of interest as well to consider isolated cases of alcoholism treatment with disulfiram, since animal experiments have shown an increase in the effect of ethylene dibromide in combination with this drug (Plotnick, 1978).

Halothane (2-bromo-2-chloro-1,1,1-trifluoroethane)

An increased frequency of spontaneous abortions among women exposed to anaesthetic gases has been reported from many countries and even paternal exposure has been suggested to play a role (see Edling, 1980, for a review and references). However, the anaesthetist is exposed to a variety of anaesthetic agents and other potentially harmful factors may operate as well: e.g., X-rays and disinfectants for surgical hand-washing.

The urine of individuals exposed to halothane has been reported to be mutagenic in bacterial test systems (McCoy et al., 1978), suggesting a potential cancer hazard. However, the mutagenic properties are not convincingly clear (Hemminki et al., 1979) and studies of sister chromatid exchanges in lymphocytes after long-term exposure of operating room personnel to waste anaesthetic gases (Husum & Wulf, 1980) and after anaesthesia in patients (Husum et al., 1981) have not indicated any mutagenic effects.

There seem to be no epidemiological studies which would permit an evaluation of carcinogenic effects due to halothane, but some studies of the mortality pattern and the cancer incidence among anaesthetists are available. A few of them indicate some excess of cancer, particularly with regard to leukemias and lymphomas (Bruce et al., 1968). Other malignancies of various types have also occurred more frequently than expected (Corbett et al., 1973) and a possible transplacental effect of anaesthetic agents has been indicated in two studies (Corbett et al., 1974; Tomlin, 1979). Nevertheless, the situation is inconclusive with regard to the existence of a cancer hazard for anaesthetists, since a prospective follow-up of the group in one of the aforementioned studies failed to disclose any further excess of cancer (Bruce et al., 1974). Doubts about some of the alleged effects of anaesthetic gases have also been expressed (Vessey, 1978), and other studies have not clearly indicated any increased cancer risk (Knill-Jones et al., 1972; Doll & Peto, 1977; Pharaoh et al., 1977). It seems unlikely that a clear-cut epidemiological evaluation of risk from halothane will be obtainable in the future, because of mixed exposures and the decreasing concentrations enforced by the usually-accepted abortion risk from anaesthetic gases (cf. Edling, 1980).

OTHER HALOGENATED HYDROCARBONS

Except for the chlorinated methyl ethers and, to some extent, epichlorohydrin, no epidemiological material relevant to cancer is to be found for the other halogenated hydrocarbons considered in this review. With regard to dibromochloropropane and allyl chloride, epidemiological interest has been focussed on the spermatotoxic (Lipshultz et al., 1980; see also Babich & Davis, 1981 for a review and further references) and neurotoxic effects (He et al., 1980).

Epichlorohydrin

An excess of chromosomal aberrations in peripheral lymphocytes has been reported in workers exposed to epichlorohydrin (3-chloro-1,2-epoxypropane) at rather low concentrations; i.e., exposure levels of 0.5-5 mg/m³ were reported in one study (Kucerova et al., 1977) and 1.9 mg/m³ in another (Picciano, 1979). Even 1 mg/m³ was thought to have had effects in a third study (Sram et al., 1980).

A mortality study from two epichlorohydrin production plants, encompassing 864 individuals with a total of 52 deaths (Enterline & Henderson, 1978; Enterline, 1982), has been reviewed (IARC, 1979b, 1982; Hemminki, 1981). Among those exposed to epichlorohydrin for 15 years or more, there were 10 respiratory cancers _versus_ 8.7 expected, and two deaths had occurred from leukemia _versus_ 0.4 expected. In view of the carcinogenic effects found in animal experiments (IARC, 1976; IARC, 1979b, 1982), the indicated excess of malignancies and the finding of chromosome aberrations among exposed workers are quite suggestive of a cancer risk to humans (IARC, 1982) so that additional studies are urgently needed.

Chloromethyl ethers

Chloromethyl methyl ether (CMME) and bis(chloromethyl)ether (BCME) have been widely used as chemical intermediates in organic synthesis and in the preparation of ion-exchange resins (IARC, 1974). Some concern has been expressed about spontaneous formation of BCME in the presence of hydrogen chloride and formaldehyde, but it seems unlikely that such a reaction would take place or be of any importance (Kallos & Solomon, 1973; Tou & Kallos, 1974; Travenius, 1982).

Both of these highly-reactive, alpha-halogenated ethers were reported to be carcinogenic in animals in the late 1960s and early 1970s (Van Duuren _et al._, 1968; Laskin _et al._, 1971; Leong _et al._, 1971), even at concentrations as low as 0.1 ppm for rats subjected to long-term exposure to BCME. Lung cancers were later observed in industrial populations for both CMME (Figueroa _et al._, 1973) and BCME (Sakabe, 1973; Thiess _et al._, 1973).

The study by Figueroa _et al._ (1973) on CMME, mixed with some BCME, found an approximately eight-fold increase of risk of lung cancer among workers in a chemical manufacturing plant. This risk estimate was based on a subcohort of 111 individuals followed for a 5-year period, and as many as four cases were observed. A total of 14 cases were identified, however, and histological confirmation was obtained in 13, of which 12 were found to be oat-cell carcinomas. Three of the 14 men were reported never to have smoked. In a more complete cohort study from the same plant, DeFonso and Kelton (1976) reported 19 cases of lung cancer against 5.6 expected. For a particular subcohort, the risk ratio was as high as 9.6 and a dose-response relation appeared to obtain.

Interestingly, the risk ratio for lung cancer among workers exposed to CMME (and some BCME) has been reported to be higher among non-smokers than among smokers (Weiss, 1976), although the latter were also at increased risk (Weiss, 1980). Weiss suggested the explanation that CMME, as well as BCME, is rapidly hydrolyzed to hydrochloric acid and formaldehyde (and methanol in the case of CMME), causing irritation of the mucous membrane, which reacts by further increasing the mucous layer of the bronchii, especially among the smokers, who also had symptoms of bronchitis to a greater extent than did the non-smokers. This interpretation would be consistent with the inverse relationship between bronchitis and lung cancer observed in many populations, although steadily overlooked (WHO, 1972; Axelson, 1984).

There are also some studies of lung cancer in workers predominantly exposed to BCME. One of these reports cited five cases in a Japanese dyestuff factory (Sakabe, 1973). Exposure to many other chemicals had also occurred, but BCME was thought to be the most likely etiological agent. Thiess et al. (1973) found six cases of lung cancer among 18 men employed in a laboratory where BCME had been used, and another two cases were identified among 50 production workers. Five of the eight lung cancers were described as oat-cell carcinomas. The majority of the subjects concerned were smokers, and the duration of exposure had been from 6 to 9 years, with induction-latency periods from first exposure to diagnosis ranging from 8 to 16 years.

EPILOGUE

The growing body of evidence showing that many halogenated hydrocarbons posess mutagenic and carcinogenic properties indicates that epidemiological studies of populations exposed to these agents are urgently needed. As can be concluded from this review, few studies have yet been undertaken. In addition, it appears to be difficult to find sufficiently large populations with reasonably pure, well-defined, exposure to the agents of interest.

Except for the findings of an obvious excess of lung cancers among workers exposed to chloromethyl ethers, no clear-cut, indisputable cancer risk has yet been shown for any of the halogenated hydrocarbons considered here. On the other hand, all the studies discussed are characterized by fairly low power to detect even a rather considerable cancer risk. This situation might improve in the future, since many of the study groups may become more informative as they get older, permitting better allowance for the induction-latency time for cancers to develop. Furthermore, some of the cross-sectional studies that have been undertaken in the past few decades concerning short-term effects (e.g., neurotoxic effects from the halogenated solvents, spermatotoxic effects from dibromochloropropane, etc.) may provide a good basis for a long-term follow-up of the subjects involved. A somewhat better appreciation of the cancer risk to humans might then be obtained.

REFERENCES

Ahmed, E.A. & Anders, M.W. (1978) Metabolism of dihalomethanes to formaldehyde and inorganic halide. Biochem. Pharmacol., 27, 2021-2025

Axelson, O. (1984) Room for a role for radon in lung cancer causation? Med. Hypotheses, 13, 51-61

Axelson, O., Andersson, K., Hogstedt, C., Holmberg, B., Molina, G. & De Verdier, A. (1978) A cohort study on trichloroethylene exposure and cancer mortality. J. Occup. Med., 20, 194-196

Axelson, O., Andersson, K., Seldén, A. & Hogstedt, C. (1984) Cancer morbidity and exposure to trichloroethylene. In: International Conference on Organic Solvent Toxicity, Stockholm, October 15-17, 1984, Abstract book. Arbete och Hälsa, 29, 126

Babich, H. & Davis, D.L. (1981) Dibromochloropropane (DBCP): A review. Sci. Total. Environ., 17, 207-221

Blair, A., Decoufle, P. & Grauman, D. (1979) Causes of death among laundry and dry-cleaning workers. Am. J. Publ. Health, 69, 508-511

Blattner, W.A., Strober, W., Muchmore, A.V., Blaese, R.M., Broder, S. & Fraumeni, J.F. (1976) Familial chronic lymphocytic leukemia. Ann. Intern. Med., 84, 554-557

Bonse, G., Urban, T., Reichert, D. & Henschler, D. (1975) Chemical reactivity, metabolic oxirane formation and biological reactivity of chlorinated ethylenes in the isolated perfused rat liver preparation. Biochem. Pharmacol., 24, 1829-1834

Bruce, D.L., Eide, K.A., Linde, H.W. & Echenhoff, J.E. (1968) Causes of death among anesthesiologists - A 20-year survey. Anesthesiology, 29, 565-569

Bruce, D.L., Eide, K.A., Smith, M.J., Seltzer, F. & Dykes, M.H.M. (1974) A prospective survey of anaesthesiologists mortality 1967-1971. Anesthesiology, 41, 71-74

Cantor, K.P., Mason, T.J. & McCabe, L.J. (1978) Associations of cancer mortality with halomethanes in drinking water. J. natl Cancer Inst., 61, 979-985

Collier, H. (1936) Methylene dichloride intoxication in industry. A report of two cases. Lancet, 1, 594-595

Corbett, T.H., Cornell, R.G., Lieding, K. & Endres, L. (1973) Incidence of cancer among Michigan nurse-anesthesists. Anesthesiology, 18, 260-263

Corbett, T.H., Cornell, G.R., Endres, J.L. & Lieding, K. (1974) Birth defects among children of nurse-anesthesists. Anesthesiology, 4, 341-344

DeFonso, L.R. & Kelton, S.C. (1976) Lung cancer following exposure to chloromethyl methyl ether. Arch. Environ. Health, 31, 125-130

Doll, R. & Peto, R. (1977) Mortality among doctors in different occupations. Br. Med. J., 1, 1433-1436

Edling, C. (1980) Anaesthetic gases as an occupational hazard - A review. Scand. J. Work Environ. Health, 6, 85-93

Enterline, P.E. & Henderson, V.L. (1978) Updated mortality in workers exposed to epichlorohydrin. Report to Shell Oil Company.

Enterline, P.E. (1982) Importance of sequential exposure in the production of epichlorhydrin and isopropanol. Ann. N.Y. Acad. Sci., 381, 344-349

Figueroa, G.W., Raszkowski, R. & Weiss, W. (1973) Lung cancer in chloromethyl-methyl ether workers. N. Engl. J. Med., 288, 1096-1097

Friedlander, B.R., Hearne, T. & Hall, S. (1978) Epidemiologic investigation of employees chronically exposed to methylene chloride. J. Occup. Med., 20, 657-666

Gottlieb, M.S. & Carr, J.K. (1982) Case-control cancer mortality study and chlorination of drinking water in Louisiana. Environ. Health Perspect., 46, 169-177

Greim, H., Bonse, G., Radwan, Z., Reichert, D. & Henschler, D. (1975) Mutagenicity in vitro and potential carcinogenicity of chlorinated ethylenes as a function of metabolic oxirane formation. Biochem. Pharmacol., 24, 2013-2017

Gu, Z.W., Sele, B., Jalbert, P., Vincent, M., Vincent, F., Marka, C., Chmara, D. & Faure, J. (1981) Induction d'echanges entre les chromatides soeurs (SCE) par le trichloroethylene et ses metabolites. Toxicol. Eur. Res., 3, 63-67

He, F., Shen, D., Guo, Y. & Lu, B. (1980) Toxic polyneuropathy due to chronic allyl chloride intoxication. A clinical and experimental study. Chinese Med. J., 93, 177-182

Hearne, T. & Friedlander, B.R. (1981) Follow-up of methylene chloride study. J. Occup. Med., 23, 660 (Letter)

Hemminki, K. (1981) Epiklorhydrin. Nordiska expertgruppen för gränsvärdesdokumentation. Stockholm, Arbete och Hälsa, 10

Hemminki, K., Sorsa, M. & Vainio, H. (1979) Genetic risks caused by occupational chemicals. Scand. J. Work Environ. Health, 5, 307-327

Husum, B. & Wulf, H.C. (1980) Sister chromatid exchanges in lymphocytes in operating personnel. Acta Anaesth. Scand., 24, 22-24

Husum, B., Wulf, H.C. & Niebuhr, E. (1981) Sister chromatid exchanges in lymphocytes after anaesthesia with halothane and enflurane. Acta Anaesth. Scand., 25, 97-98

IARC (1974) IARC Monographs on the Evaluation of the Carcinogenic Risk of Chemicals to Man, Vol. 4, Some Aromatic Amines, Hydrazine and Related Substances, N-Nitroso Compounds and Miscellaneous Alkylating Agents, Lyon, International Agency for Research on Cancer

IARC (1976) IARC Monographs on the Evaluation of the Carcinogenic Risk of Chemicals to Man, Vol. 11, Cadmium, Nickel, Some Epoxides, Miscellaneous Industrial Chemicals and General Considerations on Volatile Anaesthetics, Lyon, International Agency for Research on Cancer

IARC (1979a) IARC Monographs on the Evaluation of the Carcinogenic Risk of Chemicals to Man, Vol. 20, Some Halogenated Hydrocarbons, Lyon, International Agency for Research on Cancer

IARC (1979b) IARC Monographs on the Evaluation of the Carcinogenic Risk of Chemicals to Man, Supplement 1, Chemical and Industrial Processes Associated with Cancer in Humans, Vol. 1 to 20, Lyon, International Agency for Research on Cancer

IARC (1982) IARC Monographs on the Evaluation of the Carcinogenic Risk of Chemicals to Man, Supplement 4, Chemicals, Industrial Processes and Industries Associated with Cancer in Humans, Vol. 1 to 29, Lyon, International Agency for Research on Cancer

Johnstone, R.T. (1948) Occupational Medicine and Industrial Hygiene, Saint Louis, MO, Mosby, p. 157

Jongen, W.M.F., Alink, G.M. & Koeman, J.H. (1978) Mutagenic effect of dichloromethane on Salmonella typhimurium. Mutat. Res., 56, 245-248

Kallos, G.J. & Solomon, R.A. (1973) Investigations of the formation of bis-chloromethyl ether in simulated hydrogen chloride-formaldehyde atmosphere environments. Am. Ind. Hyg. Assoc. J., 34, 469-473

Knill-Jones, R.P., Rodrigues, L.V., Moir, D.D. & Spence, A.A. (1972) Anaesthetic practice and pregnancy: Controlled survey of women anaesthetists in the United Kingdom. Lancet, 1, 1326-1328

Kucerova, M., Zhurkov, V.S., Polikova, Z. & Ivanova, J.E. (1977) Mutagenic effect of epichlorohydrin. II. Analysis of chromosomal aberrations in lymphocytes of persons occupationally exposed to epichlorohydrin. Mutat. Res., 48, 355-360

Kuzman, R.J., Kuzman, C.M. & Buncker, C.R. (1977) Ohio drinking water source and cancer rates. Am. J. Public Health, 67, 725-729

Laskin, S., Kuschner, M., Drew, R.T., Cappiello, V.P. & Nelson, N. (1971) Tumors of the respiratory tract induced by inhalation of bis(chloromethyl)ether. Arch. Environ. Health, 23, 135-136

Leong, B.K.J., MacFarland, H.S. & Reese, Jr., W.H. (1971) Induction of lung adenomas by chronic inhalation of bis(chloromethyl)ether. Arch. Environ. Health, 22, 663-666

Lipshultz, L.I., Ross, C.E., Whorton, D., Milby, T., Smith, R. & Joyner, R.E. (1980) dibromochloropropane and its effect on testicular function in man. J. Urol., 124, 464-468

Malek, B., Kremarova, B. & Rodova, O. (1979) An epidemiological study of hepatic tumour incidence in subjects working with trichloroethylene. Pracov. Lek., 31, 124-126

McCoy, E.C., Hankel, R., Robbins, K., Rosenkrantz, H.S., Ginffrida, I.Y. & Bizzari, D.V. (1978) Presence of mutagenic substances in the urines of anesthesiologists. Mutat. Res., 53, 71

NCI (1976) Carcinogenesis Bioassay of Trichloroethylene. Natl Cancer Inst. Techn. Rep. Ser. 2, Washington DC, Department of Health, Education and Welfare

NCI (1977) Bioassay of Tetrachloroethylene for Possible Cancerogenicity. Natl Cancer Inst. Tech. Rep. Ser. 13, Washington DC, Department of Health, Education and Welfare

Novotna, E., David, A. & Malek, B. (1979) An epidemiological study of hepatic tumour incidence in subjects working with trichloroethylene. Pracov. Lek., 31, 121-123

Olson, W.A., Haberman, R.T., Weisburger, E.K., Ward, J.M. & Weisburger, J.H. (1973) Induction of stomach cancer in rats and mice by halogenated aliphatic fumigants. J. natl Cancer Inst., 51, 1993-1995

Ott, M.G., Scharnweber, H.C. & Langner, R.R. (1980) Mortality experience of 161 employees exposed to ethylene dibromide in two production units. Br. J. Ind. Med., 37, 163-168

Ott, M.G., Skory, L.K., Holder, B.B., Bronson, J.M. & Williams, P.R. (1983) Health evaluation of employees occupationally exposed to methylene chloride. Mortality. Scand. J. Work Environ. Health, 9, Suppl. 1, 8-16

Paddle, G.M. (1983) Incidence of liver cancer and trichloroethylene manufacture: joint study by industry and a cancer registry. Br. Med. J., 286, 846

Pharaoh, P.O.D., Alberman, E., Doyle, P. & Chamberlain, G. (1977) Outcome of pregnancy among women in anaesthetic practice. Lancet, 1, 34-36

Picciano, D. (1979) Cytogenetic investigation of occupational exposure to epichlorohydrin. Mutat. Res., 66, 169-173

Plotnick, H.B. (1978) Carcinogenesis in rats of combined ethylene dibromide and disulfiram. J. Am. Med. Assoc., 239, 1609

Ramsey, J.C., Park, C.N., Ott, M.G. & Gehring, P.J. (1978) Carcinogenic risk assessment: ethylene dibromide. Toxicol. Appl. Pharmacol., 47, 411-414

Ratnoff, W.C. & Gress, R.E. (1980) The familial occurrence of polycytemia vera: Report of a father and son, with consideration of the possible role of exposure to organic solvents, including tetrachloroethylene. Blood, 56, 233-236

Rodkey, F.L. & Collison, H.A. (1977) Biological oxidation of ^{14}C-methylene chloride to carbon monoxide and carbon dioxide by the rat. Toxicol. Appl. Pharmacol., 40, 33-38

Sakabe, H. (1973) Lung cancer due to exposure to bis(chloromethyl) ether. Ind. Health, 11, 145-148

Scharnweber, H.C., Spears, G.N. & Cowles, S.R. (1974) Chronic methyl chloride intoxication in six industrial workers. J. Occup. Med., 16, 112-113

Simler, M., Maure, M. & Mandard, J.C. (1964) Cancer du foie sur cirrhose au tetrachlorure de carbone. Strasbourg Med., 15, 910-917

Simmon, V.G., Kaukanen, K. & Tardiff, R.G. (1977) Mutagenic activity of chemicals identified in drinking water. In: Scott, C.D., Budges, B.A. & Sobels, F.H., eds, Progress in Genetic Toxicology, Amsterdam, Elsevier, pp. 249-258

Sram, R.J., Zudova, Z. & Kuleshov, N.P. (1980) cytogenetic analysis of peripheral lymphocytes in workers occupationally exposed to epichlorohydrin. Mutat. Res., 70, 115-120

Szmidt, M., Axelson, O. & Edling, C. (1981) Kohortstudie av freonexponerade. Acta Soc. Med. Suec; Hygiea, 90, 77

Thiess, A.M., Hey, W. & Zeller, H. (1973) Zur Toxikologie von Dichloromethyläther verdacht auf kanzerogene Wirkung auch bei Menschen. Zentralb. Arbeitsmed. Arbeitsschutz, 23, 97-102

Tola, S., Vilhunen, R., Järvinen, E. & Korkala, M.-L. (1980) A cohort study on workers exposed to trichloroethylene. J. Occup. Med., 22, 737-740

Tomlin, P.J. (1979) Health problems of anaesthetists and their families in the West Midlands. Br. Med. J., 1, 779-784

Tou, J.C. & Kallos, G.J. (1974) Study of aqueous HCl and formaldehyde mixtures for formation of bis(chloromethyl) ether. Am. Ind. Hyg. Assoc. J., 35, 419-422

Tracey, J.P. & Sherlock, P. (1968) Hepatoma following carbon tetrachloride poisoning. NY State J. Med., 68, 2202-2204

Travenius, S.Z.M. (1982) Formation and occurrence of bis(chloromethyl)ether and its prevention in the chemical industry. Scand. J. Work Environ. Health, 8, Suppl. 3 (86 pp.)

Van Duuren, B.L., Goldschmidt, B.M., Katz, C., Langseth, L., Mercado, C. & Sivak, A. (1968) Alpha-haloethers: A new type of alkylating carcinogen. Arch. Environ. Health, 16, 472-476

Vessey, M.P. (1978) Epidemiological studies of the occupational hazards of anaesthesia - a review. Anesthesiology, 33, 430-438

Weiss, C. (1967) Toxic encephalosis as an occupational hazard with methylene chloride. Zentralbl. Arbeitsmed., 17, 282-285

Weiss, W. (1976) Chloromethyl ethers, cigarettes, cough and cancer. J. Occup. Med., 18, 194-199

Weiss, W. (1980) The cigarette factor in lung cancer due to chloromethyl ethers. J. Occup. Med., 22, 527-529

WHO (1972) Health hazards of the human environment. Geneva, World Health Organization, pp. 28-29

Young, T.B., Kanarek, M.S. & Tsiatis, A.A. (1981) Epidemiologic study of drinking water chlorination and Wisconsin female cancer mortality. J. natl Cancer Inst., 67, 1191-1198

CHAPTER 2

SPECIFIC COVALENT BINDING AND GENOTOXICITY

D. Henschler

Institut für Pharmacologie und Toxikologie der Universitä Würzburg
Versbacher Strasse 9
8700 Würzburg, Federal Republic of Germany

INTRODUCTION

Halogenated aliphatic hydrocarbons have long been regarded as a pharmacological and toxicological entity. Several reasons for this may be cited: almost 150 years ago, chloroform was introduced as a prototype for anaesthesia (see Leake, 1925) and, consequently, a host of chemically-related chlorinated, fluorinated and brominated aliphatics was investigated for their anaesthetic properties by pharmacologists (for review of structure/activity relationships, see Adriani, 1979). Some of these compounds were introduced into clinical use. Invariably, all of them shared essentially the same central nervous system (CNS)-depressing activity. The process of searching for new compounds is still going on. On the other hand, many halogenated aliphatics of widely differing structures are used as volatile solvents for a variety of technical processes (again by making use of a common property, the capacity to dissolve fatty material) such as degreasing, cleaning, extracting, emulsifying, etc. A third reason comes from early toxicological work; all chlorinated aliphatics, after exposure of mammals to elevated doses, cause damage to the major parenchymous tissues, such as liver, kidney and CNS, and cause sensitization of the cardiac automatism to sympathetic nervous system stimuli. The existence of a common denominator for these acute toxic effects seemed logical.

One is not surprised, therefore, that investigators were tempted once more to assume a common mechanism of action for all chlorinated aliphatic compounds when a new toxicological property was found: genotoxicity. Vinyl chloride was the first chlorinated compound found to be carcinogenic in humans (Creech & Johnson, 1974) and experimental animals (Viola et al., 1971; Maltoni & Lefemine, 1974) and mutagenic in microbial organisms (Rannug et al., 1974; Greim et al., 1975; Bartsch et al., 1979; IARC, 1982). The mechanism of this genotoxic effect was soon discovered to be the interaction of the metabolically-formed electrophilic intermediates, chlorooxirane and its rearrangement product, chloroacetaldehyde, with the nucleic acid base, adenosine (Laib & Bolt, 1977; Green & Hathway, 1978). This again prompted some investigators to believe that other genotoxic halogenated aliphatics also act indirectly, i.e., only after metabolic activation, and form comparable adducts with the genetic material, in particular with DNA.

We will learn in the following pages that all these generalizations, both for acute and chronic effects, and particularly for genotoxic events, are unjustified. There are types of halogenated compounds which do not need metabolic activation in order to produce toxic effects on target tissues and to damage DNA and RNA, and even where metabolite formation has been demonstrated to be a prerequisite for toxicity, the mechanisms of the formation of electrophiles may vary decisively. Even the type of DNA adducts formed may be completely different within groups of chemical analogues and homologues. Before any conclusion can be drawn for a given compound, the routes and enzymatic mechanisms of biotransformation have to be studied carefully.

CHEMICAL REACTIVITY AND BIOTRANSFORMATION: GENERAL CONSIDERATIONS OF STRUCTURE AND ACTIVITY

The uniqueness of the carbon-halogen bond constitutes the determining factor for the chemical reactivity of halocarbons in general. It permits some predictions of the metabolic behaviour, and thus of the toxic properties of the reactive metabolic intermediates (Bonse & Henschler, 1976).

In general, the electron-withdrawing effect of a halogen atom interferes with the mesomeric donor effect on the first and second carbon atoms. The resulting electron deprivation in the C...C bond system adjacent to the respective halogen substituent implies completely different consequences in alkanes, alkenes and alkynes, which result in different metabolic transformations;

(a) in alkanes, the C...C bond system is labilized; consequently, the electro-negative oxygen from oxidizing enzymes will induce free radical formation,

$$-\overset{|}{\underset{|}{C}}-\overset{|}{\underset{|}{C}}-Cl \longrightarrow -\overset{|}{\underset{|}{C}}^{\bullet}+{}^{\bullet}\overset{|}{\underset{|}{C}}-Cl$$

Another possibility of metabolic conversion is, under reductive conditions, dehydrochlorination or dechlorination, with formation of the respective olefins,

$$-\underset{Cl}{\overset{|}{C}}-\underset{Cl}{\overset{|}{C}}- \xrightarrow{-HCl} {>}C{=}C{<}^{\,}_{Cl} \qquad -\underset{Cl}{\overset{|}{C}}-\underset{Cl}{\overset{|}{C}}- \xrightarrow{-Cl_2} {>}C{=}C{<}$$

(b) in alkenes, the electron-withdrawing effect results in a stabilization of the adjacent double-bond system, which favours the formation of highly reactive oxiranes. These tend to rearrange (sometimes with intramolecular chlorine migration) to carbonyl compounds,

$${>}C{=}C{<}^{Cl} \longrightarrow {>}C\overset{O}{—}C{<}^{Cl} \longrightarrow -\underset{Cl}{\overset{|}{C}}-C{\lessgtr}^{O} \quad \text{or} \quad -\overset{|}{\underset{|}{C}}-C{\lessgtr}^{O}_{Cl}$$

(c) in alkynes, the electron withdrawal by the halogen leads to a labilization of the triple bond; active oxygen then induces a C...C break, e.g.,

$$Cl-C{\equiv}C-Cl \xrightarrow{O_2} Cl_2C{=}O + CO$$

Free radicals, oxiranes, chlorinated aldehydes and acyl chlorides all constitute electrophilic species which are capable of reacting with nucleophilic sites in macromolecules. In the case of nucleic acids, stable adducts can be formed which may act as promutagens and thus induce mutations as the first step towards cancer formation in somatic cells, or as heritable disease in the germ cells.

Another metabolic activation mechanism in chlorinated alkanes may be seen in the formation of alkanols by insertion of reactive oxygen into a C-H bond, which, if it occurs at the carbon with the chlorine substitution, results in the spontaneous elimination of HCl and the formation of reactive aldehydes, e.g.,

$$-\overset{|}{\underset{|}{C}}-Cl \xrightarrow{\frac{1}{2}O_2} -\overset{|}{\underset{OH}{C}}-Cl \xrightarrow{-HCl} \rangle C=O$$

or of carbonyl chlorides, e.g.,

$$Cl-\overset{Cl}{\underset{Cl}{C}}-H \xrightarrow{\frac{1}{2}O_2} Cl-\overset{Cl}{\underset{Cl}{C}}-OH \xrightarrow{-HCl} \overset{Cl}{\underset{Cl}{\rangle}} C=O$$

Aldehydes and acyl chlorides again may form adducts at nucleophilic sites of macromolecules and thus damage DNA and RNA.

A completely different mechanism of genotoxicity prevails if the halogen atom is located in an allylic position. In this type of compound, the halogen constitutes a suitable leaving group; the cation formed is stabilized by resonance,

$$H_2C=CH-CH_2Cl \longrightarrow [H_2C \cdots CH \cdots CH_2]^{\oplus}\ Cl^{\ominus}$$

and thus may alkylate nucleophilic sites in macromolecules (for review, see Eder _et al._, 1982).

Finally, bifunctional halogenated aliphatics may be activated to genotoxic intermediates by a substitution reaction with an -SH-bearing molecule (e.g., glutathione, cysteine, etc.), which may occur spontaneously or be enzyme-catalysed,

$$Cl-\underset{|}{\overset{|}{C}}-\underset{|}{\overset{|}{C}}-Cl + GSH \longrightarrow Cl-\underset{|}{\overset{|}{C}}-\underset{|}{\overset{|}{C}}-SG$$

The resulting mustard-like compound exerts alkylating activity. This reaction is an example of the ambiguity of metabolic processes: normally, the nucleophilic -SH group of glutathione serves for the detoxification of electrophilic intermediates, whereas, in the above-mentioned case, a toxification is accomplished by a simple electrophilic-substitution reaction.

All these genotoxic activation reactions may be predicted from basic chemical knowledge and historical experience of xenobiotic metabolism. However, in the case of halocarbons with more complicated structures, two or more of these basic mechanisms may compete with each other, which renders prediction rather difficult. The different pathways of activation may, at least in the majority of cases, be explained *a posteriori* by the chemical structure(s) of the nucleic acid adducts. The identification of these is therefore mandatory if one tries to theorize about mechanisms of genotoxicity. Unfortunately, in only a few cases involving halogenated compounds have these adducts been isolated and unequivocally identified. Therefore, the following description of types of specific covalent binding in the context of enotoxic activities of halogenated alkanes and alkenes can only be fragmentary. It will focus on an analysis of genotoxicity on a molecular basis, rather than on a systematic compilation of reports on genotoxic effects of halogenated alkanes and alkenes, as observed in different biological test systems.

HALOGENATED ALKANES

Much work has been performed on the metabolism, covalent binding and mutagenicity/carcinogenicity of halo-methanes, ethanes and higher homologues. They may be bioactivated through enzymic pathways (as outlined in the preceding chapter) to electrophilic intermediates, such as free radicals, olefins (to be further activated to epoxides) and acyl halides (such as phosgene or formyl chloride), or to mustard-like adducts to glutathione in the case of bifunctional ethanes (e.g., 1,2-dichloroethane). Covalent binding to proteins and lipids is a common feature of these electrophilic species (for review, see Laib, 1982).

Binding to DNA *in vitro* has also been described for a variety of haloalkanes (DiRenzo *et al.*, 1982), whereas data on DNA binding *in vivo* remain controversial. No specific DNA adducts formed *in vivo* have been identified so far; the only unequivocal demonstration of an adduct to a macromolecule is trichloro-methylation of fatty-acid residues in microsomal lipids from carbon tetrachloride (Link et al., 1984). Therefore, the relevance of DNA binding of halo-alkanes for genotoxic effects cannot be evaluated on a chemical basis at present, and will not be dealt with in detail in this chapter.

HALOGENATED ALKENES

Halogenated alkenes may be subdivided, according to the mechanisms of genotoxicity, into two categories:

(a) those with halogen substitution in a vinylic position, such as vinyl chloride or trichloroethylene (these are not genotoxic *per se* but have to be metabolically activated to electrophilic intermediates).

(b) compounds with an allylic halogen substitution, such as allyl chloride or allyl bromide, which possess alkylating properties and thus may display direct genotoxic activities, i.e., without metabolic activation, or may be enzymically converted to α,β-unsaturated carbonyl compounds which can damage nucleic acid bases *via* different mechanisms.

Halogenated ethylenes

As outlined above, the major metabolic oxidative pathway of halogenated ethylenes is epoxidation; the activity of mixed function oxidases in this process is now well established:

mercapturic acid
Cl
O_2, NADPH
MFO
O
Cl
H_2O
EH
OH OH
Cl
C_2 acids
C_1 acid, CO
DNA binding
O
Cl
O
Cl

The halogenated oxiranes may undergo a variety of secondary reactions:

(a) addition to glutathione (or similar low molecular weight nucleophiles), both enzymatically (glutathione transferases GST) and non-enzymatically to form, *via* secondary reactions, mercapturic acid derivatives which are the major urinary excretion products;

(b) hydrolysis, both spontaneously or catalysed by appropriate enzymes (epoxide hydrolases EH), to vicinal diols. If more than one halogen is substituted to the diol, the molecule is rendered very unstable and will undergo a variety of secondary reactions to either C_2 carbonyl compounds or, under C...C breakage, to formate and carbon oxides;

(c) rearrangement (RA) after C...O heterolysis to carbonyl compounds, sometimes accompanied by intramolecular chlorine migration, to molecules of lower energy state, either halogenated aldehydes or acyl halides;

(d) direct reaction with nucleophilic sites in nucleic acid bases (NA) by electrophilic-substitution reactions. NA adducts of identical structure may also be formed by the rearrangement products, halogenated aldehydes.

Reactions (a)-(c) constitute deactivation pathways, whereas (d) represents the toxification mechanism. If one performs genotoxicity studies with halogenated ethylenes, it will depend on the relative rates of formation of the oxirane(s) and of secondary reactions (a)-(d) whether or not (and, if so, to what extent) NA binding will occur. The relative rates may differ in intact organisms and in in-vitro experiments, a fact which complicates extrapolations of genetic risk from short-term tests to humans and other mammals.

The rate of formation of oxiranes by mixed-function oxidases is greatly influenced by the number and site of halogen substitutions. In general, high numbers and asymmetry will imbalance the C...C double bond system. On the other hand, the bulky chlorine or bromine residues may confer steric protection and thus lower the metabolic transformation rate.

All these molecular considerations may elucidate the difficulties encountered in making qualitative and quantitative predictions of the genotoxic activities of halogenated ethylenes. Another modifying factor which has often been forgotten is the chemical structure and the stability of the adducts formed, which will be stressed in the following detailed description of individual compounds.

Vinyl chloride, vinyl bromide, vinyl fluoride

Vinyl chloride (VC) has been the most extensively investigated halogenated aliphatic hydrocarbon (IARC, 1982). The formation of VC epoxide by mixed function oxidases and its rearrangement to chloroacetaldehyde are well established (Green & Hathway, 1975; Hefner et al., 1975; Bonse et al., 1975; Bolt et al., 1976):

$$H_2C{=}CHCl \xrightarrow[P450]{O_2,\ NADPH} H_2C\overset{O}{-}CHCl \longrightarrow H{-}\underset{Cl}{\overset{H}{C}}{-}C(=O){-}H$$

Chloroacetaldehyde may be partially reduced by alcohol dehydrogenase to chloroethanol, or oxidized by aldehyde oxidase (Aldox) to chloroacetic acid; the intermediate formation of the vicinal diol from the epoxide is suggested by the demonstration of glycol aldehyde as one of the secondary products (Guengerich et al., 1979):

$$H_2C\overset{O}{-}CHCl \longrightarrow H{-}\underset{H}{\overset{Cl}{C}}{-}C(=O){-}H \xrightarrow{ADH} H{-}\underset{H}{\overset{Cl}{C}}{-}CH_2OH$$

$$H{-}\underset{H}{\overset{Cl}{C}}{-}C(=O){-}H \xrightarrow{Aldox} H{-}\underset{H}{\overset{Cl}{C}}{-}C(=O){-}OH$$

$$H_2C\overset{O}{-}CHCl \xrightarrow{H_2O} H{-}\underset{H}{\overset{OH}{C}}{-}\underset{H}{\overset{OH}{C}}{-}Cl \xrightarrow{-HCl} H{-}\underset{H}{\overset{OH}{C}}{-}C(=O){-}H$$

The major pathway of detoxification of both the oxirane and chloroacetaldehyde goes through conjugation with glutathione and ensuing reactions to form excretable mercapturic acids. Well-established end-products are: 2-hydroxyethylcysteine, its N-acetyl product and thiodiglycolic acid (Green & Hathway, 1975; Hefner et al., 1975).

Three types of covalent binding of reactive VC metabolites have been described: to proteins, to DNA and RNA, and to the SH-group of coenzyme A (Bolt et al., 1982). Under in-vivo conditions, the more reactive epoxide will bind preferentially to DNA bases, whereas the rearrangement product, chloroacetaldehyde, will bind primarily to proteins. Some of the oxirane can react, immediately after its formation, with the P-450 of monooxygenases, thus partly deactivating the processing enzyme (Guengerich et al., 1981; Guengerich & Strickland, 1977).

The reaction mechanism of VC oxirane (or chloroacetaldehyde) with DNA bases has been found to be cyclisation with an endocyclic and an exocyclic nitrogen of, e.g., adenine (Green & Hathway, 1975; Laib & Bolt, 1977, 1978)

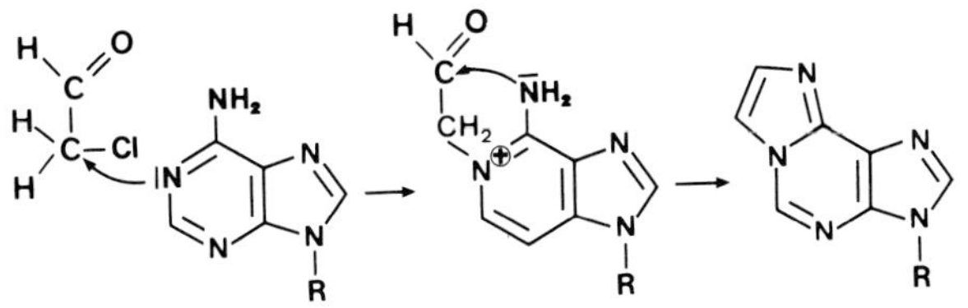

or with the N^7 and O^6-position of guanine (Osterman-Golkar et al., 1977), the product being in equilibrium with the open oxo form (Laib et al., 1981; Scherer et al., 1981),

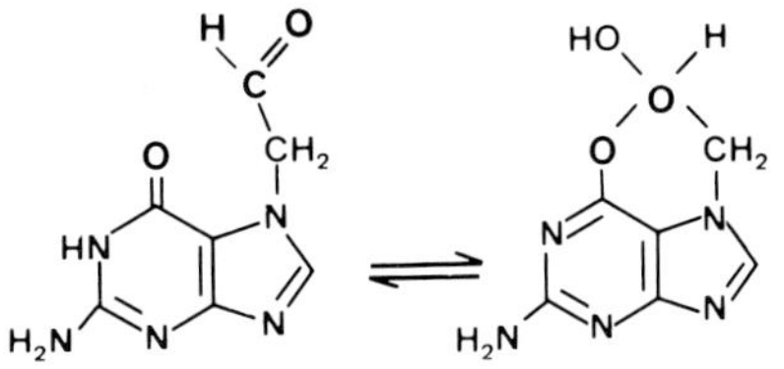

In addition, $3,N^4$ ethenoxytosine and N^2-3-ethenoguanine have been described as adducts *in vitro* ,

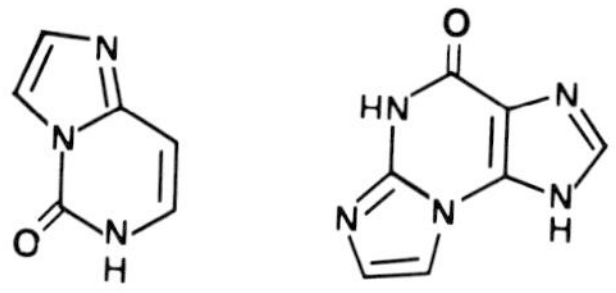

Under in-vivo conditions, 7-oxoethylguanine seems to be the major, if not the only, DNA-alkylation product (Laib *et al.*, 1981).

There is good evidence for the ethenoderivative formation being the cause for the mutagenic activity of VC: the co-planar position of the imidazole ring interferes with translation and transcription mechanisms of DNA- and RNA-bases by miscoding (Hathway, 1981).

The validity of the proposed activation and genotoxic mechanisms is stressed by a variety of additional findings. Microsomal incubation of VC leads to alkylating species which can be trapped by model nucleophiles (Göthe *et al.*, 1974; Barbin *et al.*, 1975). Chlorooxirane is definitely mutagenic *in vitro* (Huberman *et al.*, 1975; Rannug *et al.*, 1976) and carcinogenic *in vivo* (Zajdela *et al.*, 1980), chloroacetaldehyde being much less effective *in vitro* and *in vivo* (van Duuren *et al.*, 1979).

Vinyl bromide (VB) also binds covalently to proteins (Bolt *et al.*, 1978) and forms DNA- and RNA-adducts identical to those obtained after exposure to VC (Ottenwälder *et al.*, 1978). VB is mutagenic after metabolic activation (Bartsch *et al.*, 1979) and carcinogenic in rats (Huntingdon, 1978).

Vinyl fluoride (VF) has been less intensively studied. However, it may produce preneoplastic liver nodules after inhalation of VC in newborn rats (Bolt *et al.*, 1981), the order of potency in this type of short-term carcinogenicity testing has been found to be vinyl chloride > vinyl fluoride > vinyl bromide (Bolt *et al.*, 1982b). All these findings indicate that the same mechanisms underly the genotoxicity of the three analogues, with some expected quantitative differences.

Vinylidene chloride, vinylidene fluoride

Vinylidene chloride (VDC) differs from the other members of the family of chlorinated ethylenes in that it exerts strong acute toxic effects on liver and kidney cells. Due to the extremely uneven electron distribution, the metabolization rate of VDC is high, and the stability of the presumed intermediate, 1,1-dichlorooxirane, is so low that the compound cannot be synthesized (Bonse *et al.*, 1975). Nevertheless, the metabolites identified in several whole-animal studies (Jones & Hathway, 1978; McKenna *et al.*, 1978; Reichert *et al.*, 1979) are indicative of oxirane formation and rearrangement to chloroacetyl chloride. So far, the following metabolites have been identified; carboxymethyl cysteine, N-acetyl carboxymethyl cysteine, thiodiglycolic acid, thioglycolic acid, dithioglycolic acid, chloroacetic acid and carbon dioxide.

A different pathway also suggests oxirane formation, the (assumed) reaction of VDC with phosphotidyl ethanolamine, with the lipid component subsequently splitting off, and substitution of the residual chlorine by S-methyl to form methylthioacetyl-aminoethanol (Reichert *et al.*, 1979):

O_2, NADPH
P450

$CH_2Cl{-}C(=O){-}Cl$

$CH_2Cl{-}C(=O){-}NH{-}CH_2{-}CH_2{-}O{-}$phosphatidyl

$CH_2Cl{-}C(=O){-}NH{-}CH_2{-}CH_2OH$

$CH_3{-}S{-}CH_2{-}C(=O){-}NH{-}CH_2{-}CH_2OH$

This metabolic pathway may explain the acute toxic activity of VDC *via* destruction of lipid membranes by the attack on the major lipid component, phosphotidyl ethanolamine.

No nucleic acid adducts from VDC have been identified so far, although covalent DNA binding does occur. Despite the similarity of VC and VDC metabolism and metabolites, conceivable etheno derivatives of nucleic acid bases from VDC are expected to differ from those from VC, because the binding reaction through the two chlorine atoms will carry an oxygen into the adduct, e.g.,

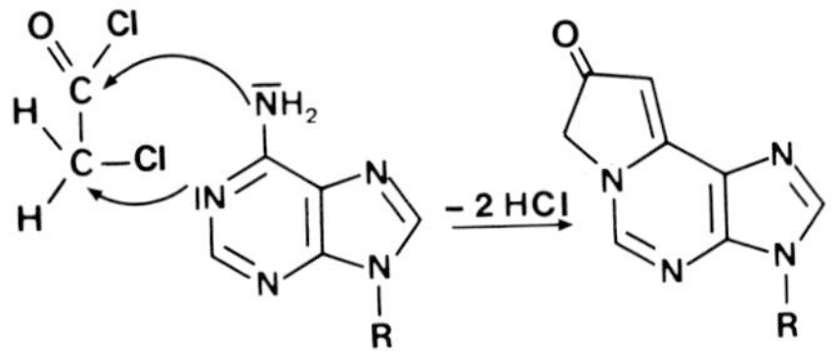

This might have different influences on translation and transcription mechanisms (Hathway, 1977).

Thus it is not unexpected that VC and VDC have distinct, indirect mutagenic properties (Greim et al., 1975; Bartsch et al., 1975). Carcinogenicity studies are still controversial: positive results (Lee et al., 1978; Maltoni & Tovoli, 1979) stand against negative findings (Reitz et al., 1980). The latter may be due to the high, acute and "non-specific" toxicity of VDC.

Vinylidene fluoride has been found to produce preneoplastic liver foci in postnatally exposed rats (Stöckle et al., 1979), which indicates activation mechanisms comparable to those discussed for VDC, but which occur at much lower rates.

1,2-Dichloroethylenes, cis- and trans-

No data are available on specific binding of these two chloroethylenes. Metabolites identified are derivatives of dichloroacetaldehyde (Bonse et al., 1975; Costa & Ivanetich, 1982), which indicates oxidative metabolic activation through dichlorooxiranes and rearrangement to dichloroacetaldehyde. Mutagenicity studies have been negative with E. coli K 12, indicating a complete deactivation of the oxiranes in the metabolic processing.

Trichloroethylene

Despite the fact that no specific binding or genotoxicity has been found with this compound so far, it will be dealt with in more detail because considerable controversy concerning the compound's carcinogenic potential is still being carried on in the literature; in addition, trichloroethylene has been most extensively studied with regard to its metabolism.

Oxirane formation in the metabolic conversion of trichloroethylene (tri) was postulated as early as 1945 by Powell, because the identification of 1,1,1-trichloro compounds, trichloroethanol and trichloroacetic acid, as the major urinary excretion products could only be explained by intramolecular chlorine migration in the (labile) oxirane transition state;

$$Cl_2C{=}CHCl \xrightarrow[P\,450]{O_2,\ NADPH} Cl_2C\overset{O}{—}CClH \rightarrow CCl_3{-}CHO \;\begin{cases} \rightarrow CCl_3{-}CH_2OH \\ \rightarrow CCl_3{-}COOH \end{cases}$$

In fact, the situation is more complicated, since it has been shown that trichlorooxirane will rearrange thermally (unlike the in-vivo situation) to dichloroacetyl chloride,

$$Cl_2C\overset{O}{—}CClH \;\begin{cases} \xrightarrow{thermal} CH_2Cl{-}COCl \\ \xrightarrow{metabol.} CCl_3{-}CHO \end{cases}$$

The reasons for the thermal behaviour have been explained elsewhere (Henschler, 1977). Also, an explanation has been provided for the (almost exclusive) in-vivo transformation to trichloroacetaldehyde; a Lewis acid-catalysed Cl-shift through the trivalent iron in P-450 (Bonse & Henschler, 1976). An alternative explanation has also been proposed, which postulates the intermediate formation of a tri-adduct to P-450 haeme-iron, the existence of which has not, however, been demonstrated (Miller & Guengerich, 1982).

It has also been shown that tri-oxirane decomposes extremely rapidly in the presence of water to a variety of C_2- and C_1-units, which suggests the following reaction mechanism (Henschler et al., 1979);

$$\text{Cl}_2\text{C}\overset{\text{O}}{—}\text{C(Cl)H} \xrightarrow{H_2O} \text{Cl–C(OH)(Cl)–C(OH)(Cl)–H} \xrightarrow{-2HCl} \text{Cl(O=)C–C(=O)H} \xrightarrow[-HCl]{H_2O} \text{HO(O=)C–C(=O)H}$$

$$\text{Cl–C(OH)(Cl)–C(OH)(Cl)–H} \rightarrow \text{Cl–C(OH)(Cl)–H} \xrightarrow{H_2O,\ -2\,HCl} \text{HO–C(=O)–H}$$

$$\text{Cl–C(OH)(Cl)–C(OH)(Cl)–H} \rightarrow \text{Cl–C(=O)–H} \rightarrow \text{CO} + \text{HCl}$$

If these reactions were to take place in vivo , carbon monoxide and formic acid would be expected as final metabolites. Recent careful metabolism studies of tri after gavage in rats and mice have shown, however, that these products are not formed, even after extremely high single doses; nor have mercapturic acids been detected (Dekant et al., 1984). These findings are indicative of a protective mechanism which hinders the escape of trichloro-oxirane from the hydrophobic premise of the oxidizing enzyme system, P-450, by inducing complete rearrangement to 1,1,1-trichloroacetaldehyde before the highly electrophilic oxirane gets access to DNA or RNA, or to soluble, low molecular weight nucleophiles, such as glutathione. Only some covalent binding to phosphotidyl ethanolamine in the lipid fraction of the microsomal multi-enzyme complex does occur and results in the formation of the recently-discovered, novel metabolite, 2-hydroxyacetyl-ethanolamide (Dekant et al., 1984):

$$\text{Cl}_2\text{C}\overset{\text{O}}{—}\text{C(Cl)H} \rightarrow \rightarrow \rightarrow \text{HO–CH}_2\text{–C(=O)–NH–CH}_2\text{–CH}_2\text{–OH}$$

Only after extremely high oral doses to mice, is some dichloroacetic acid formed, which may be explained by an overloading of the above-mentioned protective mechanism (Hathway, 1980).

Incubation of ^{14}C-tri with activating enzyme systems, or addition of trichlorooxirane, may lead to the formation of adducts to DNA or RNA (Banerjee & van Duuren, 1978; Laib *et al.*, 1979). These have, however, never been identified; they are not identical with those formed with vinyl chloride (Laib *et al.*, 1979). All attempts to identify DNA- or RNA-adducts from tri *in vivo* have been unsuccessful so far and, in fact, covalent binding as such could not be demonstrated unequivocally (Parchman & Magee, 1980; Bergman, 1983).

All these findings cast serious doubts on the supposed genotoxic activity of tri. In accordance with this, most mutagenicity experiments with purified samples of tri have been negative (Henschler *et al.*, 1977; Bartsch *et al.*, 1979). A few positive findings may be due to the presence of impurities, such as epichlorohydrin or epoxibutane, which are added as stabilizers to conventional technical samples of tri (Henschler *et al.*, 1977). Furthermore, simple chemical consideration of the presumptive DNA adducts from trichlorooxirane reveal that these should be expected to be rather unstable, due to the residual chlorine residue(s).

Trichloroethylene has been reported to exert carcinogenic activity in a certain hypersensitive strain of mouse (NCI, 1976). In the light of the metabolism described above and of mutagenicity studies, it must be concluded that this weak carcinogenic effect cannot be the result of a genotoxic potential of tri. In fact, other carcinogenicity studies under realistic exposure conditions in rats, mice and hamsters gave no indication of tumour formation (Henschler *et al.*, 1979, 1984).

Tetrachloroethylene

This perchlorinated ethylene derivative lacks any C-H bond and is only slowly metabolized to oxygenated compounds, the chemical structure of which points again to epoxidation as the essential metabolic step (Bonse *et al.*, 1975);

$$Cl_2C{=}CCl_2 \xrightarrow[P450]{O_2,\ NADPH} Cl_2C\overset{O}{—}CCl_2 \longrightarrow CCl_3{-}C({=}O)Cl \xrightarrow[-HCl]{H_2O} CCl_3{-}C({=}O)OH$$

$$CCl_3{-}C({=}O)Cl \longrightarrow R{-}CO{-}CCl_3$$

The intra-molecular rearrangement to trichloroacetylchloride produces an acylating species which leads to some covalent, but easily hydrolysable, binding to proteins (Bonse et al., 1975). Hydrolysis of the oxirane to the vicinal diol and subsequent dehydrochlorination leads to the excretable metabolite, oxalic acid (Yllner, 1961; Pegg et al., 1979; Dekant & Henschler, unpublished). Covalent binding of metabolites to proteins has been described (Bolt & Link, 1980; Costa & Ivanetich, 1980; Schumann et al., 1980), but no DNA binding could be detected *in vivo* (Schumann et al., 1980). These findings indicate the existence of a protective mechanism similar to that which has been described for trichloroethylene. In accordance with this, no mutagenicity was found in microbial tester organisms (Greim et al., 1975; Bartsch et al., 1979), whereas the tetrachlorooxirane is definitely mutagenic in *Salmonella typhimurium* (Kline et al., 1982). Carcinogenicity testing with a technical sample of tetrachloroethylene at extremely high oral doses revealed positive results in a hypersensitive strain of mouse (NCI, 1977), whereas inhalation studies with rats were negative (Rampy et al., 1977).

In conclusion, the present state of knowledge concerning tri- and tetrachloroethylenes warrants caution in the application of simple rules of metabolic formation of electrophilic intermediates as a means of providing proof of genotoxic potential. Rather, all steps of metabolism and type of covalent binding, in particular the stability of the adducts formed with macromolecules, have to be taken into consideration before predictions of mutagenic and carcinogenic risks are released.

Hexachlorobutadiene

The metabolism of this interesting compound of major practical interest has been only partially elucidated. Two metabolites have been unequivocally identified (Reichert, 1982; Schütz & Reichert, 1984);

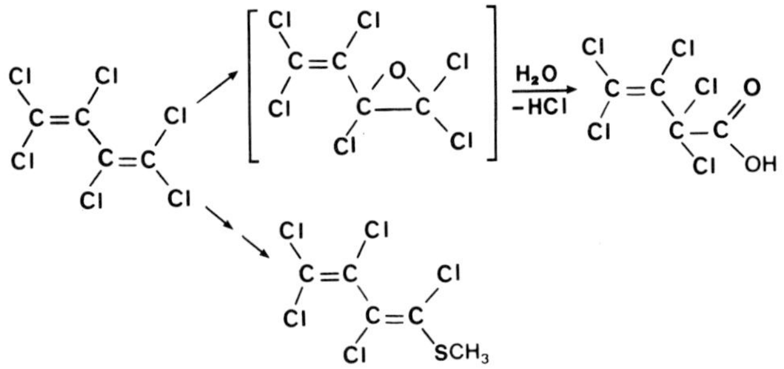

The oxidative pathway, with epoxidation and rearrangement to the acyl chloride and subsequent hydrolysis, follows well-established rules for this type of compound, whereas the formation of a thioether, obviously through an enzymic conjugation reaction (glutathione transferase), constitutes a novel pathway. Both pentachloro-1-butenic acid and pentachloro-1-methylthio-butadiene are direct mutagens (Reichert et al., 1984). Hexachlorobutadiene is a strong nephrocarcinogen in rats (Kociba et al., 1977). DNA binding after metabolic activation does occur (Oesch & Wolf, 1984), but the chemical structure of the adducts formed has not yet been identified.

Halogenated allyl compounds

Although chemically closely related to halogenated vinyl compounds, the allyl derivatives show completely different metabolic behaviour. They are direct alkylating substances which may act through three different alkylating mechanisms (Eder et al., 1982b);

(a) S_N-1-reactivity *via* a resonance-stabilized allyl cation (for equation, see p. 24).

(b) S_N-2-reactivity and S_N-2'-reactivity (bi-molecular reaction),

$$Nu| + CH_2{=}CH{-}CH_2{-}X \longrightarrow [Nu{-}CH_2{-}CH{=}CH_2]^{\oplus}\ X^{\ominus}$$

(c) free radical mechanisms,

$$CH_2{=}CH{-}CH_2{-}X \xrightarrow{-X^{\bullet}} [CH_2{=}CH{-}\dot{C}H_2 \rightleftarrows {}^{\bullet}CH_2{-}CH{=}CH_2]$$

Therefore, typical allyl halides display direct mutagenic activity (Eder et al., 1980), whereas addition of metabolizing enzymes to the test systems abolishes the mutagenic activity, which means that epoxidation is not involved in the genotoxicity of this type of compound. Alkylating and mutagenic potential increase within the group of halogens in the order Cl>Br>I. Further halo or alkyl substitutions at positions other than the allylic increase, due to inductive and mesomeric influences, the direct alkylating and genotoxic activity. Structure-activity relationships of a large group of halogenated allyl compounds have been worked out (for review, see Eder et al., 1982a).

An interesting case is encountered with 2,3-dichloro-1-propene, which may in part alkylate directly *via* the above-described mechanisms, but, in addition, may be epoxidised,

$$ClHC{=}CHCH_2Cl \longrightarrow ClHC(O)CHCH_2Cl \longrightarrow ClCH_2{-}CO{-}CH_2Cl$$

and further rearrange to the highly-reactive and genotoxic dichloroacetone.

So far, the following DNA-adducts from allyl bromide have been identified _in vitro_ and _in vivo_ ; O^6-allyl guanine, 7-allyl guanine, N^2-allyl guanine, 3-allyl guanine and N^6-allyl guanine (Eder & Sebeikat, 1982).

REFERENCES

Adriani, J. (1979) _The chemistry and physics of anaesthesia_, 2nd Ed., Springfield, IL, C.C. Thomas Publ.

Banerjee, S. & van Duuren, B.L. (1978) Covalent binding of the carcinogen trichloroethylene to hepatic microsomal proteins and to exogenous DNA _in vitro_ . _Cancer Res._, _38_, 776-780

Barbin, A., Brésil, H., Croisy, A., Jacquignon, P., Malaveille, C., Montesano, R. & Bartsch, H. (1975) Liver microsome-mediated formation of alkylating agents from vinyl bromide and vinyl chloride. _Biochem. Biophys. Res. Commun._, _67_, 596-603

Bartsch, H., Malaveille, C., Montesano, R. & Tomatis, L. (1975) Tissue-mediated mutagenicity of vinylidene chloride and 2-chlorobutadiene in _Salmonella typhimurium_. _Nature_, _255_, 641-643

Bartsch, H., Malaveille, C., Barbin, A. & Planche, G. (1979) Mutagenic and alkylating metabolites of halo-ethylenes, chlorobutadienes and dichlorobutenes produced by rodent or human liver tissues. _Arch. Toxicol._, _41_, 249-277

Bergman, K. (1983) Interactions of trichloroethylene with DNA in vitro and with RNA and DNA of various mouse tissues in vivo. _Arch. Toxicol._, _54_, 181-193

Bolt, H.M., Kappus, H., Buchter, A. & Bolt, W. (1976) Disposition of [1,2-^{14}C]vinyl chloride in the rat. _Arch. Toxicol._, _35_, 153-162

Bolt, H.M., Filser, J.G. & Hinderer, R.K. (1978) Rat liver microsomal uptake and irreversible protein binding of [1,2-^{14}C]vinyl bromide. _Toxicol. Appl. Pharmacol._, _44_, 481-489

Bolt, H.M. & Link, B. (1980) Zur Toxikologie von Perchloräthylen. Verh. Dtsch. Ges. Arbeitsmed., 20, 463-468

Bolt, H.M., Laib, R.J. & Klein, K.P. (1981) Formation of preneoplastic hepatocellular foci by vinyl fluoride in newborn rats. Arch. Toxicol., 47, 71-73

Bolt, H.M., Filser, J.G. & Laib, R.J. (1982a) Covalent binding of haloethylenes. In: Snyder, R., Parke, Kocsis, Gollow, Gibson, Witmer, eds, Biological reactive intermediates, II. Part A., New York & London, Plenum Press, pp. 667-676

Bolt, H.M., Laib, R.J. & Filser, J.G. (1982b) Reactive metabolites and carcinogenicity of halogenated ethylenes. Biochem. Pharmacol., 31, 1-4

Bonse, G. & Henschler, D. (1976) Chemical reactivity, biotransformation, and toxicity of polychlorinated aliphatic compounds. CRC Crit. Rev. Toxicol., 4, 395-409

Bonse, G., Urban, T., Reichert, D. & Henschler, D. (1975) Chemical reactivity, metabolic oxirane formation and biological reactivity of chlorinated ethylenes in the isolated perfused rat liver preparation. Biochem. Pharmacol., 24, 1829-1834

Costa, A.K. & Ivanetich, K.M. (1980) Tetrachloroethylene metabolism by the hepatic microsomal cytochrome P-450 system. Biochem. Pharmacol., 29, 2863-2869

Costa, A.K. & Ivanetich, K.M. (1982) The 1,2-dichloroethylenes: their metabolism by hepatic cytochrome P-450 in vitro. Biochem. Pharmacol., 31, 2093-2102

Creech, J.L. Jr. & Johnson, M.N. (1974) Angiosarcoma of liver in the manufacture of polyvinyl chloride. J. Occup. Med., 16, 150-151

Dekant, W., Metzler, M. & Henschler, D. (1984) Novel metabolites of trichloroethylene through dechlorination reactions in rats, mice and humans. Biochem. Pharmacol. (in press)

Di Renzo, A.B., Gandolfi, A.J. & Sipes, I.G. (1982) Microsomal bioactivation and covalent binding of aliphatic halides to DNA. Toxicol. Lett., 11, 243-252

van Duuren, B.L., Goldschmidt, B.M., Loewengart, G., Smith, A.C., Melchionne, S., Seidman, L. & Roth, D. (1979) Carcinogenicity of halogenated olefinic and aliphatic hydrocarbons in mice. J. natl Cancer Inst., 63, 1433-1439

Eder, E. & Sebeikat, D. (1982) In vivo alkylating of DNA by allyl bromide. Naunyn Schmiedeberg's Arch. Pharmacol., Suppl. 319, 380

Eder, E., Neudecker, T., Lutz, D. & Henschler, D. (1980) Mutagenic potential of allyl and allylic compounds. Structure-activity relationship as determined by alkylating and direct in vitro mutagenic properties. Biochem. Pharmacol., 29, 993-998

Eder, E., Neudecker, T., Lutz, D. & Henschler, D. (1982a) Correlation of alkylating and mutagenic activities of allyl and allylic compounds: standard alkylation test vs. kinetic investigation. Chem.-Biol. Interact., 38, 303-315

Eder, E., Henschler, D. & Neudecker, T. (1982b) Mutagenic properties of allylic and α,β-unsaturated compounds: consideration of alkylating mechanisms. Xenobiotica, 12, 831-848

Göthe, R., Calleman, C.J., Ehrenberg, L. & Wachtmeister, C.A. (1974) Trapping with 3,4-dichlorobenzenethiol of reactive metabolites formed in vitro from the carcinogen vinyl chloride. Ambio, 3, 234-239

Green, T. & Hathway, D.E. (1975) The biological fate in rats of vinyl chloride in relation to its oncogenicity. Chem.-Biol. Interact., 11, 545-562

Green, T. & Hathway, D.E. (1978) Interactions of vinyl chloride with rat-liver DNA in vivo. Chem.-Biol. Interact., 22, 211-224

Greim, H., Bonse, G., Radwan, Z., Reichert, D. & Henschler, D. (1975) Mutagenicity in vitro and potential carcinogenicity of chlorinated ethylenes as a function of metabolic oxirane formation. Biochem. Pharmacol., 24, 2013-2017

Guengerich, F.P. & Strickland, T.W. (1977) Metabolism of vinyl chloride: destruction of the heme of highly purified liver microsomal cytochrome P-450 by a metabolite. Mol. Pharmacol., 13, 993-1004

Guengerich, F.P., Crawford, W.M., Jr & Watanabe, P.G. (1979) Activation of vinyl chloride to covalently bound metabolites: roles of 2-chloroethylene oxide and 2-chloroacetaldehyde. Biochemistry, 18, 5177-5182

Guengerich, F.P., Mason, P.S., Stott, W.S., fox, T.R. & Watanabe, P.G. (1981) Roles of 2-haloethylene oxides and 2-haloacetaldehydes derived from vinyl bromide and vinyl chloride in irreversible binding to protein and DNA. Cancer Res., 41, 4391-4398

Hathway, D.E. (1977) Comparative mammalian metabolism of vinyl chloride and vinylidene chloride in relation to oncogenic potential. Environ. Health Perspect., 21, 55-59

Hathway, D.E. (1980) Consideration of the evidence for mechanisms of 1,1,2-trichloroethylene metabolism, including new identification of its dichloroacetic acid and trichloroacetic metabolites in mice. Cancer Lett., 8, 263-269

Hathway, D.E. (1981) Mechanisms of vinyl chloride carcinogenicity/mutagenicity. Br. J. Cancer, 44, 597-600

Hefner, R.E., Jr., Watanabe, P.G. & Gehring, P.J. (1975) Preliminary studies of the fate of inhaled vinyl chloride monomer (VCM) in rats. Ann. N.Y. Acad. Sci., 246, 135-148

Henschler, D., Eder, E., Neudecker, T. & Metzler, M. (1977) Carcinogenicity of trichloroethylene: fact or artifact? Arch. Toxicol., 37, 233-236

Henschler, D., Hoos, W.R., Fetz, H., Dallmeier, E. & Metzler, M. (1979) Reactions of trichloroethylene epoxide in aqueous systems. Biochem. Pharmacol., 28, 543-548

Henschler, D., Elsässer, H.M., Romen, W. & Eder, E. (1984) Carcinogenicity study of trichloroethylene, with and without epoxide stabilisers, in mice. J. Cancer Res. Clin. Oncol. (in press)

Huberman, E., Bartsch, H. & Sachs, L. (1975) Mutation induction in Chinese hamster V79 cells by two vinyl chloride metabolites, chloroethylene oxide and 2-chloroacetaldehyde. Int. J. Cancer, 16, 639-644

Huntingdon Research Center (1978) 18-Month sacrifice report (vinyl bromide). HCR Report 7511-253, June 26, New York.

IARC (1982) IARC Monographs on the Evaluation of the Carcinogenic Risk of Chemicals to Humans, Suppl. No. 4, Chemicals, industrial processes and industries associated with cancer in humans (IARC Monographs, Volumes 1 to 29), Lyon, International Agency for Research on Cancer, pp. 260-260

Jones, B.K. & Hathway, D.E. (1978) the biological fate of vinylidene chloride in rats. Chem.-Biol. Interact., 20, 27-41

Kline, S.A., McCoy, E.C., Rosenkranz, H.S. & van Duuren, B.L. (1982) Mutagenicity of chloroalkene epoxides in bacterial systems. Mutat. Res., 101, 115-125

Kociba, R.J., Keyes, D.G., Jersey, G.C., Ballard, J.J., Dittenbar, D.A., Quast, J.F., Wade, C.E., Humiston, C.G. & Schwetz, B.A. (1977) Results of a two year chronic toxicity study with hexachlorobutadiene in rats. Am. Ind. Hyg. Assoc. J., 38, 589-602

Laib, J.R. (1982) Specific covalent binding and toxicity of aliphatic halogenated xenobiotics. In: Reviews on Drug Metabolism and Drug Interactions, Vol. 4, No. 1, London, Freund Publishing House Ltd., pp. 1-48

Laib, R.J. & Bolt, H.M. (1977) Alkylation of RNA by vinyl chloride metabolites in vitro and in vivo: formation of 1,N^6-ethenoadenosine. Toxicology, 8, 185-195

Laib, R.J. & Bolt, H.M. (1978) Formation of 3,N^4-ethenocytidine moieties in RNA by vinyl chloride metabolites in vitro and in vivo. Arch. Toxicol., 39, 235-240

Laib, R.J., Stöckle, G., Bolt, H.M. & Kunz, W. (1979) Vinyl chloride and trichloroethylene: comparison of alkylating effects of metabolites and induction of preneoplastic enzyme deficiencies in rat liver. J. Cancer res. Clin. Oncol., 94, 139-147

Laib, R.J., Gwinner, L.M. & Bolt, H.M. (1981) DNA alkylation by vinyl chloride metabolites: etheno-derivatives or 7-alkylation of guanine? Chem.-Biol. Interact., 37, 219-231

Leake, C.D. (1925) Historical development of surgical anaesthesia. Sci. Mon., 20, 304-308

Lee, C.C., Bhandari, J.C., Winston, J.M., House, W.B., Dixon, R.L. & Woods, J.S. (1978) Carcinogenicity of vinyl chloride and vinylidene chloride. J. Toxicol. Environ. Health, 4, 15-30

Link, B. & Dürk, H. & Frank, H. (1982) Trichloromethylation and crosslinking of membrane lipids: an exploration of carbon tetrachloride toxicity. Naunyn-Schmiedeberg's Arch. Pharmacol., Suppl. 325, 103

Maltoni, C. & Lefemine, G. (1974) Carcinogenicity bioassays on vinyl chloride. I. Research plan and early results. Environ. Res., 7, 387-405

Maltoni, C. & Tovoli, D. (1979) First experimental evidence of the carcinogenic effects of vinylidene chloride. Med. Lav., 70, 363-368

McKenna, M.J., Zempel, J.A., Madrid, E.O., Braun, W.H. & Gehring, P.J. (1978) Metabolism and pharmacokinetic profile of vinylidene chloride in rats following oral administration. Toxicol. Appl. Pharmacol., 45, 821-835

Miller, R.E. & Guengerich, F.P. (1982) Oxidation of trichloroethylene by liver microsomal cytochrome P-450: evidence for chlorine migration in a transition state not involving trichloroethylene oxide. Biochemistry, 21, 1090-1097

NCI Carcinogenesis bioassay of trichloroethylene (1976) Cas No. 79-01-6. NCI Tech. Rep. Ser., No. 2, DHEW Publication No. 76-802, Washington, DC, US Dept. Health, Education and Welfare

NCI Bioassay of tetrachloroethylene for possible carcinogenicity (1977) NCI-GG-TR-13, DHEW Publication NIH-77-813, Washington, DC

Oesch, F. & Wolf, C.R. (1984) Potential of microsomal glutathione transferases as mediators of reactions which lead to toxicity or carcinogenicity and evidence for multiple forms. Carcinogenesis (in press)

Osterman-Golkar, S., Hultmark, D., Segerbäck, D., Calleman, C.J., Göthe, R. & Ehrenberg, C.A. (1977) Alkylation of DNA and proteins in mice exposed to vinyl chloride. Biochem. Biophys. Res. Commun., 63, 259-266

Ottenwälder, H., Laib, R.J., Filser, J.G. & Bolt, H.M. (1978) Disposition and metabolic activation of vinyl bromide in rats. 7th Int. Congr. Pharmacol., Abstract 719, Paris

Parchman, L.G. & Magee, P. (1980) Production of $^{14}CO_2$ from trichloroethylene in rats and mice and a possible interaction of a trichloroethylene metabolite with DNA. Abstracts, No. 153, 19th SOT Meeting, March 9-13, 1980, Washington, DC

Pegg, D.G., Zempel, J.A., Braun, W.H. & Watanabe, P.G. (1979) Disposition of tetrachloro(^{14}C)ethylene following oral and inhalation exposure in rats. Toxicol. Appl. Pharmacol., 51, 465-474

Rampy, L.W., Quast, J.P., Leong, B.K.J. & Gehring, P.J. (1977) Results of long-term inhalation toxicity studies on rats of 1,1,1-trichloroethane and perchloroethylene formulations. Toronto, Canada, 1st Int. Congr. Toxicol. Abstract, p. 27

Rannug, U., Johansson, A., Ramel, C. & Wachtmeister, C.A. (1974) the mutagenicity of vinyl chloride after metabolic activation. Ambio, 3, 194-198

Rannug, U., Göthe, R. & Wachtmeister, C.A. (1976) The mutagenicity of chloroethylene oxide, chloroacetaldehyde, 2-chloroethanol and chloroacetic acid, conceivable metabolites of vinyl chloride. Chem.-Biol. Interact., 12, 251-263

Reichert, D. (1982) Metabolic activation of hexachloro(1,3)butadiene by mixed function oxygenase system. Naunyn Schmiedeberg's Arch. Pharmacol., Suppl. 321, 113

Reichert, D., Werner, H.W., Metzler, M. & Henschler, D. (1979) Molecular mechanism of 1,1-dichloroethylene toxicity: excreted metabolites reveal different pathways of reactive intermediates. Arch. Toxicol., 42, 159-169

Reichert, D., Neudecker, T. & Schütz, S. (1984) Mutagenicity of hexachlorobutadiene, perchlorobutenoic acid and perchlorobutenoic acid chloride. Mutat. Res. (in press)

Reitz, R.H., Watanabe, P.G., McKenna, M.J., Quast, J.F. & Gehring, P.J. (1980) Effects of vinylidene chloride on DNA synthesis and DNA repair in the rat and mouse: a comparative study with dimethylnitrosamine. Toxicol. Appl. Pharmacol., 52, 357-370

Scherer, E., van der Laken, C.J., Gwinner, L.M., Laib, R.J. & Emmelot, P. (1981) Modification of deoxyguanosine by chloroethylene oxide. Carcinogenesis, 2, 671-677

Schütz, S. & Reichert, D. (1984) Identification of novel conjugation product in hexachlorobutadiene metabolism. Naunyn-Schmiedeberg's Arch. Pharmacol., Suppl. 325, 67

Schumann, A.M., Quast, J.F. & Watanabe, P.G. (1980) the pharmacokinetics and macromolecular interactions of perchloroethylene in mice and rats as related to oncogenicity. Toxicol. Appl. Pharmacol., 55, 207-219

Stöckle, G., Laib, R.J., Filser, J.G. & Bolt, H.M. (1979) Vinylidene fluoride: metabolism and induction of preneoplastic hepatic foci in relation to vinyl chloride. Toxicol. Lett., 3, 337-342

Uehleke, H., Tabarelli-Poblawski, S., Bonse, G. & Henschler, D. (1977) Spectral evidence for 2,2,3-trichlorooxirane formation during microsomal trichloroethylene oxidation. Arch. Toxicol., 37, 95-105

Viola, P.L., Bigotti, A. & Caputo, A. (1971) Oncogenic response of rat skin, lungs and bones to vinyl chloride. Cancer Res., 31, 516-522

Yllner, S. (1961) Urinary metabolites of ^{14}C-tetrachloroethylene in mice. Nature, 191, 820

Zajdela, F., Croisy, A., Barbin, A., Malaveille, C., Tomatis, L. & Bartsch, H. (1980) Carcinogenicity of chloroethylene oxide, an ultimate reactive metabolite of vinyl chloride, and bis(chloromethyl)ether after subcutaneous administration and in initiation-promotion experiments in mice. Cancer Res., 40, 352-356

OCCURRENCE

CHAPTER 3

HALOGENATED ALIPHATIC HYDROCARBONS: USES AND ENVIRONMENTAL OCCURRENCE

L. Fishbein

Department of Human and Health Services
Food and Drug Administration
National Center for Toxicological Research
Jefferson, Arkansas 72079, USA

INTRODUCTION

In recent decades much concern has arisen with regard to the toxicological and environmental effects of a spectrum of halogenated hydrocarbons, primarily the persistent organochlorine pesticides and related derivatives; e.g., DDT, heptachlor, endrin, dieldrin, chlordane, Mirex, Kepone, toxaphene, DBCP (dibromochloropropane), chlorinated biphenyls (PCBs) and chlorinated trace impurities, such as chlorinated dibenzo-p-dioxins (PCDDs) [2,3,7,8-tetrachlorodibenzo-p-dioxin (TCDD)] and chlorinated dibenzofurans (PCDFs) (IARC, 1974b, 1975, 1977, 1978, 1979, 1982; Winteringham, 1977; Kimbrough, 1979).

This concern has now been extended to practically all of the major halogenated hydrocarbons, particularly the lower molecular weight, volatile C_1-C_3 chlorinated alkanes and alkenes, numerous members of which are carcinogenic and/or mutagenic, are produced in extremely large amounts and are extensively employed as intermediates and as major end products. Production figures suggest a growth rate for the halocarbons industry in excess of 6% per annum. Halocarbons are used as solvents (in extraction of components from foods and drugs) and in adhesives, industrial solvent blends, paint and coatings, as fumigants, degreasing agents, dry-cleaning fluids, aerosol propellants, anaesthetics, refrigerants, textile processing agents, anti-knock scavengers, flame-retardants and extinguishants, cutting-fluids, synthetic feedstuffs and as intermediates in the production of other chemicals and of textiles and plastics (IARC, 1974a,b, 1975, 1977, 1979, 1982; Fishbein, 1976, 1979a-d; Rosenkranz, 1977; Weisburger, 1977; Archer, 1979a,b; Henschler, 1980; Stolzenberg <u>et al.</u>, 1980; Marier, 1982).

A number of these agents (e.g., 1,1',2-trichloroethane, carbon tetrachloride, methylene chloride, trichloroethylene and tetrachloroethylene, Freons 11, 12, 114) have been employed in a wide range of commonly-available household items, such as degreasing agents, paint and varnish removers and brush cleaners, and can thus be an important source of air pollutants in the indoor environment (Bridbord *et al.*, 1975).

In addition, solvents such as trichloroethylene, tetrachloroethylene, 1,1,1-trichloroethane (methyl chloroform) and methylene chloride, as well as the anaesthetic gas halothane, have been abused (Garriott & Petty, 1980; Ramsey & Flanagan, 1982).

Of more recent and increasing concern is the presence of a large number of halogenated hydrocarbons, including a number of mutagenic and/or carcinogenic derivatives, in many of the global water sources as a result of industrial outfall or disinfection of water supplies by chlorination (e.g., trihalomethanes, haloacetonitriles) (NAS, 1977, 1980).

Additional potential exposure results from slow dissipation of the volatile halocarbons (e.g., Freons 11, 12, 144, methylchloride, methylene chloride, chloroform, trichloroethylene, methyl chloroform) into the global atmosphere and marine environments (Pearson & McConnel, 1975; Lovelock, 1977; Lahl *et al.*, 1981; Pearson, 1982; Miller, 1983).

The lipophilic nature of the volatile halogenated alkanes and alkenes raises an additional concern inasmuch as these agents will accumulate in the fatty tissues of biological organisms. The presence of these pollutants (e.g., chloroform, tri- and tetrachloroethylene) in samples of marine organisms has been reported (Pearson & McConnel, 1975; Dickson & Riley, 1976; Fishbein, 1979a,b; Ofstad *et al.*, 1981).

Prevalent disposal practices include deep-well disposal, burning the wastes in land-based incinerators and burning them at sea (Miller, 1983).

The general class of halogenated alkanes and alkenes therefore give cause for concern on both toxicological and environmental grounds, due to their extensive production volumes, use patterns and dissipative nature, and to the broad potential exposure of large segments of the population, including worker, consumer and general public.

PRODUCTION AND USES

Chlorinated hydrocarbon derivatives are produced by a number of basic processes, such as direct-addition chlorination, hydrochlorination, oxychlorination, dehydrochlorination and chlorinolysis. Aliphatic hydrocarbon feedstocks for chlorination reactions include methane (natural gas), ethane, ethylene, propylene and propane (natural gas); methanol and various waste steams may also be used (EPA, 1974; Ahlstrom & Steele, 1979; Anthony, 1979; Archer, 1979a,b; DeShon, 1979a,b; Fishbein, 1979a-d; Keil, 1979; McNeill, 1979; Pearson, 1982). Typical manufacturing processes for C_1 and C_2 chlorohydrocarbons (e.g., methyl chloride, methylene chloride, chloroform, carbon tetrachloride, ethylchloride, 1,1-dichloroethane, 1,2-dichloroethane, vinyl chloride, vinylidene chloride, trichloroethylene, tetrachloroethylene, 1,1,2-trichloroethane, 1,1,1-trichloroethane and tetrachloroethane) are shown in Figure 1 (Archer, 1979a).

FIG. 1. MANUFACTURING PROCESSES FOR C_1 AND C_2 CHLOROHYDROCARBONS (Archer, 1979a)

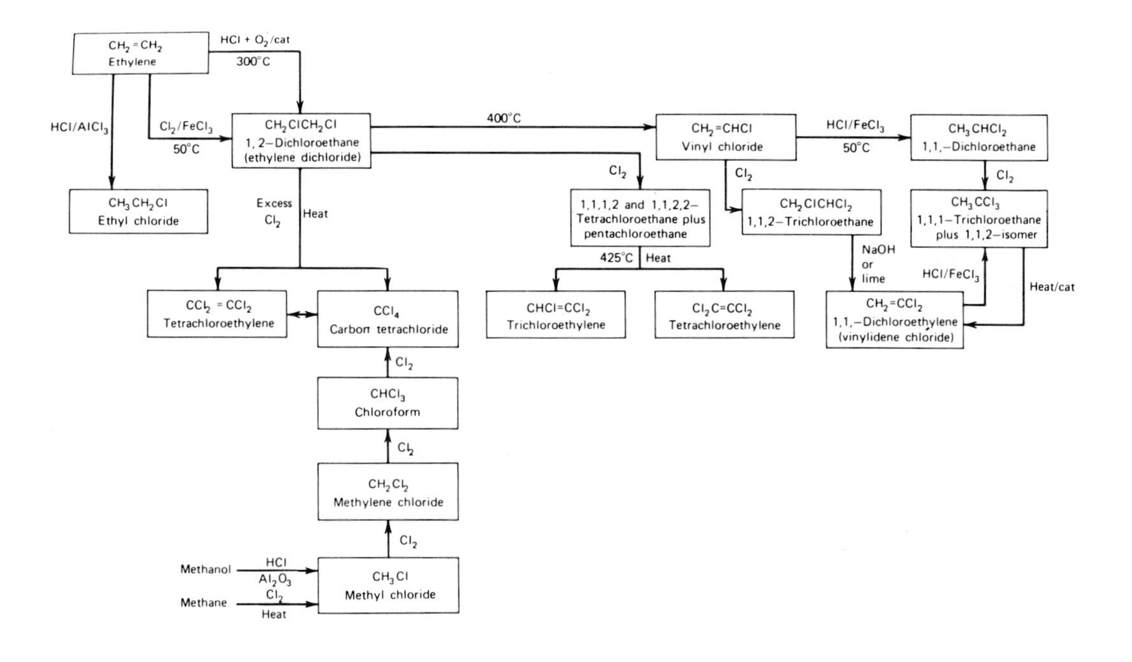

Progressive chlorination of a hydrocarbon molecule yields both liquids and solids of increasing nonflammability, density and viscosity, as well as enhanced solvent power for a large number of inorganic and organic materials. In addition, water solubility, dielectric constant and specific heat exhibit a progressive decrease with increasing chlorine content. All chlorinated hydrocarbons undergo pyrolysis at elevated temperatures, with the formation of hydrogen chloride. While olefinic chlorinated derivatives are oxidized in the presence of UV light to yield hydrogen chloride, phosgene and chlorinated acetyl chlorides, saturated aliphatic chlorine derivatives are generally quite stable to oxidation.

Inhibitors, such as antioxidants, acid acceptors and metal stabilizers, are added to minimize degradation. For example, alcohols and amines are often added to an oxidation-sensitive solvent (e.g., 1,1,2-trichloroethane) to minimize this mode of degradation. Epoxides have been added as acid acceptors to neutralize hydrogen chloride, which is a major product of degradation of the chlorinated hydrocarbons (Archer, 1979a). Table 1 lists the global production and uses of the halogenated aliphatic hydrocarbons of major commercial importance. The production volumes listed are those of the major producers, e.g., United States, Western Europe and Japan (no figures are available from Eastern Europe or China). The figures are rounded off to 1978-1980 levels (Pearson, 1982).

The 1982 U.S. production capacities for a number of major chlorinated derivatives are listed in Table 2 (SRI, 1983). A number of C_2 chlorinated solvents (primarily trichloroethylene, perchloroethylene and 1,1,1-trichloroethane), although still consumed in major amounts, have exhibited considerable variations in production in recent years, due mainly to increasing concern over possible adverse effects on humans and the environment. World capacity for production of these three solvents is about 3.1 million metric tons. Western Europe has approximately 50% of the total capacity, while the United States and Japan have 40% and 10%, respectively. In 1980 the world industry operated at only about 55% of capacity and produced some 1.7 million metric tons of these solvents (40% perchloroethylene, 25% trichloroethylene and 35% 1,1,1-trichloroethane (SRI, 1982). Of the total world capacity for ethylene dichloride (EDC) in 1981, the U.S. held about 50% and Western Europe about 26% (SRI, 1981).

Table 1. Production and uses of the major halogenated aliphatic hydrocarbons of commercial importance (Pearson, 1982)

Name	Synonym	World production (million tonnes/year)	Major uses
Chloromethane	Methyl chloride	0.4	Intermediate, solvent
Dichloromethane	Methylene chloride	0.5	Solvent
Trichloromethane	Chloroform	0.25	Intermediate, solvent
Tetrachloromethane	Carbon tetrachloride	1.0	Intermediate, fumigant
Bromomethane	Methyl bromide	0.02	Fumigant
Chloroethene	Vinyl chloride	10.00	Monomer for plastics, intermediate
1,1-Dichloroethene (asym)	Vinylidene chloride	0.2	Monomer for plastics
Trichloroethene	Trichloroethylene	0.6	Solvent
Tetrachloroethene	Perchloroethylene	1.1	Solvent
Chloroethane	Ethyl chloride	0.4	Solvent, intermediate
1,2-Dichloroethane	Ethylene dichloride	13.0	Intermediate
1,1-Dichloroethane	Ethylidene chloride	0.5	Intermediate
1,1,1-Trichloroethane	Methyl chloroform	0.6	Solvent
1,2-Dibromoethane	Ethylene dibromide	0.25	Antiknock scavenger
3-Chloropropene-1	Allyl chloride	0.5	Intermediate
3-Chloropropane, 1-2 epoxide	Epichlorohydrin	0.5	Monomer for resin

Table 2. U.S. production capacity of a number of major chlorinated derivatives in 1982 (SRI, 1983)

Compound	Production (million pounds)
Methyl chloride	690
Chloroform	515
Carbon tetrachloride	1 213
Methylene chloride	960
1,2-Dichloroethane	20 295
1,1,1-Trichloroethane	1 300
Trichloroethylene	320
Tetrachloroethylene	1 085
Epichlorohydrin	640
Fluorocarbons (including F-11, F-12, F-22)	1 166

Methyl chloride is used mainly in the production of tetramethyl lead (as an antiknock agent), silicones, synthetic rubber and methylcellulose, and as a general methylating agent. Its utility as a refrigerant and in extractant applications is of secondary importance. Approximately 31 000 individuals are estimated by the National Institute of Occupational Safety and Health (NIOSH) to be exposed to methyl chloride (Clement Associates, 1977; Ahlstrom & Steele, 1979; Fishbein, 1979b).

Methylene chloride demand has grown in the past decade, competing with trichloroethylene and perchloroethylene largely because of increased recognition of their toxicity. Methylene chloride is extensively used for solvent degreasing, as a paint remover, in aerosol sprays, as a solvent for the extraction of naturally-occurring, heat-sensitive substances, as a substitute for trichloroethylene for decaffeinating coffee, as a process solvent in the pharmaceutical industry in the manufacture of steroids, antibiotics and vitamins. Its use in the urethane foam industry as an auxilliary blowing agent for flexible foams and as a substitute for fluorocarbons in aerosol products is increasing. Other applications include low-pressure refrigerants and as a low-temperature, heat-transfer medium (Anthony, 1979; Fishbein, 1979b). It is estimated that from 1976 on, about 70 000 U.S. employees were potentially exposed to a working environment containing methylene chloride (NIOSH, 1976).

Chloroform was one of the first organic chemicals produced on a large scale in the U.S., with production commencing in 1900 (DeShon, 1979a). Chloroform is used primarily in the manufacture of chlorodifluoromethane refrigerants (F-22) and aerosol propellants, and as a raw material for polytetrafluoroethylene resins (e.g., PTFE, Teflon). Chloroform was used chiefly as an anaesthetic and in pharmaceutical preparations prior to World War II, and in toiletries (e.g., mouthwashes, dentifrices), hair tinting and permanent-waving formulations untill recently. It has also been used as a fumigant and insecticide. NIOSH estimates that 40 000 persons are exposed occupationally to chloroform in the United States (NIOSH, 1976b; DeShon, 1979a; Fishbein, 1979b).

Carbon tetrachloride was one of the first organic chemicals to be produced on a large scale, with production commencing in Germany in the 1890's (DeShon, 1979b). Carbon tetrachloride is principally employed for the production of chlorofluoromethanes, e.g., dichlorofluoromethane (CF_2Cl_2) (F-12) and trichlorofluoromethane ($CFCl_3$) (F-11) for refrigeration, aerosol and blowing-agent markets. Carbon tetrachloride was formerly used extensively for metal degreasing and as a dry-cleaning fluid, fabric-spotting fluid, fire-extinguisher fluid, grain fumigant and reaction medium (DeShon, 1979b; Fishbein, 1979b).

Ethylene dichloride (EDC; 1,2-dichloroethane) is the principal alkane halide of industrial utility. Production of EDC in the U.S. increased from about 510 million pounds in 1955 to almost 8 billion pounds in 1972. this 16-fold increase was mostly due to increased vinyl chloride monomer (VCM) production, for which EDC is one of the basic raw materials. Current U.S. consumption of EDC is about 10 billion pounds and it is the 15th largest-volume U.S. commercial chemical. The worldwide capacity for EDC production is approximately 51 billion pounds. Estimated U.S. domestic consumption in 1976 was as follows: 86% as an intermediate for vinyl chloride; about 3% each as an intermediate for 1,1,1,-trichloroethane, methyl chloroform and ethylene-imines; 2% each as an intermediate for perchloroethylene, 1,1-dichloroethane and trichloroethylene; and as a lead scavenger for motor fuels. The yearly growth-rate of EDC production in the United States has been 9% from 1969 to 1979. The use of EDC as an additive in tetraethyl lead antiknock mixtures will decrease due to increased marketing of unleaded gasolines (Archer, 1979b; Fishbein, 1979c, 1980). Baier (1978) estimated that approximately 2 million workers in the U.S. may receive some exposure to EDC, with perhaps 200 000 receiving a substantial exposure primarily due to its use as a solvent in textile cleaning and metal degreasing and in certain adhesives, as well as component of fumigants. However, NIOSH estimated in 1976 that approximately 18 000 people are potentially exposed to EDC in their working environments (NIOSH, 1976c).

The U.S. production of ethylene dibromide (EDB; 1,2-dibromoethane) increased from an estimated 64 million pounds in 1940 to a peak of 332, primarily due to increased consumption of gasoline containing EDB as an additive, which has always been its greatest single use (NIOSH, 1977). Although the recent U.S. production of EDB was approximately 250 million pounds, down 32% from the peak of 332 million pounds, it remains the leading bromine-based chemical. The outlook is for further sharp declines in demand, possibly by as much as 75% by 1980, because of diminishing use of lead additives in gasoline and, in recent months, the restriction or elimination of its use in pesticides by the U.S. Environmental Protection Agency. EDB was a component of more than 100 pesticides registered with the EPA and was primarily used as a soil fumigant on a variety of vegetable, fruit and grain crops, and for disinfecting fruits, vegetables, grains, tobacco and seeds in storage (EPA, 1970). An estimated 5 million pounds of EDB were used in the U.S. in 1975 as a fumigant.

NIOSH estimated in 1977 that approximately 9 000 employees (chiefly manufacturing, formulating and fumigating) are potentially exposed to EDB in the workplace and that 650 000 gasoline station attendants are also exposed (NIOSH, 1977b). No estimate can currently be given concerning the number of motorists who are potentially exposed to EDB during "self-serve" operations at the gas pump.

Methyl chloroform (1,1,1-trichloroethane) is used in the U.S. principally (about 70%) as a cold solvent for degreasing metals and cleaning electrical and electronic equipment. The commercial metal-cleaning grades contain added inhibitors (e.g., epoxides, glycol diesters, nitroaliphatic hydrocarbons, 1,4-dioxane, morpholine or a variety of alcohols) that make usage acceptable for all common metals, including aluminum. It is also widely employed as a solvent for various greases, oils, tars and waxes and a broad spectra of organic material (Archer, 1979b; Fishbein, 1979b). A number of consumer aerosol products also contain methyl chloroform; these include cleaners (oven and spot removers, containing 25 to 70% methyl chloroform) waxes and polishes, automotive and specialty products (Aviado _et al._, 1976). NIOSH estimated that approximately 100 000 workers are exposed to methyl chloroform (Fishbein, 1979b). Historically, methyl chloroform has been a common substitute for carbon tetrachloride (Aviado _et al._, 1976). It is also being increasingly substituted for trichloroethylene in solvent applications, largely because the latter is a suspect carcinogen. Methyl chloroform also contributes less to smog formation and is relatively inert in the troposphere, compared to trichloroethylene (Fishbein, 1979b).

Trichloroethylene (TCE, 1,1,2-trichloroethylene) production began in Germany in 1920 and in the United States in 1925 and was stimulated by improvements in metal degreasing techniques during the 1920's and by the growth of dry-cleaning estalbishments during the 1930's. Although the market expanded steadily until 1970 (with TCE accounting for 82% of all the chlorinated solvents used in vapor degreasing), the utility of trichloroethylene has been under increasing attack due to its toxic properties and its danger as an atmospheric pollutant. Approximately 90% of TCE (345 million pounds) consumed in the United States in 1974 was for vapor degreasing and cold cleaning of fabricated metal parts and 6% (25 million pounds) was employed as a chain terminator for polyvinyl chloride production. Previously, it was widely employed as a solvent in the textile industry, as an ingredient in printing inks, lacquers, varnishes and adhesives, in the dry-cleaning of fabrics and as an extractant in food processing (e.g., for decaffeinating coffee). A pharmaceutical grade of trichloroethylene has been employed as a general anaesthetic in surgical, dental and obstetrical procedures. It should be noted that since TCE is slowly decomposed (autooxidized) by air to give oxidation products which are acidic and corrosive, stabilizers, (e.g., epichlorohydrin and butylene oxide) are added to all commercial grades of trichloroethylene to scavenge any free HCl and $AlCl_3$ (Fishbein, 1979a; McNeil, 1979). Largely because of its solvent properties, TCE has been incorporated in a number of consumer products (e.g., cleansers for automobiles, spot removers, rug cleaners, disinfectants and deodorants) (Lloyd et al., 1975). The number of U.S. workers exposed to trichloroethylene has been estimated by Lloyd et al. (1975) to be about 283 000.

Tetrachloroethylene (perchloroethylene) was manufactured before World War I in the United Kingdon and Germany, and in the United States since 1925. the major application of tetrachloroethylene is in dry-cleaning, with approximately 80% of all dry-cleaners using this solvent as their primary cleaning agent. It's estimated use pattern (%) in the United States in 1976 was dry-cleaning (66), textile processing (13), metal degreasing (13), fluorocarbon manufacture (3) and miscellaneous (5). Tetrachloroethylene is also used as a solvent in the manufacture of rubber solutions, paint removers, printing inks, fats, oils, silicones and sulfur and as a heat-transfer medium (Fishbein, 1979a; Keil, 1979). NIOSH estimated in 1978 that some 500 000 workers were at risk of exposure to tetrachloroethylene and also noted that over 20 000 dry-cleaning establishments and a large number of other industries use or manufacture tetrachloroethylene (Anon, 1978b).

Since their introduction approximately 40 years ago as refrigerants, and later as propellants for aerosol products, the fluorocarbons (chlorofluoromethanes) have generally been considered to possess an extremely low order of biological activity (Clayton, 1967) and, as a consequence, their production and diversity of use have increased enormously. Between 1960 and 1970, world production was estimated to have grown exponentially, with a doubling time of 3.5 years (Lovelock et al., 1973; Crutzen, 1974). These chemicals (mostly known as "Freons"), unlike carbon tetrachloride, have no known natural sources or sinks in the troposphere and possess relative chemical inertness and high volatility.

Fluorocarbons are made commercially by a number of procedures including: a) the electrolysis of hydrocarbons in anhydrous hydrogen fluoride; b) reaction of acetylene or olefins and hydrogen fluoride, or chlorocarbons and anhydrous hydrogen fluoride, in the presence of an antimony fluoride catalyst (Smart, 1980).

The major fluorocarbons of environmental as well as toxicological concern are trichlorofluoromethane and dichlorofluoromethane, which have been used almost exclusively as aerosol propellant gases (approximately 50-60%) and refrigerants (approximately 20-30%), but have also found use as blowing agents, solvents and fire extinguishers. Other fluorocarbons (e.g., F-22) are used as feedstocks for fluorocarbon resins (Fishbein, 1979b).

World production of trichlorofluoromethane (F-11) and dichlorodifluoromethane (F-12) (excluding the USSR and Eastern Europe) was 1.7 billion pounds in 1973, which represents an increase of 11% compared with 1972. Approximately half of that production and use occurred in the United States. The total

U.S. production of fluorocarbons has been doubling every 5 to 7 years since the early 1950's. The U.S. production of F-11 and F-12 in 1974 was estimated to have been approximately 860 million pounds and it is believed that the market will continue to grow by about 5 to 6% a year (Lovelock et al., 1973). By 1975 it was estimated that approximately 13.8 billion pounds of F-11 and F-12 had been produced in the world (Council on Environmental Quality, 1976). Approximately 3 billion aerosol cans containing F-11 and F-12 were sold yearly in the U.S. for use with the following products (in millions of units, for 1972): personal products (e.g., shaving creams, cosmetics, perfumes, deodorants, anti-perspirants, hair-sprays), 1 490; household products (e.g., window cleaners, air freshners, oven cleaners, furniture polishes), 699; coatings and finishing products (e.g., paint spray), 270; insect sprays, 135; lubricants and degreasers, 100; automotive products, 76. Miscellaneous uses have included mold releases, silicone sprays and as additives in some foods (Council on Environmental Quality, 1975).

It was estimated in 1975 that approximately 4 000 people were directly engaged in fluorocarbon production, sales and research by 6 companies at 15 plants (Council on Environmental Quality, 1975).

Dibromochloropropane (DBCP; 1,2-dibromo-3-chloropropane) is produced commercially by the addition of bromine to allyl chloride and is used mainly as a soil fumigant and nematocide and as an intermediate in organic synthesis. About 12 million pounds of DBCP were used in 1972. In 1969, U.S. production was 3.9 million kg. DBCP is also produced in the Benelux, France, Italy, Spain and Switzerland and the annual production in these countries was estimated to be 3-30 million kg annually (IARC, 1977). In 1971, 1.6 million kg of DBCP were employed on crops in the U.S. and in 1975, 285 thousand kg were used in California alone (Fishbein, 1979c). Approximately two to three thousand employees in manufacturing and formulating facilities have been exposed to DBCP. About 80 formulators have had labels registered with EPA for the approximately 160 products which contain DBCP (OSHA, 1977).

Haloethers, primarily alpha chloromethyl ethers, are alkylating agents which are giving rise to increasing concern due to the establishment of a causal relationship betweeen occupational exposure to bis(chloromethyl)ether (BCME) and chloromethylmethylether (CMME) and lung cancer in the United States, the Federal Republic of Germany and Japan. These compounds have been widely used in industry as chloromethylation agents for the preparation of anion-exchange resins, the formation of water repellants and other textile-treating agents, the manufacture of polymers and as solvents for polymerization reactions. Anion-exchange resins (modified polystyrene resins which are chloromethylated, then treated with a tertiary amine or with a polyamine) have been produced in the United States, France, Federal Republic of Germany, German Democratic Republic, Italy, The Netherlands, United Kingdom, USSR and Japan (no data are available on the quantities produced) (Fishbein, 1979c; IARC, 1974).

BCME can be produced by saturating a solution of paraformaldehyde in cold sulfuric acid with hydrogen chloride. BCME has been primarily produced in the U.S. as a chemical intermediate. CMME can be produced by reaction of methanol, formaldehyde and anhydrous hydrogen chloride. It should be noted that commercial grades of CMME can be contaminated with 1-8% BCME (Fishbein, 1979c; IARC, 1974).

Allyl chloride (3-chloro-1-propene) is the most important of all commercial allyl compounds and has been used as a monomer in the production of various plastics and resins that are used per se, or incorporated into surface coatings, adhesives, etc. A number of commercially-important compounds, such as glycerol, epichlorohydrin and allyl alcohol, are made directly from allyl chloride. The production of allyl chloride in the U.S. in 1973 was estimated to be about 300 million pounds. NIOSH estimates that approximately 5 000 workers are potentially exposed to allyl chloride during its manufacture or use (Fishbein, 1979a).

LOSSES TO THE ENVIRONMENT

It is generally acknowledged that even in a well-operated, integrated, production unit, some losses of both original reactants and products will always occur. Such inputs can be significant for compounds which are mainly employed as intermediates for further synthesis, e.g., methyl chloride, chloroform, carbon tetrachloride, vinyl chloride, vinylidene chloride, epichlorohydrin and chloroprene. It has been estimated that losses of between 0.5 and 2% of raw material and finished product occur, depending on the age of the plant and the nature of the process (EPA, 1973, 1974; Pearson, 1982). Another principal source of halocarbon input to the environment involves those materials which are extensively employed as solvents or fumigants. These include methylene chloride, methyl bromide, trichloroethylene, tetrachloroethylene, methyl chloroform, chloroform, carbon tetrachloride, ethyl chloride, 1,2-dichloroethane (EDC) and 1,2-dibromoethane (EDB).

The U.S. Environmental Protection Agency (EPA) estimated that about 163 million pounds of EDC entered the environment in 1974 from its use in the U.S. alone (Anon, 1978a). Best estimates of current emissions indicate that approximately 11 000 - 44 000 metric tons of EDC are emitted annually from production and process facilities (EPA, 1979). Losses of EDC to the environment arise principally from vapours released during primary production or during end-product manufacture, during dispersive applications and as a contaminant in waste water and in waste solids, principally EDB tars - a complex mixture consisting chiefly of chlorinated aliphatic hydrocarbons, e.g., approximately 33% EDC, 1,1,2-trichloroethane and about 0.06% VCM monomer (Jensen et al., 1975). It should be noted that the composition of EDC tar varies not only from one factory to another, but also from time to time within the same factory. The EDC content of EDC tars from VCM production is estimated at about 60 million pounds annually. In the U.S., disposal of EDC tar is usually accomplished by burial in landfills or by incineration. Because of its volatility, there is a probability that EDC in buried wastes may eventually leak into the atmosphere (EPA, 1979; Fishbein, 1980). Dispersive uses of EDC are generally considered to result in the release of all the EDC to the environment. Use of EDC as a solvent or fumigant accounts for the release of approximately 5 000 metric tons annually, while auto emissions and the use of EDC as an intermediate in the synthesis of other organic compounds are estimated to release about 4 000 metric tons annually. The use of EDC as a grain fumigant may account for annual emissions of an additional 500 metric tons (EPA, 1979).

The chief sources of 1,2-dibromoethane (EDB) emissions are from automotive sources via evaporation from the fuel tank and carburetor of cars operated on leaded fuel. Emissions from these sources were estimated to range from 2 to 25 mg/day for 1972-1974 model cars in the U.S. (EPA, 1975). (The concentration of EDB in tetraethyllead anti-knock mixtures varies and can be present in amounts of approximately 18% by weight (2.8 g/L); aviation gasoline anti-knock mixtures can contain about 36% EDB by weight (IARC, 1977).

Emissions of commercial organic solvent vapors into the atmosphere have been increasing dramatically in the last decade. The loss of trichloroethylene and tetrachloroethylene to the global environment in 1973 was estimated to be over 1 million tons each (McConnell et al., 1975). Emissions of these chlorocarbons occur principally from 3 sources; production, transportation and consumption. Estimated emission from trichloroethylene production are 57 pounds emitted/ton produced (Garner & Dzierlenca, 1976; EPA, 1977). The major source of emission of trichloroethylene arose from its use as a solvent in open-top vapor degreasers (EPA, 1977a). Assuming that 55% of the vapor-degreasing operations in 1974 used trichloroethylene, the total U.S. emissions would have been approximately 121 000 tons, or roughly 70% of the total amount of trichloroethylene used in metal-cleaning operations (EPA, 1979; SRI, 1975).

Losses of carbon tetrachloride to the global environment are considerable. It was estimated by Singh et al. (1976) that, by 1973, accumulated world-wide emissions amounted to about 2.5 million metric tons and that this had grown at a rate of 60 000 metric tons/year for at least 30 years. McConnell et al. (1975) estimated the 1974 loss of carbon tetrachloride to the global environment to be of the order of 1 million tons. The occurrence of carbon tetrachloride in the atmosphere cannot be accounted for from direct product emission data (Lillian et al., 1975; Singh et al., 1976; Neeley, 1977).

The major source of fluorocarbon release to the atmosphere arises from their use as propellants; the annual loss was estimated to be in excess of 650 million pounds in the early 1970's (McCarthy, 1974). Lesser amounts are released from foaming agents, refrigerants, fire extinguishers and solvents. Approximately 1% of fluorocarbons are lost during production and 1% during transportation and storage, amounting to approximately 10 million pounds/year in each case, in the U.S. (Council on Environmental Quality, 1975). Atmospheric emission rates for trichlorofluoromethane alone have increased from an average of 0.14 billion pounds per year between 1961 and 1965 to 0.51 billion pounds per year between 1971 and 1972, with the United States and Canada accounting for 44% of world emissions (Lillian et al., 1975). The yearly world-wide emission of dichlorodifluoromethane into the troposphere in 1974 was estimated to be nearly 1 billion pounds (Lillian et al., 1975). Fluorocarbons released at ground level are estimated to take about 10 years to reach stratospheric heights at which they are photolyzed. Fluorocarbons may remain in the atmosphere for 40-150 years and concentrations can be expected to reach 10 to 30 times present levels (Molina & Roland, 1974; Roland & Molina, 1974).

It was suggested in 1975 that fluorocarbon release to the environment to date may have resulted in a reduction in average ozone concentration of between 0.5 and 1% (possibly as large as 2%) and eventually may result in as much as 1.3 to 3% reduction in the equilibrium ozone concentration (Council on Environmental Quality, 1975). Model calculations predict that if release of fluorocarbons were to continue at the 1972 rate, a maximum reduction of about 1% in the equilibrium ozone concentration would be expected after several decades (Council on Environmental Quality, 1975).

Table 3 depicts the U.S. production and estimated annual release rate of 17 halocarbons based on 1972 U.S. Tariff Commission reports (Stephenson, 1977).

An important additional source of halocarbons that has engendered increasing concern results from the use of chlorine for the disinfection of drinking water. This has lead to a significant increase in the level of trihalomethanes (THM), the most important of which are chloroform, bromoform, bromodichloromethane and chlorodibromomethane. The levels detected range from 1 μg/L to 100 μg/L and above (as total THM), of which typically 80% will be chloroform (EPA, 1977b; NAS, 1977, 1980).

For a population of 1 million with a water use of 100 L/day/person and a mean chloroform level of 20 μg/L, the annual production of chloroform would amount to about 7 000 tons (Pearson, 1982). A suggested additional source of chloroform can arise from the extensive use of chlorine in the bleaching of paper pulp. A conversion efficiency as low as 6% in the bleaching process would supply a global source of chloroform (via a haloform reaction) of the magnitude of 3×10^5 tons/year (Yung et al., 1976).

An additional source of halocarbons can arise from their formation in the environment. For example, decomposition of tetrachloroethylene could provide atmospheric chloroform via the photolysis of dichloroacetylchloride (Yung et al., 1976). Moreover, chloroform, as well as carbon tetrachloride, may arise naturally via the reaction between chlorine and methane in the atmosphere (Lovelock et al., 1973; McConnell et al., 1975). Methyl chloride has been suggested to be a major product of the combustion of agricultural waste and of slash-and-burn land clearance (Palmer, 1976). It has been estimated that as much as 5×10^6 tonnes/year of methyl chloride could be emitted from all fires throughout the world (Pearson, 1982). It has also been suggested that large quantities of methyl chloride (as high as 40 000 tons per annum) are produced as a result of the activities of marine algae, possibly via interchange reactions between chloride and methyl iodide, methyl bromide or dimethyl sulfide, all of which have been associated with the metabolism of specific algae (Pearson, 1982).

Table 3. Annual production and estimated annual release rate of seventeen halocarbons in the United States[a] (Stephenson, 1977)

Compound	Production (millions of pounds)	Estimated annual release rate (millions of pounds)
Methyl chloride	453.5	16.7
Ethyl chloride	575.5	34.6
Methyl bromide	24.6	22.4
Methylene chloride	471.3	366.9
Chloroform	234.7	38.7
Carbon tetrachloride	967.7	60.0
1,2-Dichloroethane	8 600.0	458.0
1,2-Dibromoethane	315.5	304.0
1,1,1-Trichloroethane (methyl chloroform)	440.7	284.5
Trichloroethylene	426.7	429.5
Tetrachloroethylene (perchloroethylene)	734.2	562.0
Chloroprene	402.0	6.0
Vinylidene chloride	60.0	0.9
Hexachlorobutadiene	8.0	7.3
Allyl chloride	295.0	4.4
Dichlorodifluoromethane (F-12)	439.2	445.8
Trichlorofluoromethane (F-11)	299.6	274.1

[a] Based on 1972 U.S. Tariff Commission Reports

A summary estimate of total annual sources of a number of the major chlorinated aliphatic compounds is given in Table 4 (Pearson, 1982).

Table 4. Estimated total inputs to the environment (kilo-tons per annum) of some chlorinated aliphatic compounds (Pearson, 1982).

Compound	Lost after use	Production losses	Production by-products	Water chlorination	Natural origin	Total
Methyl chloride	-	4	-	-	5 000+	5 000+
Methylene chloride	500	-	-	-	-	500
Chloroform	5	5	-	10?	?	20+
Carbon tetrachloride	10	20	-	-	?	50+
Vinyl chloride	-	200	-	-	?	200
Vinylidene chloride	-	2	-	-	-	2
Trichloroethylene	-	600	-	-	-	600
Perchloroethylene	-	1 100	-	-	-	1 100
Ethyl chloride	-	15	-	-	-	15
Ethylene dichloride (EDC)	-	1 200	-	-	-	1 200
Methyl chloroform	-	600	-	-	-	600
Hexachloroethane	-	5	-	-	-	5
Allyl chloride	-	5	-	-	-	5
Chloroprene	-	5	-	-	-	5
Propylene dichloride	-	-	50	-	-	50
Hexachlorobutadiene	-	-	10	-	-	10

REFERENCES

Ahlstrom, R.C. & Steele, J.M., (1979) Methyl chloride. In: Kirk-Othmer Encyclopedia of Chemical Technology, Vol. 5, 3rd ed., New York, Wiley-Interscience, pp. 677-685

Anon (1978a) NCI finds ethylene dichloride to be carcinogenic. Toxic Mater. News, 5, 284-285

Anon (1978b) NIOSH issues intelligence bulletin citing carcinogenicity of perchloroethylene. Chem. Regul. Reporter, 1, 1539

Anthony, T. (1979) Methylene chloride. In: Kirk-Othmer, Encyclopedia of Chemical Technology, Vol. 5, 3rd ed., New York, Wiley-Interscience, pp. 686-693

Archer, W.L. (1979a) Chlorocarbons and chlorohydrocarbons - Survey. In: Kirk-Othmer Encyclopedia of Chemical Technology, Vol. 5, 3rd ed., New York, Wiley-Interscience, pp. 668-676

Archer, W.L. (1979b) Other chloroethanes. In: Kirk-Othmer Encyclopedia of Chemical Technology, Vol. 5, 3rd ed., New York, Wiley-Interscience, pp. 772-742

Aviado, D.M., Simaan, J.A., Zakhari, S. & Ulsamer, A.G. (1976) Methyl chloroform and Trichloroethylene in the Environment, Cleveland OH, CRC Press

Baier, E.J. (1978) Statement on Ethylene Dichloride before the Subcommittee on Oversight and Investigations House Committee on Interstate and Foreign Commerce, Washington DC, January 23

Bridbord, K., Brubaker, P.E., Gay, B., Jr & French, J.G. (1975) Exposure to halogenated hydrocarbons in the indoor environment. Environ. Health Perspect., 11, 215-220

Clayton, J.W. (1967) Fluorocarbon toxicity and biological action. Fluor. Chem. Rev., 1, 197-252

Clement Associates (1977) Information Dossiers on Substances Designated by TSCA Interagency Testing Committee. Report on Contract NSF-C-ENV-77-15417, Washington, DC

Council on Environmental Quality (1976) Fluorocarbons and the Environment, Council on Environmental Quality, Federal Council for Science and Technology, Washington, DC (June)

Crutzen, P.J. (1974) Estimates of possible variations in total ozone due to natural causes and human activities. Ambio, 3, 201-210

DeShon, H.D. (1979a) Chloroform. In: Kirk-Othmer Encyclopedia of Chemical Technology, Vol. 5, 3rd ed., New York, Wiley-Interscience, pp. 693-703

DeShon, H.D. (1979b) Carbon Tetrachloride. In: Kirk-Othmer Encyclopedia of Chemical Technology, Vol. 5, 3rd ed., New York, Wiley-Interscience, pp. 704-714

Dickson, A.H. & Riley, J.P. (1976) The distribution of short-chain halogenated aliphatic hydrocarbons in some marine organisms. Mar. Pollut. Bull., 7, 167-169

EPA (1970) U.S. Environmental Compendium of Registered Pesticides, Washington DC, U.S. Environmental Protection Agency, pp. III-E-0.1 - III-E-9.5

EPA (1973) Major Organics Products, Development Document Effluent Limitation Guidelines, EPA 440/1-73/009, Washington DC, U.S. Environmental Protection Agency

EPA (1974) Organic chemicals Industry, Phase II, Development Document Effluent Limitation Guidelines, Washington DC, U.S. Environmental Protection Agency

EPA (1975) Sampling and Analysis of Selected Toxic Substances, Washington DC, U.S. Environmental Protection Agency

EPA (1977a) Status Assessment of Toxic Chemicals, No. 13, Trichloroethylene, Cincinnati OH, Washington DC, U.S. Environmental Protection Agency

EPA (1977b) Organic Compounds Identified in Drinking Water in the United States, Cincinnati OH, Washington DC, U.S. Environmental Protection Agency (April 1)

EPA (1979) Ethylene dichloride, Environmental Criteria and Assessment Office, Research Triangle Park, NC, U.S. Environmental Protection Agency (June 22)

Fishbein, L. (1976) Industrial mutagens and potential mutagens. I. Halogenated derivatives. Mutat. Res., 32, 267-308

Fishbein, L. (1979a) Potential halogenated industrial carcinogenic and mutagenic chemicals. I. Halogenated unsaturated hydrocarbons. Sci. Total Environ., 11, 111-161

Fishbein, L. (1979b) Potential halogenated industrial carcinogenic and mutagenic chemicals. II. Halogenated saturated hydrocarbons. Sci. Total Environ., 11, 163-195

Fishbein, L. (1979c) Potential halogenated industrial carcinogenic and mutagenic chemicals. III. Alkane halides, alkanols and ethers. Sci. Total Environ., 11, 223-257

Fishbein, L. (1979d) Potential Industrial Carcinogens and Mutagens, Amsterdam, Elsevier Scientific Publ., pp. 165-265

Fishbein, L. (1980) Production, uses and environmental fate of ethylene dichloride and ethylene dibromide. In: Ames, B., Infante, P. & Reitz, R., eds, Banbury Report No. 5, Ethylene Dichloride: A Potential Health Risk, Cold Spring Harbor Laboratory, Bar Harbor, NY, pp. 227-238

Garner, D.N. & Dzierlenca, P.S. (1976) Organic Chemical Producers Data Base, Vol. II, Final Report for Contract No. 68-02-1319, Task Number 15, Austin TX, Radian Corporation (August)

Garriott, A.J. & Petty, C.S. (1980) Death from inhalant abuse: toxicological and pathological evauation of 34 cases. Clin. Toxicol., 16, 305-315

Henschler, D. (1977) Metabolism and mutagenicity of halogenated olefins - a comparison of structure and activity. Environ. Health Perspect., 21, 61-64

IARC (1974a) Monographs on the Evaluation of the Carcinogenic Risk of Chemicals to Man, Vol. 4, Some Aromatic Amines, Hydrazine and Related Substances, N-Nitroso Compounds and Miscellaneous Alkylating Agents, Lyon, International Agency for Research on Cancer, pp. 231-245

IARC (1974b) Monographs on the Evaluation of the Carcinogenic Risk of Chemicals to Man, Vol. 5, Some Organochlorine Pesticides, Lyon, International Agency for Research on Cancer

IARC (1975) Monographs on the Evaluation of the Carcinogenic Risk of Chemicals to Man, Vol. 8, Some Aromatic Azo Compounds, Lyon, International Agency for Research on Cancer

IARC (1977) Monographs on the Evaluation of the Carcinogenic Risk of Chemicals to Man, Vol. 15, Some Fumigants, the Herbicides 2,4-D and 2,4,5-T, Chlorinated Dibenzodioxins, and Miscellaneous Industrial Chemicals, Lyon, International Agency for Research on Cancer, pp. 44-102; 139-147; 195-209

IARC (1978) Monographs on the Evaluation of the Carcinogenic Risk of Chemicals to Man, Vol. 18, Polychlorinated Biphenyls and Polybrominated Biphenyls, Lyon, International Agency for Research on Cancer

IARC (1979) Monographs on the Evaluation of the Carcinogenic Risk of Chemicals to Man, Vol. 20, Some Halogenated Hydrocarbons, Lyon, International Agency for Research on Cancer, pp. 45-65; 83-96; 129-154; 283-301; 327-348; 379-571

IARC (1982) Monographs on the Evaluation of the Carcinogenic Risk of Chemicals to Man, Suppl. 4, Chemicals, Industrial Processes and Industries Associated with Cancer in Humans, IARC Monographs, Vols 1-29, Lyon, International Agency for Research on Cancer

ITC (1979) Synthetic Organic Chemicals, United States Production and Sales, 1978, U.S. ITC Publication 1001, Washington DC, International Trade Commission

Jensen, S., Lange, R., Berge, G., Plark, K.H. & Renberg, K. (1975) On the chemistry of EDC-tar and its biological significance in the sea. Proc. R. Soc. Lond., 189, 333-343

Keil, S.L. (1979) Tetrachloroethylene. In: Kirk-Othmer Encyclopedia of Chemical Technology, Vol. 5, 3rd ed., New York, Wiley-Interscience, pp. 754-762

Kimbrough, R.D. (1979) The carcinogenic and other chronic effects of persistent halogenated organic compounds. Ann. N.Y. Acad. Sci., 320, 415-418

Lahl, U., Cetinkaya, M., von Düszeln, J., Stachel, B., Thiemann, W., Gabel, B., Kozicki, R. & Podbielski, A. (1981) Health risks from volatile halogenated hydrocarbons. Sci. Total Environ., 20, 171-189

Lillian, D., Singh, H.B., Appelby, L., Lobban, R., Arnts, R., Gumpert, R., Hague, R., Toumey, J., Kazazis, J., Antell, M., Hansen, D. & Scott, B. (1975) Atmospheric fates of halogenated compounds. Environ. Sci. Technol., 9, 1042-1048

Lloyd, J.W., Moore, R.M. & Breslin, P. (1975) Background information on trichloroethylene. J. Occup. Med., 17, 603-610

Iovelock, J.E. (1977) Halogenated hydrocarbons in the atmosphere. Ecotoxicol. Environ. Saf., 1, 399-406

Lovelock, J.E., Maggs, R.J. & Wade, R.J. (1973) Halogenated hydrocarbons in and over the Atlantic. Nature, 241, 194-196

Marier, J.R. (1982) Halogenated hydrocarbon environmental pollution: The special case of halogenated anesthetics. Environ. Res., 28, 212-239

McCarthy, R.L. (1974) Fluorocarbons in the Environment. Presented at American Geophysical Union Meeting, San Francisco, CA

McConnell, G., Ferguson, O.M. & Pearson, C.R. (1975) Chlorinated hydrocarbons in the environment. Endeavor, 34, 13-18

McNeil, W.C. (1979) Trichloroethylene. In: Kirk-Othmer Encyclopedia of Chemical Technology, Vol. 5, 3rd ed., New York, Wiley-Interscience, pp. 745-753

Miller, S. (1983) Chlorinated hydrocarbons wastes. Environ. Sci. Technol., 17, 290A-291A

Molina, M.J. & Rowland, F.S. (1974) Stratospheric sink for chlorofluoromethanes: chlorine catalyzed destruction of ozone. Nature, 249, 810-812

NAS (1977) Drinking Water and Health, Vol. I., Washington DC, National Academy of Sciences

NAS (1980) Drinking Water and Health, Vol. II., Washington DC, National Academy of Sciences

Neeley, S.B. (1977) Material balance analysis of trichlorofluoromethane and carbon tetrachloride in the atmosphere. Sci. Total Environ., 8, 267-274

NIOSH (1976a) Criteria Document - Methylene Chloride. Rockville MD, National Institute for Occupational Safety and Health

NIOSH (1976b) Criteria document - Chloroform. Rockville MD, National Institute for Occupational Safety and Health (March 15)

NIOSH (1976c) Criteria for a Recommended Standard: Occupational Exposure to Ethylene dichloride (1,2-Dichloroethane). Rockville MD, National Institute for Occupational Safety and Health

NIOSH (1977a) Criteria for a Recommended Standard: Occupational Exposure to Ethylene Dibromide. Rockville MD, National Institute for Occupational Safety and Health

NIOSH (1977b) Criteria for a Recommended Standard: Occupational Exposure to Ethylene Dibromide. Washington DC, U.S. Government Printing Office

Ofstad, E.B., Drangsholt, H. & Carlberg, G.E. (1981) Analysis of volatile halogenated organic compounds in fish. Sci. Total Environ., 20, 205-215

OSHA (1977) OSHA Emergency Temporary Standard for Occupational Exposure to Dibromochloropropane. Fed. Regist., 42, 45536 (Sept. 9)

Palmer, T.Y. (1976) Combustion sources of atmospheric chlorine. Nature, 263, 44-46

Pearson, C.R. (1982) C_1 and C_2 Halocarbons. In: Hutzinger, O., ed., Handbook of Environmental Chemistry, Vol. 3, Part B, Anthropogenic Compounds. Berlin, Heidelberg, New York, Springer-Verlag, pp. 69-88

Pearson, C.R. & McConnel, G. (1975) Chlorinated C_1 and C_2 hydrocarbons in the marine environment. Proc. R. Soc. Lond. (Biol.), 189, 305-332

Ramsey, J.D. & Flanagan, R.J. (1982) Detection and identification of volatile organic compounds in blood by headspace gas chromatography as an aid to the diagnosis of solvent abuse. J. Chromatogr., 240, 423-444

Rosenkranz, H.S. (1977) Mutagenicity of halogenated alkanes and their derivatives. Environ. Health Perspect., 21, 79-84

Rowland, M.S. & Molina, M.J. (1974) Chlorofluoromethanes in the Environment, Atomic Energy Commission Report No. 1974-1. Iwting, C.A., University of California

Singh, H.B., Fowler, D.P. & Peyton, T.O. (1976) Atmospheric carbon tetrachloride: Another man-made pollutant. Science, 192, 1231-1234

Smart, B.E. (1980) Fluorinated aliphatic compounds. In: Kirk-Othmer Encyclopedia of Chemical Technology, Vol. 10, 3rd ed., New York, Wiley-Interscience, pp. 856-870

SRI (1975) Chemical Economics Handbook. Menlo Park, CA, Stanford Research Institute (November)

SRI (1981) Chemical Industries Division Newsletter. Stanford Research International, Menlo Park, CA, p. 4 (Nov/Dec)

SRI (1982) Chemical Industries Division Newsletter. Stanford Research International, Menlo Park, CA, p. 5 (Jan/Feb)

SRI (1983) 1983 Directory of Chemical Producers-United States. Menlo Park, CA, pp. 492, 503, 575, 584, 627, 730, 733, 772, 949

Stephenson, M.E. (1977) An approach to the identification of organic compounds hazardous to the environment and human health. Ecotoxicol. Environ. Saf., 1, 39-48

Stolzenberg, S.J. & Hine, C.H. (1980) Mutagenicity of 2- and 3-carbon halogenated compounds in the Salmonella/mammalian-microsome test. Environ. Mutagen., 2, 59-66

Weisburger, E.K. (1977) Carcinogenicity studies on halogenated hydrocarbons. Environ. Health Perspect., 21, 7-16

Winteringham, F.P.W. (1977) Comparative ecotoxicology of halogenated hydrocarbon residues. Ecotoxicol. Environ. Saf., 1, 407-425

Yung, Y.L., McElroy, M.B. & Wolfsy, S. (1976) Reply to comment by A. Appleby, D. Lillian & H.B. Singh on Atmospheric Halocarbons: A Discussion with Emphasis on Chloroform. Geophys. Res. Lett., 3, 238

CHAPTER 4

FORMATION OF TRIHALOMETHANES IN DRINKING WATER

D.T. Williams

Environmental Health Directorate
Health and Welfare Canada
Ottawa, Ontario, Canada

INTRODUCTION

Following John Snow's deduction, in 1854, that a polluted water supply was responsible for a severe cholera epidemic, Mills and Reincke, in 1893, showed that replacement of polluted drinking water by a pure supply greatly improved the overall health of the community. This was confirmed, in 1903, by Hazen, who reported that for each person saved from death by typhoid, three other persons would be saved from death from other causes associated with a polluted water supply (White, 1972). Chlorine disinfection of potable water was therefore introduced in order to eliminate waterborne diseases caused by pathogenic organisms and it is now the most commonly used disinfection method world-wide. However, even today, the single most important cause of waterborne infectious disease in the USA has been identified as the failure to maintain adequate chlorination of the water supply (Craun, 1979). Other benefits of chlorine use in water treatment are control of taste and odour problems, bleaching of colour, removal of iron and manganese, destruction of hydrogen sulfide and improvement in coagulation, flocculation and filtration (White, 1972).

Despite all of these advantages, the use of chlorine as a potable water disinfectant is now being re-evaluated. This is primarily due to the realization that the chlorine also reacts with organic materials naturally present in the water to form halogenated compounds, particularly trihalomethanes. Although there had been occasional reports of the presence of chloroform in drinking water, it was not until 1974 that Rook (1974) and Bellar _et al._ (1974) clearly demonstrated that trihalomethanes were being formed during the treatment process. Since that time, numerous studies have been carried out to determine the levels of trihalomethanes in potable water, the chemistry related to their formation and their potential health hazards.

TRIHALOMETHANE CHEMISTRY

Chlorine chemistry in water

When chlorine gas is dissolved in water, hydrolysis rapidly occurs to give an equilibrium with hypochlorous acid which, itself, partially ionizes to hypochlorite ion *via* a rapid reversible process.

$$Cl_2 + H_2O \rightleftharpoons H^+ + Cl^- + HOCl$$

$$HOCl \rightleftharpoons H^+ + OCl^-$$

The reactions occur within a few tenths of a second at 18°C and within a few seconds at 0°C (White, 1978). These three forms of free available chlorine (Cl_2, HOCl, OCl^-) exist together in equilibrium in aqueous solution, with their relative proportions governed by pH, ionic strength and temperature. Table 1 shows their distribution at various pH values at 15°C and a chloride content equal to 350 mg/L (Morris, 1978). If sodium or calcium hypochlorite are used as the source of chlorine, the hypochlorite ion rapidly establishes equilibrium with hypochlorous acid and, at the same temperature and pH, results in the same composition of the aqueous solution as the use of chlorine gas (Jolley & Carpenter, 1983).

Table 1. Distribution of aqueous chlorine species, 15°C

		Fraction of oxidizing chlorine		
pH	pCl	$Cl_2 (\times 10^6)$	HOCl	OCl^-
5	2	360	0.997	0.003
6	2	36	0.975	0.025
7	2	2.9	0.797	0.203
8	2	0.10	0.280	0.720
9	2	0.001	0.038	0.962
7.8	0.3	11	0.382	0.618

Other transient species such as H_2OCl^+, Cl^+ and Cl_3^- are not likely to be of significance under conditions found in potable water treatment, except in the immediate vicinity of the injection of strong solutions of chlorine or hypochlorite, where the pH change may be very significant (Jolley & Carpenter, 1983). Morris (1978) has also pointed out that to determine the major reactive species one must consider the specific reactivity as well as the concentration of each of the forms. Although the specific reactivities will depend upon the substrate with which reaction is occurring, Morris (1978) has estimated net relative reactivities in dilute aqueous solution at pH 7 (Table 2). Even though these estimates are not in any sense quantitative, Morris feels that HOCl should be regarded as the major reactive species in dilute aqueous solutions between pH 5 and pH 9.

Table 2. Estimated net reactivities of forms of active chlorine; pH 7, 15°C

Species	Estimated specific reactivity	Fraction of total Cl	Net relative reactivity
Cl_2	10^3	3×10^{-6}	0.003
HOCl	1	0.80	0.80
OCl^-	10^{-4}	0.20	0.00002
H_2OCl^+	10^5	10^{-8}	0.001

Jolley and Carpenter (1983) state in more general terms that the nature and concentrations of reactive chlorine species and chlorine-producing oxidant species formed in chlorinated waters are a function of the chlorine dosage and chemical composition of the water (e.g. pH, temperature, ammonia concentration, salinity, organic and inorganic constituents, and sunlight). The usual reactant behaviour of aqueous chlorine solutions with organic carbon gives rise to oxidation, addition or substitution reactions which are consistent with HOCl acting as an electrophilic agent. The formation of trihalomethanes mainly involves substitution reactions.

Halogen substitution reactions

Rook (1974) postulated that the trihalomethanes were being formed *via* the classical haloform reaction (Fuson & Bull, 1934), wherein aqueous hypochlorites react with methyl ketones, under acid or base-catalysed conditions, to give trihalomethanes. In dilute aqueous solutions at pH > 5, the base-catalysed reaction predominates (Fig. 1) and the mechanism is believed to involve initial proton abstraction from the α-carbon to give a carbanion, which is then subjected to electrophilic attack by HOCl. After three chlorine atoms have been substituted, hydrolysis takes place to give the trihalomethane.

FIG. 1. THE HALOFORM REACTION

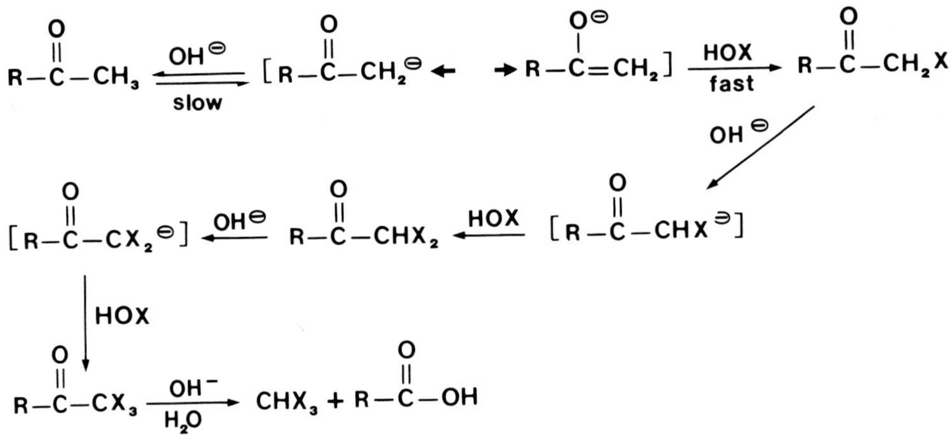

$$R-\overset{\displaystyle O}{\overset{\|}{C}}-CH_3 \underset{\text{slow}}{\overset{OH^{\ominus}}{\rightleftharpoons}} [R-\overset{\displaystyle O}{\overset{\|}{C}}-CH_2^{\ominus} \leftrightarrow R-\overset{\displaystyle O^{\ominus}}{\overset{|}{C}}=CH_2] \xrightarrow[\text{fast}]{HOX} R-\overset{\displaystyle O}{\overset{\|}{C}}-CH_2X$$

$$R-\overset{\displaystyle O}{\overset{\|}{C}}-CH_2X \xrightarrow{OH^{\ominus}} [R-\overset{\displaystyle O}{\overset{\|}{C}}-CHX^{\ominus}] \xrightarrow{HOX} R-\overset{\displaystyle O}{\overset{\|}{C}}-CHX_2 \xrightarrow{OH^{\ominus}} [R-\overset{\displaystyle O}{\overset{\|}{C}}-CX_2^{\ominus}]$$

$$[R-\overset{\displaystyle O}{\overset{\|}{C}}-CX_2^{\ominus}] \xrightarrow{HOX} R-\overset{\displaystyle O}{\overset{\|}{C}}-CX_3 \xrightarrow[H_2O]{OH^-} CHX_3 + R-\overset{\displaystyle O}{\overset{\|}{C}}-OH$$

Subsequent studies have shown that while the classical haloform reaction can explain much of the trihalomethane production, it cannot fully explain all of the production of trihalomethanes and other chlorinated organic compounds and it is most likely that a number of other complex reactions are occurring (Rook, 1977; Peters et al., 1980; Tomita et al., 1982; Boyce & Hornig, 1983).

The redox potential of hypochlorous acid (E_o=1.49 V) is greater than that of hypobromous acid (E_o=1.33 V) and hypoiodous acid (E_o=0.99 V) and, therefore, in the presence of hypochlorite, bromide and iodide ions will be oxidized to hypobromous acid and iodine, respectively (Rook et al., 1978; Dore et al., 1982). These can then participate in the substitution reactions to give trihalomethanes with more than one type of halogen in the molecule. The equilibria for Br_2/HOBr and I_2/HOI are significantly different from that for Cl_2/HOCl. There is a much greater percentage of molecular bromine and iodine at neutral pH, and hypobromous acid and hypoiodous acid ionize far less than hypochlorous acid at a given pH. The kinetics of their reactions will, therefore, differ significantly from those of chlorine (Cooper et al., 1983; Jolley & Carpenter, 1983).

Bunn et al. (1975) showed that chlorination of Missouri River water, to which potassium bromide and potassium iodide had been added, resulted in the formation of all ten possible trihalomethanes (Fig. 2). Addition of potassium fluoride was not expected to, and did not, give rise to fluorine-containing trihalomethanes. Rook et al. (1978) also showed that, when hypochlorite and hypobromite reacted in combination, the ratio of bromine to chlorine in the trihalomethanes formed was much higher than expected from the ratio of bromide ion originally present to chlorine added. This can be explained by the fact that when hypobromite acts as an oxidizing agent it is reduced to bromide ion, which can then be rapidly reconverted to hypobromite by chlorine/hypochlorite present in the system. Therefore, in the presence of excess chlorine, bromine is effectively removed from the reaction system only by substitution reactions. Thus chlorine may be considered to be involved principally in oxidation reactions, while bromine acts preferentially as a substituting agent (Rook _et al._, 1978; Luong _et al._, 1982). Other workers have shown that increasing bromide ion concentration results not only in an increased percentage of brominated trihalomethanes, but also in an increase in the total trihalomethane concentration and in their rate of formation (Trussel & Umphres, 1978; Minear & Bird, 1980; Oliver, 1980).

FIG. 2. STRUCTURAL FORMULAE OF THE TRIHALOMETHANES

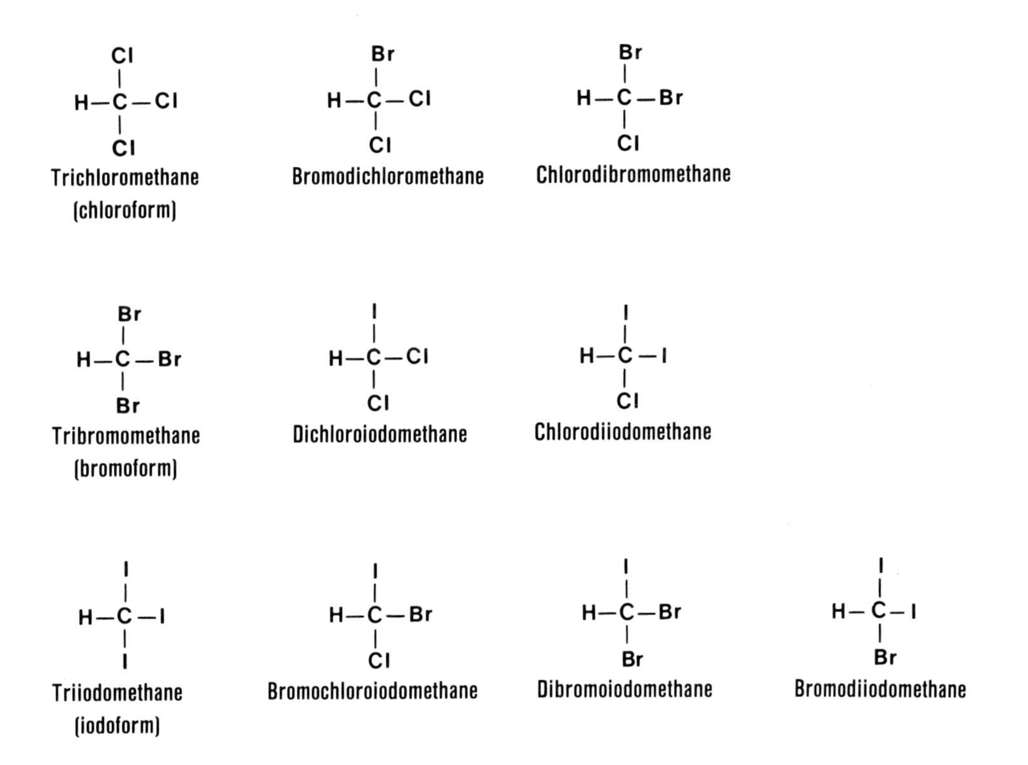

Despite the fact that dichloroiodomethane was detected frequently in a United States national survey (Brass et al., 1977), there has been little further qualitative or quantitative study of the occurrence of iodine-containing trihalomethanes (Thomas et al., 1980). Further work is needed to define the extent of iodide participation in trihalomethane formation in drinking water.

Trihalomethane precursors

Rook (1974, 1977) first suggested that trihalomethanes were formed by reaction of chlorine/hypochlorite with the humic/fulvic acids present in the raw water and this postulate has been supported by other workers (Stevens et al., 1976; Peters et al., 1980). However, little is known of the chemical structure of aquatic humic material, although chemical degradation studies (Schnitzer & Khan, 1972; Christman et al., 1978) have suggested that the core structure includes polyhydroxy phenols and phenolic acids. Model compound studies with dihydroxy benzenes, hydroxy benzoic acids and 1,3-diketones (Rook, 1977; Norwood et al., 1980; Dore et al., 1982; Tomita et al., 1982; Boyce & Hornig, 1983) have shown that 1,3-diketones and meta dihydroxy aromatic compounds give high yields of trihalomethanes under mild reaction conditions. Other dihydroxy aromatic compounds give low yields of trihalomethanes, except under strongly basic conditions.

The reaction of chlorine with various molecular weight fractions of humic/fulvic acids has been investigated (Trussel & Umphres, 1978; Schnoor et al., 1979; Oliver & Visser, 1980; Ishikawa, 1982) and it has been shown that there are large variations in trihalomethane production with molecular weight and that the low molecular weight humic/fulvic acids give the highest percentage yield of trihalomethanes. However, even this represents only a small fraction of the chlorinated by-products, most of which are polar non-volatile halogenated compounds (Rook, 1976; Johnson et al., 1982; Fleischacker & Randtke, 1983; Miller & Uden, 1983).

Although humic/fulvic acids are typically the major natural constituents (60-80%) of water, there are a considerable number of other compounds present which can also act as precursors for trihalomethane production (Newell, 1976; Bedding et al., 1982). Morris and Baum (1978), using model pyrollic compounds, have shown that compounds such as chlorophyll, indole derivatives and acetogenins can act as trihalomethane precursors. Other precursors have been identified as algae and algal by-products (Briley et al., 1980; Crane et al., 1980; Hoehn et al., 1980; Oliver & Shindler, 1980), halogenated phenols and anilines (Hirose & Okitsu, 1982), tannic, vanillic and gallic acids (Youssefi et al., 1978) and even polyelectrolytes used in water treatment (Kaiser & Lawrence, 1977).

Many attempts have been made to correlate trihalomethane levels with precursor concentrations and it has been shown in model systems that increasing the concentration of humic/fulvic acids causes a proportional increase in trihalomethane concentration, if other conditions are kept the same (Stevens et al., 1976; Oliver & Lawrence, 1979). However, only crude relationships have been found with different raw water sources, and although high trihalomethane levels are usually associated with high levels of dissolved organic carbon, the correlation of these two parameters is relatively weak (r=0.59, Glaze & Rawley, 1979; r=0.74, Singer et al., 1976; r=0.65, Singer et al., 1981). This is not entirely surprising, since no two sources of humic/fulvic acids are exactly the same and, although the profiles of organic compounds formed on chlorination are qualitatively similar, it has been shown that there are quantitative differences which vary with the source of the aqueous humic acids (Miller & Uden, 1983). However, Oliver and Thurman (1983) have shown that the trihalomethane potential of aquatic fulvic acids is well correlated with colour (r=0.82 at pH 7). They also feel this indicates that aromatic rings conjugated with olefins, ketonic groups or other aromatic rings are more likely trihalomethane precursor structures than are resorcinol-type structures.

The effect of pH on trihalomethane formation

The classic haloform reaction is base-catalysed and it is not unexpected, therefore, that an increase in pH causes an increase in trihalomethane levels (Rook, 1976). However, this increase is due not merely to an increase in the rate of reaction, but also to a change in mechanism which allows additional compounds to form trihalomethanes (Stevens et al., 1976; Fleischacker & Randtke, 1983). It has been shown that the total amount of organohalogen compounds formed remains relatively constant with change in pH (Oliver, 1978; Miller & Uden, 1983). However, at low pH (< 5) the trihalomethane levels are low and non-volatile organohalogen compounds are high in concentration. As the pH increases, this ratio reverses and at high pH (> 9) the non-volatile organohalogen compounds are low in concentration and the trihalomethane levels are high (Oliver, 1978; Fleischacker & Randtke, 1983; Miller & Uden, 1983). It has also been shown that chlorination of organic intermediates at neutral pH produces intermediates that hydrolyse to trihalomethanes when the pH is raised following dechlorination (Morris & Baum, 1978; Peters et al., 1980). However, this two-step process gives lower trihalomethane levels than would immediate chlorination at the higher pH (Peters et al., 1980; Fleischacker & Randtke, 1983). The combined total of free trihalomethane and intermediates is not greatly affected by change in pH (Peters et al., 1980). However, only the free trihalomethane levels have been considered in most surveys and health evaluations; these can increase by a factor of two or three by increasing the pH from 7 to 9 before chlorination (Trussel & Umphres, 1978; Glaze & Rawley, 1979).

Effect of temperature on trihalomethane formation

Stevens et al. (1976) have shown that the rate of formation and the yield of trihalomethanes increases with increase in temperature and that approximately twice as much trihalomethanes are found, in a given time, at 25°C than at 3°C. The activation energy of chloroform formation has been calculated to be between 20 and 30 kJ/mol (Peters et al., 1980). Therefore, in general, the levels of trihalomethanes in samples collected in the summer are higher than in samples collected in the winter, but it has been shown that this is not entirely due to increased water temperatures, and other factors need to be considered (Singer et al., 1981; Otson et al., 1981). The total amount of organohalogen compounds is also increased at higher temperatures (Oliver, 1980; Peters et al., 1980; Fleischacker & Rundtke, 1983). Because of differences in the rates of reaction, low temperatures can increase the relative proportion of brominated trihalomethanes (Oliver, 1980).

Effect of chlorine dosage on trihalomethane formation

The overall reactions of chlorine with organic compounds in water can be considered to occur in three phases. There is an immediate chlorine demand, mainly inorganic, when only traces of trihalomethanes are formed, followed by a rapid increase in trihalomethane levels with increase in chlorine dosage (Stevens et al., 1976; Trussel & Umphres, 1978). When all of the precursors have reacted, the trihalomethane levels either reach a plateau or increase slowly with time. This last phase is indicative of the fact that hydrolysis of trihalomethane intermediates can be slow and possibly dependent on factors such as steric hindrance and mesomeric and inductive effects (Peters et al., 1980). Although an excess of chlorine will not increase the total levels of trihalomethanes, it can increase the rate of reaction and will favour the formation of the less-brominated trihalomethanes (Trussel & Umphres, 1978). Levels of non-volatile organohalogen compounds also increase with increase in chlorine dosage and can represent more than 75% of the organic chlorine (Fleischacker & Randtke, 1983; Miller & Uden, 1983).

Laboratory studies have shown that trihalomethane levels are strongly positively correlated with chlorine dosage and demand (Rook, 1976; Stevens et al., 1976; Trussel & Umphres, 1978). Multivariate analysis of survey data has also shown that chlorine dosage and demand are the independent variables that can explain the greatest percentage of trihalomethane production (Singer et al., 1976; Williams et al., 1980; Otson et al., 1981). Because of the complexity of the reaction mechanisms and the variation in precursors in different waters, it is not possible to derive a general rate equation linking chlorine demand with trihalomethane formation. However, it might be possible to derive an approximate rate equation for a specific source of waters (Peters et al., 1980).

In the presence of ammonia, low doses of chlorine will react to form combined chlorine, which can then react to form non-volatile organochlorine compounds and trihalomethanes, particularly those containing bromine, although the levels are lower than with free chlorine (Luong et al., 1982; Fleischacker & Randtke, 1983). Kajino and Yagi (1980) have reported that dichloramine reacts with humic/fulvic acids to give trihalomethanes, but monochloramine does not. However, Fleischacker and Randtke (1983) feel that monochloramine hydrolyses back to ammonia and free chlorine and that the free chlorine then reacts to form trihalomethanes.

WATER ANALYSIS AND TREATMENT

Sampling and analytical methods

In any evaluation or comparison of survey data it is important to be aware of the sampling and analytical methods used in the survey. This is particularly essential for trihalomethanes in drinking water, since the trihalomethane levels can vary significantly, depending on sample handling, storage and choice of analytical method. Unless the residual free chlorine is removed at the time the sample is collected, reaction between chlorine and trihalomethane precursors can continue during sample storage, so that trihalomethane values may be obtained that are higher than those present at the time of sampling. Consequently, in most surveys the residual free chlorine has been eliminated at the time of sampling (Arguello et al., 1979; Brett & Calverley, 1979; Williams et al., 1980). However, some workers have chosen not to do this in order to simulate trihalomethane levels in the distribution system (Symons et al., 1975; Stevens & Kopfler, 1976).

Because of the additional time available for reaction, trihalomethane levels for samples taken in the distribution system are normally higher than for those taken at the treatment plant (Brett & Calverley, 1979; Williams et al., 1980). Since chlorinated intermediates are present in the sample even after removal of free residual chlorine, increases in pH or temperature can cause increases in trihalomethane levels (Peters et al., 1980; Fleischacker & Randtke, 1983). Such changes are normally avoided during sample storage, but if the samples are analysed by direct injection into a heated gas chromatograph injection port, additional trihalomethanes can be formed (Nicholson et al., 1977; Pfaender et al., 1978). Surveys in which this method of analysis has been used give values representative of potential trihalomethane levels, rather than those actually present in the samples (Smillie, 1977). Most surveys have used a method of analysis which measures the actual trihalomethane content of the water samples. The two most common procedures involve either solvent extraction (Henderson et al., 1976) or purge-and-trap (Novak et al., 1973; Bellar & Lichtenberg, 1974) techniques, followed by gas chromatography using halogen-specific detectors. These two techniques have been shown to give similar results for trihalomethane analyses (Dressman et al., 1979; Glaze & Rawley, 1979; Otson et al., 1979).

Trihalomethane levels in drinking water

Because of the large number of factors which can influence the formation of trihalomethanes, significant temporal and seasonal variations of trihalomethane levels are not entirely unexpected. In a study of Houston, USA, tap water it was found that trihalomethane levels varied, in a seven day period, from a low of 69 µg/L to a high of 132 µg/L and that rapid changes of as much as 30 to 40 µg/L could occur in a four-hour period (Smith et al., 1980). Significant day-to-day and within-day variations in trihalomethane levels were also reported for Ottawa-Hull, Canada, drinking water (Otson et al., 1981). In the same study (Otson et al., 1981), over a one-year period, chloroform levels ranged from a low of 7 µg/L in the winter to a high of 187 µg/L in the summer. Other studies have also reported significantly higher trihalomethane levels in summer samples than in winter samples (Arguello et al., 1979; Singer et al., 1981; Cech et al., 1982), although the levels were not always greatly different (Brett & Calverley, 1979). The results of trihalomethane surveys which represent only single samples per site or have been carried out during one period of the year should, therefore, be interpreted with caution. The values reported in these surveys may not be indicative of the average trihalomethane levels present in the drinking water supplies.

The most extensive surveys have been carried out in the USA, including two national surveys (Table 3) (Symons et al., 1975; Brass et al., 1977) and a large number of regional or city surveys (Rawley, 1979; Allgeier et al., 1980; Singer et al., 1981) which reported trihalomethane levels in the same range as the national surveys. Canadian surveys have reported a similar range of trihalomethane values in drinking water (Smillie, 1977; Williams et al., 1980).

Trihalomethane levels in European drinking waters have tended to be somewhat lower than North american values (Brett & Calverley, 1979; Chambon et al., 1983; Dick, 1981; Gabel et al., 1981; Norin et al., 1981; Quaghebeur & De Wulf, 1980). However, very high trihalomethane levels, up to 1.3 mg/L, have been reported in Finnish potable waters (Hemminki et al., 1983) obtained from water sources high in humic content. A survey of drinking water from twelve countries showed low levels of trihalomethanes in these samples (Trussel et al., 1980).

Table 3. Trihalomethane levels in USA drinking water (μg/L)

Trihalomethane	NOMS survey		NORS survey	
	Median	Range	Median	Range
Chloroform	22	NF-200	21	NF-311
Bromodichloromethane	6	NF-72	6	NF-116
Chlorodibromomethane	2	NF-114	1	NF-100
Bromoform	LD	NF-137	5	NF-92

NF = not found; LD = less than detection limit

Reduction of trihalomethane levels

Since the realization that trihalomethanes are formed whenever chlorine is used in drinking water treatment, considerable attention has been paid to ways to minimize trihalomethane levels in the finished water. This can be achieved by the use of two approaches, either separately or in combination. The first approach is to prevent or minimize the formation of trihalomethanes and the second approach is to remove the trihalomethanes once they are formed.

Improvements in raw water quality either by a change of source or by control of algal growth in lake or reservoir waters can be considered where feasible. Lowering the chlorine dosage, or delaying the use of chlorination until some of the trihalomethane precursors have been removed by coagulation-filtration-sedimentation, has also been investigated and found to reduce trihalomethane levels (Bolton, 1977; Young & Singer, 1979). Reductions in trihalomethane precursor levels of 65% (Semmens & Field, 1980) and 29-51% (Ohio River, 1980) have been reported. Because trihalomethane precursors are not defined organic chemicals, but a mixture of compounds that vary from one location to another, the potential for removing these materials also varies with location (AWWA, 1982). Provided that the microbiological quality of the finished water is maintained, the removal of precursors and/or change in the point of chlorination should not result in any additional health risks. Corrosion problems may, however, occur in the distribution system (Kuhn & Sontheimer, 1981).

Considerable studies have also been carried out to evaluate the replacement of chlorine with other disinfectants, such as chloramines, chlorine dioxide and ozone (AWWA, 1982). Ozone and chlorine dioxide are strong disinfectants, equal to or better than free chlorine, but chloramines are weaker disinfectants and may not provide adequate disinfection (Federal Register, 1979). Significant reductions (>80%) in trihalomethane levels can be obtained by replacing free chlorine by chloramines (Brodtmann et al., 1980; Norin et al., 1981; AWWA, 1982). Treatment with free chlorine for a few minutes, to ensure adequate disinfection, followed by ammonia addition to form chloramines, also reduces trihalomethane levels significantly (> 75%) (Norman et al., 1980).

Pure chlorine dioxide gives rise to little, if any, trihalomethanes when used in potable water treatment (NRC, 1980; Ohio River, 1980; Norin et al., 1981). Even when some chlorine is present in the chlorine dioxide, the levels of trihalomethanes formed are much lower than those formed when only chlorine is used as the disinfectant (Hubbs et al., 1980; Vogt & Regli, 1981; AWWA, 1982). The addition of ammonia immediately after addition of the chlorine dioxide-chlorine mixture results in even lower trihalomethane levels (Hubbs et al., 1980).

When ozone is used alone, virtually no trihalomethanes are formed (AWWA, 1982), although some bromoform may be formed in waters with a high bromide content under certain pH and ozone dosage conditions (Haag & Holgné, 1983). However, since ozone does not produce a residual disinfectant, chlorine is often added as the final treatment step to prevent aftergrowth problems in the distribution system (Vogt & Regli, 1981). In these cases, trihalomethane levels have been reported to be both higher and lower than with chlorine alone (Riley et al., 1978; Rook et al., 1978). Whether trihalomethane precursors are removed or created by ozonation depends on ozone dosage and pH (Riley et al., 1981) and on the unique characteristics of each water supply (Veenstra et al., 1983).

Other disinfectants, such as bromine chloride, bromine and iodine, have been considered, but have been used only rarely (NRC, 1980).

The major disadvantage to using alternative disinfectants to control trihalomethane production is that they produce organic by-products which are, at present, ill-defined. Whether these by-products are more or less safe for consumers than those produced by chlorination needs further evaluation (AWWA, 1982).

The second major approach to reducing trihalomethane levels is to remove the trihalomethanes after they are formed, and a number of methods have been extensively investigated (AWWA, 1982).

Aeration of the finished water has been demonstrated to be a feasible technical approach for removal of chloroform, but it is less efficient for the brominated trihalomethanes (AWWA, 1982). It also does not remove non-volatile intermediates, which can subsequently form trihalomethanes, and it may create an air pollution problem.

Powdered activated carbon added during the treatment process does not effectively remove chloroform, although it is somewhat better for bromoform.

Granular activated carbon (GAC) used in filter beds has been extensively investigated for the removal of taste and odour, specific organic compounds and, more recently, trihalomethanes (McCreary & Snoeyink, 1977). Trihalomethanes are effectively removed by GAC until breakthrough occurs, when the carbon has to be replaced or regenerated. The lifetime of the GAC before saturation depends on a wide variety of factors, but is typically several months. Trihalomethane precursors are less effectively removed by GAC and if chlorine is added to the treated water after the GAC, trihalomethanes can be formed.

Synthetic resins have also been evaluated for removal of trihalomethanes, but (except for ambersorb XE-340) are not effective. The Ambersorb resin, which was specifically designed to adsorb low molecular weight halogenated compounds, appears to remove trihalomethanes more effectively than GAC in preliminary experiments. However, full-scale plant studies have not yet been carried out (AWWA, 1982).

REFERENCES

Allgeier, G.D., Mullins, R.L., Wilding, D.A., Zogorski, J.S. & Hubbs, S.A. (1980) Trihalomethane levels at selected water utilities in Kentucky, USA. In: Afghan, B.K. & Mackay, D., eds, Hydrocarbons and Halogenated Hydrocarbons in the Aquatic Environment, New York, Plenum Press, pp. 473-490

Arguello, M.D., Chriswell, C.D., Fritz, J.S., Kissinger, L.D., Lee, K.W., Richard, J.J. & Svec, H.J. (1979) Trihalomethanes in water: a report on the occurrence, seasonal variation in concentrations, and precursors of trihalomethanes. J. Am. Water Works Assoc.. 71, 504-508

AWWA (1982) Treatment Techniques for Controlling Trihalomethanes in Drinking Water, Colorado, American Water Works Association

Bedding, N.D., McIntyre, A.E., Perry, R. & Lester, J.N. (1982) Organic contaminants in the aquatic environment. 1. Sources and occurrence. Sci. Total Environ., 25, 143-167

Bellar, T.A. & Lichtenberg, J.J. (1974) Determining volatile organics at microgram-per-litre levels by gas chromatography. J. Am. Water Works Assoc., 66, 739-744

Bellar, T.A., Lichtenberg, J.J. & Kroner, R.C. (1974) The occurrence of organohalides in chlorinated drinking water. J. Am. Water works Assoc., 66, 703-706

Bolton, C.M. (1977) Cincinnati research in organics. J. Am. Water works Assoc., 69, 405-406

Boyce, S.D. & Hornig, J.F. (1983) Reaction pathways of trihalomethane formation from the halogenation of dihydroxyaromatic model compounds for humic acid. Environ. Sci. Technol., 17, 202-211

Brass, H.J., Feige, M.A., Halloran, T., Mello, J.W., Munch, D. & Thomas, R.F. (1977) The National Organic Monitoring Survey: Samplings and analyses for purgeable organic compounds. In: Pojasek, R.B., ed., Drinking Water Quality Enhancement Through Source Protection, Michigan, Ann Arbor Science, pp. 393-416

Brett, R.W. & Calverley, R.A. (1979) A one-year survey of trihalomethane concentration changes within a distribution system. J. Am. Water Works Assoc., 71, 515-520

Briley, K.F., Williams, R.F., Langley, K.E. & Sorber, C.A. (1980) Trihalomethane production from algal precursors. In: Jolley, R.L., Brungs, W.A. & Cumming, R.B., eds, Water Chlorination, Environmental Impact and Health Effects, Vol. 3, Michigan, Ann Arbor Science, pp. 117-129

Brodtmann, N.V., Koffskey, W.E. & De Marco, J. (1980) Studies of the use of combined chloramine (monochloramine) as a primary disinfectant of drinking water. In: Jolley, R.L., Brungs, W.A. & Cumming, R.B., eds, Water Chlorination, Environmental Impact and Health Effects, Vol. 3, Michigan, Ann Arbor Science, pp. 777-788

Bunn, W.W., Haas, B.B., Deane, E.R. & Kleopfer, R.D. (1975) Formation of trihalomethanes by chlorination of surface water. Environ. Lett., 10, 205-213

Cech, I., Smith, V. & Henry, J. (1982) Spatial and seasonal variations in concentration of trihalomethanes in drinking water. Pergamon Ser. Environ. Sci., 7, 19-38

Chambon, P., Taveau, M., Morin, M., Chambon, R. & Vial, J. (1983) Survey of trihalomethane levels in Rhone-Alps water supplies - estimates on the formation of chloroform in wastewater treatment plants and swimming pools. Water Res., 17, 65-69

Christman, R.F., Johnson, J.D., Hass, J.R., Pfaender, F.K., Liao, W.T., Norwood, D.L. & Alexander, H.J. (1978) Natural and model aquatic humics: reactions with chlorine. In: Jolley, R.L., Gorchev, H. & Hamilton, D.H., eds, Water Chlorination, Environmental Impact and Health Effects, Vol. 2, Michigan, Ann Arbor Science, pp. 15-28

Cooper, W.J., Meyer, L.M., Bofill, C.C. & Cordal, E. (1983) Quantitative effects of bromine on the formation and distribution of trihalomethanes in groundwater with a high organic content. In: Jolley, R.L., Brungs, W.A., Cotruvo, J.A., Cumming, R.B., Mattice, J.S. & Jacobs, V.A., eds, Water Chlorination, Environmental Impact and Health Effects, Vol. 4, Michigan, Ann Arbor Science, pp. 285-296

Crane, A.M., Kovacic, P. & Kovacic, E.D. (1980) Volatile halocarbon production from the chlorination of marine algal byproducts, including D-mannitol. Environ. Sci. Technol., 14, 1371-1374

Craun, G.F. (1979) Waterborne disease outbreaks in the United States. J. Environ. Health, 41, 259-265

Dick, T.A. (1981) The control of potential health risks related to drinking water in the U.K. Sci. Total Environ., 18, 317-334

Dore, M., Merlet, N., De Laat, J. & Goichon, J. (1982) Reactivity of halogens with aqueous micropollutants: a mechanism for the formation of trihalomethanes. J. Am. Water Works Assoc., 74, 103-107

Dressman, R.G., Stevens, A.A., Fair, J. & Smith, B. (1979) Comparison of methods for determination of trihalomethanes in drinking water. J. Am. Water Works Assoc., 71, 392-396

Federal Register (1979) 44, 68624-68707

Fleischacker, S.J. & Randtke, S.J. (1983) Formation of organic chlorine in public water supplies. J. Am. Water Works Assoc., 75, 132-138

Fuson, R.C. & Bull, R.A. (1934) The haloform reaction. Chem. Rev., 15, 275-309

Gabel, B., Lahl, U., Batjer, K., Cetinkaya, M., V. Duszeln, J., Kozicki, R., Podbielski, A., Stachel, B. & Thiemann, W. (1981) Volatile halogenated compounds in drinking waters of the FRG. Sci. Total Environ., 18, 363-366

Glaze, W.H. & Rawley, R.A. (1979) A preliminary survey of trihalomethane levels in selected East Texas water supplies. J. Am. Water Works Assoc., 71, 509-515

Haag, W.R. & Holgné, J. (1983) Ozonation of bromide-containing waters: Kinetics of formation of hypobromous acid and bromate. Environ. Sci. Technol., 17, 261-267

Henderson, J.E., Peyton, C.R. & Glaze, W.H. (1976) A convenient liquid-liquid extraction method for the determination of halomethanes in water at the parts-per-billion level. In: Keith, L.H., ed., Identification and Analysis of Organic Pollutants in Water, Michigan, Ann Arbor Science, pp. 105-111

Hemminki, K., Vainio, H., Sorsa, M. & Salminen, S. (1983) An estimation of the exposure of the population in Finland to suspected chemical carcinogens. J. Environ. Sci. Health, C, 1, 55-95

Hirose, Y. & Okitsu, T. (1982) Formation of trihalomethanes by reaction of halogenated phenols or halogenated anilines with sodium hypochlorite. Chemosphere, 11, 81-87

Hoehn, R.C., Barnes, D.B., Thompson, B.C., Randall, C.W., Gizzard, T.J. & Shaffter, P.T.B. (1980) Algae as sources of trihalomethane precursors. J. Am. Water Works Assoc., 72, 344-350

Hubbs, S.A., Goers, M. & Siria, J. (1980) Plant-scale examination and control of a chlorine dioxide chloramination process at the Louisville Water Company. In: Jolley, R.L., Brungs, W.A. Cumming, R.B., eds, Water Chlorination, Environmental Impact and Health Effects, Vol. 3, Michigan, Ann Arbor Science, pp. 769-775

Ishikawa, T. (1982) Precursors of trihalomethanes in aquatic environments. Eisei Kagaku, 28, 10-15

Johnson, J.D., Christman, R.F., Norwood, D.L. & Millington, D.S. (1982) Reaction products of aquatic humic substances with chlorine. Environ. Health Perspect., 46, 63-71

Jolley, R.L. & Carpenter, J.H. (1983) A review of the chemistry and environmental fate of reactive oxidant species in chlorinated water. In: Jolley, R.L., Brungs, W.A., Cotruvo, J.A., Cumming, R.B., Mattice, J.S. & Jacobs, V.A., eds, Water Chlorination, Environmental Impact and Health Effects, Vol. 4, Michigan, Ann Arbor Science, pp. 3-47

Kaiser, K.L.E. & Lawrence, J. (1977) Polyelectrolytes: potential chloroform precursors. Science, 196, 1205-1206

Kajino, M. & Yagi, M. (1980) Formation of trihalomethanes during chlorination and determination of halogenated hydrocarbons. In: Afghan, B.K. & Mackay, D., eds, Hydrocarbons and Halogenated Hydrocarbons in the Aquatic Environment, New York, Plenum Press, pp. 491-501

Kuhn, W. & Sontheimer, H. (1981) Treatment: improvement or deterioration of water quality. Sci. Total Environ., 18, 219-233

Luong, T.V., Peters, C.J. & Perry, R. (1982) Influence of bromide and ammonia upon the formation of trihalomethanes under water-treatment conditions. Environ. Sci. Technol., 16, 473-479

McCreary, J.J. & Snoeyink, V.L. (1977) Granular activated carbon in water treatment. J. Am. Water Works Assoc., 69, 437-444

Miller, J.W. & Uden, P.C. (1983) Characterization of nonvolatile aqueous chlorination products of humic substances. Environ. Sci. Technol., 17, 150-157

Minear, R.A. & Bird, J.C. (1980) Trihalomethanes: impact of bromide ion concentration on yield, species distribution, rate of formation and influence of other variables. In: Jolley, R.L., Brungs, W.A. & Cumming, R.B., eds, Water Chlorination, Environmental Impact and Health Effects, Vol. 3, Michigan, Ann Arbor Science, pp. 151-160

Morris, J.C. (1978) The chemistry of aqueous chlorine in relation to water chlorination. In: Jolley, R.L., ed., Water Chlorination, Environmental Impact and Health Effects, Vol. 1, Michigan, Ann Arbor Science, pp. 21-35

Morris, J.C. & Baum, B. (1978) Precursors and mechanisms of haloform formation in the chlorination of water supplies. In: Jolley, R.L., Gorchev, H. & Hamilton, D.H., eds, Water Chlorination, Environmental Impact and Health Effects, Vol. 2, Michigan, Ann Arbor Science, pp. 29-46

Newell, I.L. (1976) Naturally occurring organic substances in surface waters and effect of chlorination. New Engl. Water works Assoc. J., 90, 315-340

Nicholson, A.A., Meresz, O. & Lemyk, B. (1977) Determination of free and total potential haloforms in drinking water. Anal. Chem., 49, 814-819

Norin, H., Renberg, L., Hjort, J. & Lundblad, P.O. (1981) Factors influencing formation of trihalomethanes in drinking water with special reference to Swedish conditions. Chemosphere, 10, 1265-1273

Norman, T.S., Harms, L.L. & Looyenga, R.W. (1980) The use of chloramines to prevent trihalomethane formation. J. Am. Water works Assoc., 72, 176-180

Norwood, D.L., Johnson, J.D., Christman, R.F., Hass, J.R. & Bobenrieth, M.J. (1980) Reactions of chlorine with selected aromatic models of aquatic humic material. Environ. Sci. Technol., 14, 187-190

Novak, J., Zluticky, J., Kubelka, V. & Mostecky, J.(1973) Analysis of organic constituents present in drinking water. J. Chromatogr., 76, 45-50

NRC (1980) Drinking Water and Health, Vol. 2, National Research Council, Washington, National Academy Press

Ohio River Valley Water Sanitation Commission (1980) Water treatment process modifications for trihalomethane control and organic substances in the Ohio River. EPA-600/2-80-028, Cincinnati, USEPA

Oliver, B.G. (1978) Chlorinated non-volatile organics produced by the reaction of chlorine with humic materials. Can. Res., 11, 21-22

Oliver, B.G. (1980) Effect of temperature, pH and bromide concentration on the trihalomethane reaction of chlorine with aquatic humic material. In: Jolley, R.L., Brungs, W.A. & Cumming, R.B., eds, Water Chlorination, Environmental Impact and Health Effects, Vol. 3, Michigan, Ann Arbor Science, pp. 141-150

Oliver, B.G. & Lawrence, J. (1979) Haloforms in drinking water: a study of precursors and precursor removal. J. Am. Water works Assoc., 71, 161-163

Oliver, B.G. & Shindler, D.B. (1980) Trihalomethanes from the chlorination of aquatic algae. Environ. Sci. Technol., 14, 1502-1505

Oliver, B.G. & Thurman, E.M. (1983) Influence of aquatic humic substance properties on trihalomethane potential. In: Jolley, R.L., Brungs, W.A. & Cumming, R.B., eds, Water Chlorination, Environmental Impact and Health Effects, Vol. 4, Michigan, Ann Arbor Science, pp. 231-241

Oliver, B.G. & Visser, S.A. (1980) Chloroform production from the chlorination of aquatic humic material: the effect of molecular weight, environment and season. Water Res., 14, 1137-1141

Otson, R., Williams, D.T. & Bothwell, P.D. (1979) A comparison of dynamic and static head space and solvent extraction techniques for the determination of trihalomethanes in water. Environ. Sci. Technol., 13, 936-939

Otson, R., Williams, D.T., Bothwell, P.D. & Quon, T.K. (1981) Comparison of trihalomethane levels and other water quality parameters for three treatment plants on the Ottawa River. Environ. Sci. Technol., 15, 1075-1080

Peters, C.J., Young, R.J. & Perry, R. (1980) Factors influencing the formation of haloforms in the chlorination of humic materials. Environ. Sci. Technol., 14, 1391-1395

Pfaender, F.K., Jonas, R.B., Stevens, A.A., Moore, L. & Hass, J.R. (1978) Evaluation of direct aqueous injection method for analysis of chloroform in drinking water. Environ. Sci. Technol., 12, 438-441

Quaghebeur, D. & De Wulf, E. (1980) Volatile halogenated hydrocarbons in Belgian drinking waters. Sci. Total Environ., 14, 43-52

Riley, T.L., Mancy, K.H. & Boettner, E.A. (1979) The effect of preozonation on chloroform production in the chlorine disinfection process. In: Jolley, R.L., Gorchev, H. & Hamilton, D.H., eds, Water Chlorination, Environmental Impact and Health Effects, Vol. 2, Michigan, Ann Arbor Science, pp. 593-603

Rook, J.J. (1974) Formation of haloforms during chlorination of natural waters. Water Treat. Exam., 23, 234-243

Rook, J.J. (1976) Haloforms in drinking water. J. Am. Water Works Assoc., 68, 168-172

Rook, J.J. (1977) Chlorination reactions of fulvic acids in natural waters. Environ. Sci. Technol., 11, 478-482

Rook, J.J., Gras, A.A., Van der Heijden, B.G. & De Wee, J. (1978) Bromide oxidation and organic substitution in water treatment. J. Environ. Sci. Health, Part A, 13, 91-116

Schnitzer, M. & Khan, S.U. (1972) Humic Substances in the Environment, New York, Marcel Dekker

Schnoor, J.L., Nitzschke, J.L., Lucas, R.D. & Veenstra, J.N. (1979) Trihalomethane yields as a function of precursor molecular weight. Environ. Sci. Technol., 13, 1134-1138

Semmens, M.J. & Field, T.K. (1980) Coagulation: Experiences in organic removal. J. Am. Water Works Assoc., 72, 476-483

Singer, P.C., Lawler, D.L. & Babcock, D.B. (1976) Notes and comments. J. Am. Water Works Assoc., 68, 452

Singer, P.C., Barry, J.J., Palen, G.M. & Scrivner, A.E. (1981) Trihalomethane formation in North Carolina drinking waters. J. Am. Water Works Assoc., 73, 392-401

Smillie, R.D. (1977) Chloroform levels in Ontario drinking water. Water Pollut. Control, 115, 8-9

Smith, V.L., Cech, I., Brown, J.H. & Bogdan, G.F. (1980) Temporal variations in trihalomethane content of drinking water. Environ. Sci. Technol., 14, 190-196

Stevens, A.A. & Kopfler, F.C. (1976) Analyzing drinking water. Chem. Eng. News, 54, 58-71

Stevens, A.A., Slocum, C.J., Seeger, D.R. & Robeck, G.G. (1976) Chlorination of organics in drinking water. J. Am. Water Works Assoc., 68, 615-620

Symons, J.M., Bellar, T.A., Carswell, J.K., De Marco, J., Kropp, K.L., Robeck, G.G., Seeger, D.R., Slocum, C.J., Smith, B.L. & Stevens, A.A. (1975) National organics reconnaissance survey for halogenated organics. J. Am. Water Works Assoc., 67, 634-647

Thomas, R.F., Weisner, M.J. & Brass, H.J. (1980) The fifth trihalomethane: dichloroiodomethane, its stability and occurrences in chlorinated drinking water. In: Jolley, R.L., Brungs, W.A. & Cumming, R.B., eds, Water Chlorination: Environmental Impact and Health Effects, Vol. 3, Michigan, Ann Arbor Science, pp. 161-168

Tomita, M., Manabe, H., Honma, K. & Hamada, A. (1982) Studies on trihalomethane formation from model compounds of aqueous chlorination. Eisei Kagaku, 28, 21-27

Trussel, R.R. & Umphres, M.D. (1978) The formation of trihalomethanes. J. Am. Water Works Assoc., 70, 604-612

Trussel, A.R., Cromer, J.L., Umphres, M.D., Kelley, P.E. & Moncur, J.G. (1980) Monitoring of volatile halogenated organics: a survey of twelve drinking waters from various parts of the world. In: Jolley, R.L., Brungs, W.A. & Cumming, R.B., eds, Water Chlorination, Environmental Impact and Health Effects, Vol. 3, Michigan, Ann Arbor Science, pp. 39-53

Veenstra, J.N., Barber, J.B. & Khan, P.A. (1983) Evaluation of ozone pretreatment of surface waters prior to chlorination for reduction in trihalomethane formation. In: Jolley, R.L., Brungs, W.A. & Cumming, R.B., eds, Water Chlorination, Environmental Impact and Health Effects, Vol. 4, Michigan, Ann Arbor Science, pp. 455-465

Vogt, C. & Regli, S. (1981) Controlling trihalomethanes while attaining disinfection. J. Am. Water Works Assoc., 73, 33-40

White, G.C. (1972) Handbook of Chlorination, New York, Van Nostrand Reinhold, Co.

White, G.C. (1978) Disinfection of Wastewater and Water for Reuse, New York, Van Nostrand Reinhold Co.

Williams, D.T., Otson, R., Bothwell, P.D., Murphy, K.L. & Robertson, J.L. (1980) Trihalomethane levels in Canadian drinking water. In: Afghan, B.K. & Mackay, D., eds, Hydrocarbons and Halogenated Hydrocarbons in the Aquatic Environment, New York, Plenum Press, pp. 503-512

Young, J.S. & Singer, P.C. (1979) Chloroform formation in public water supplies: A case study. J. Am. Water Works Assoc., 71, 87-95

Youssefi, M., Zenchelsky, S.T. & Faust, S.D. (1978) Chlorination of naturally occurring organic compounds in water. J. Environ. Sci. Health, A13, 629-637

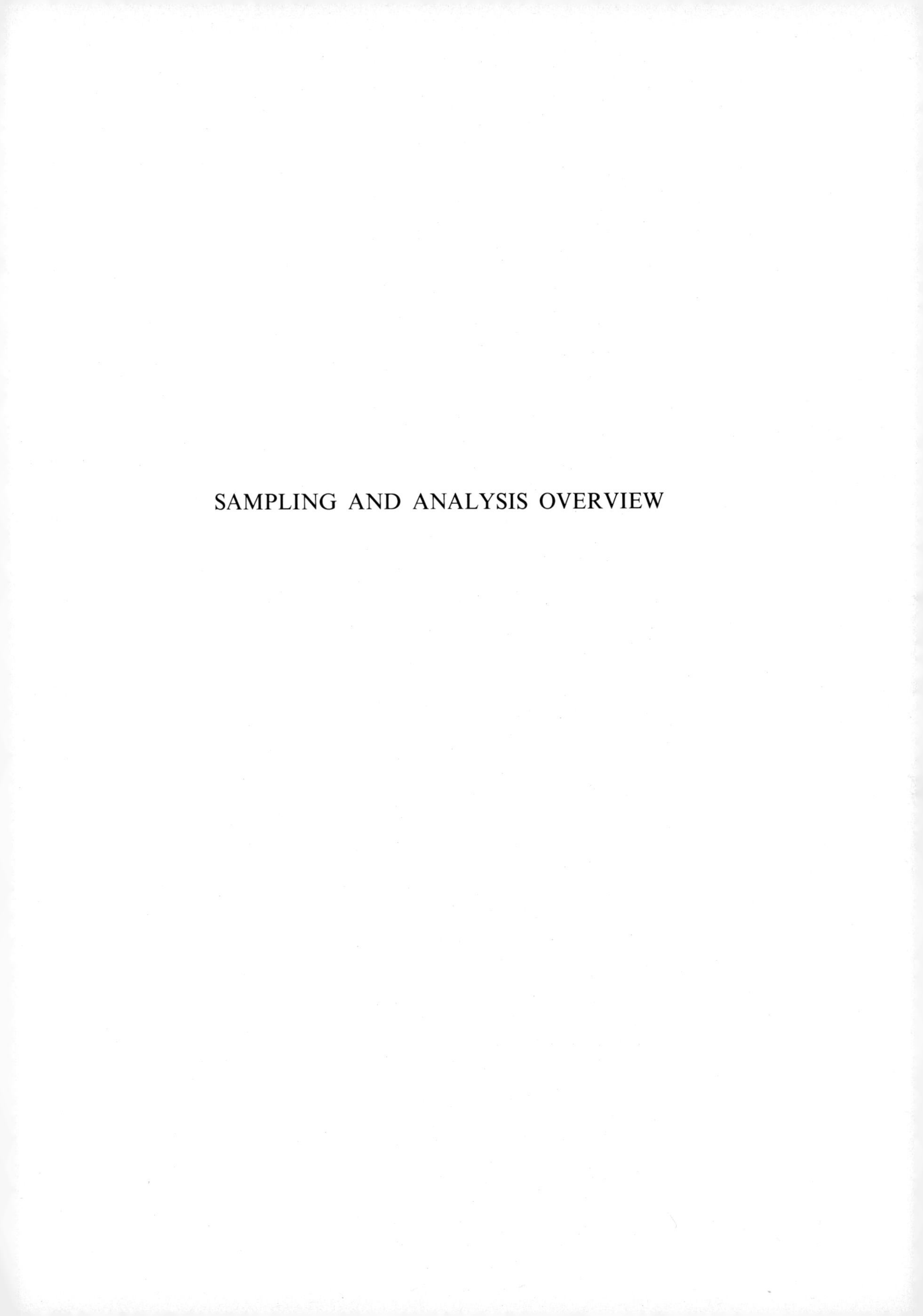

SAMPLING AND ANALYSIS OVERVIEW

CHAPTER 5

HALOCARBONS IN INDOOR ENVIRONMENTS

L. Fishbein

Department of Health and Human Services
Food and Drug Administration
National Center for Toxicological Research
Jefferson, Arkansas 72079, USA

INTRODUCTION

In contrast to pollution in workplace, urban and rural environments, pollution in residences, offices and public buildings has not been extensively investigated. It is becoming increasingly recognized, however, that indoor exposure can constitute an important fraction of the total exposure to many pollutants (Bridbord et al., 1975; WHO, 1979; GAO, 1980; Abraham et al., 1981; Hollowell & Miksch, 1981; National Research Council, 1981; Seifert & Abraham, 1982; Spengler & Sexton, 1983).

Concern about potential public health problems due to indoor air pollution is primarily based on evidence that many urban residents typically spend more than 90% of their time indoors, and that concentrations of some contaminants are often higher indoors that outdoors (Bridbord et al., 1975; National Research Council, 1981; Seifert & Abraham, 1982; spengler & Sexton, 1983). For example, although there is little epidemiological evidence concerning the health effects of indoor pollutants, indoor concentrations of some of them exceed the levels cited for primary ambient-air quality standards (National Research Council, 1981). Even if indoor pollutant concentrations are low, they may make a substantial contribution to time-weighted exposures (National Research Council, 1981; Spengler & Sexton, 1983).

The situation, moreover, can be aggravated in homes that are close to extensive toxic waste sites (e.g., Love Canal) (Pellizzari & Bunch, 1979; Pellizzari et al., 1979; Barkley et al., 1980). In addition, personal exposure to some pollutants is not characterized adequately by outdoor measurements.

CLASSES OF INDOOR POLLUTANTS AND SOURCES OF EMISSIONS

The principal indoor contaminants which are associated with health or irritation effects are generally recognized to be tobacco smoke, radon decay products, carbon monoxide, nitrogen dioxide, formaldehyde, asbestos fibres, microorganisms and aeroallergens (National Research Council, 1981; Spengler & Sexton, 1983). It is recognized that indoor pollution in homes, offices and public buildings is created principally by the activities of the occupants and a variety of other factors, including fuel burning, smoking, use of consumer products, emanations from building materials and the intrusion of outdoor pollutants. The growing need for energy conservation by enhanced insulation, and perhaps reduced ventillation rates, might result in higher pollutant concentrations in the future (National Research Council, 1981; Seifert & Abraham, 1982; Spengler & Sexton, 1983). A broad spectrum of organic vapors can arise from a large variety of consumer products, including aerosol sprays, solvents, resin products, pesticides, adhesives, cosmetics, cleaning products and from cooking (Bridbord et al., 1975; National Research Council, 1981; Spengler & Sexton, 1983).

PROBLEMS IN SAMPLING AND MEASUREMENT

It is generally recognized that indoor air pollutants are generated or released in concentrations and mixtures that are often episodic and generally vary with time and from one space to another. Most investigators will agree that the distribution of indoor air quality is extremely difficult to categorize and quantify on a geographic basis, since it is determined by a number of complex dynamic relationships, including occupant activity, highly-variable seasonal and structural characteristics, outdoor air quality, emission from indoor sources and air-infiltration rates. It is considered likely that these factors can vary within each metropolitan and suburban area, as well as within each neighbourhood (National Research Council, 1981; Spengler & Sexton, 1983).

Additionally, substantial problems exist with regard to sampling and measurement of indoor pollutants. For example, the measurement requires a sampling protocol that must take account of the spatial and temporal profile of several pollutants, as well as of air diffusion and ventilation characteristics. The detection of low levels of contaminants in admixture, furthermore, requires adequate sensitivity, selectivity and freedom from interference. In many cases, the instruments which might characterize short-term and long-term indoor pollutant concentrations rapidly and economically do not exist, and those which are available are often of limited utility (National Research Council, 1981).

SOURCES OF HALOCARBONS IN INDOOR AIR

Tobacco smoke

There is increasing concern about the absorption of constituents of tobacco smoke by non-smokers (Weiss et al., 1983; Hoffmann et al., 1983, 1984). Our well-established knowledge of the health hazards of active smoking underlines the need to investigate the effects of exposure to "sidestream smoke" (passive smoking) in populations at potential risk, including early childhood (Greenberg et al., 1984).

It has been estimated by Replace (1981) that cigarette smokers liberate an estimated 2.25 million metric tons of gaseous and inhalable particulate matter into the indoor environment each year.

The composition of cigarette smoke is tremendously complex and about 200 of 3 800 individual compounds identified in tobacco smoke have now been quantitatively determined in both "mainstream" and "sidestream" smoke (Johnson et al., 1973; Schmeltz et al., 1975; Dube & Green, 1982; Klus & Kuhn, 1982; U.S. Surgeon General, 1982; Hoffman et al., 1984). Tobacco smoke is the source of a broad spectrum of chemicals which include relatively simple aliphatic and aromatic derivatives, as well as complex structures, e.g., polycyclic aromatic hydrocarbons, a number of which are carcinogenic and/or mutagenic. While a large research effort has been dedicated to the analysis of tobacco products, relatively little is known about the substances to which non-smokers are exposed through the presence of sidestream smoke, the major contributor to indoor pollution. It is generally agreed that both sampling and analysis have presented problems.

Although almost all constituents are far more concentrated in mainstream than in sidestream smoke, it is recognized that, given the approximately 24 to 1 disparity in burning time (i.e., the sidestream smoke is produced during 96% of the total smoking time) and the difference in combustion conditions, sidestream smoke can be enriched in many compounds (National Research Council, 1981; Spengler & Sexton, 1983).

Tobacco contains traces of chlorinated pesticides (Guthrie & Bowery, 1967), some unknown chlorinated organic compounds (Wynder & Hoffmann, 1967; Stedman, 1968) and up to 3% of inorganic chloride (Tso, 1982), which may be the source of some of the halogenated organic compounds detected in tobacco smoke.

Of the halogenated, low molecular weight compounds which have been reported in tobacco smoke (e.g., methyl chloride, methylene chloride, chloroform, trichloroethylene, tetrachloroethylene (Johnson et al., 1973; Holzer et al., 1976; Harsch, 1977; Sloan et al., 1977; Kadaba et al., 1978; NAS, 1978; Dirinck et al., 1980) and vinyl chloride (Hoffmann et al., 1976)), methyl chloride is considered to be the most important (Johnson et al., 1973; Kadaba et al., 1978; NAS, 1978). Methyl chloride has been found in mainstream smoke at levels that range from 100-700 μg/cigarette. Depending on the type and content of tobacco, the ratio of methyl chloride in sidestream to mainstream smoke is 1.7-3.3 (Johnson et al., 1973). The amount of methyl chloride in cigarette smoke is determined to a large extent by the concentration of inorganic chloride in tobacco (Johnson & Smith, 1970).

It has been estimated by the U.S. National Academy of Sciences (NAS, 1978) that tobacco smoking world-wide results in about 20 million kilograms of chloromethane entering the atmosphere annually. Elevated levels of methyl chloride can occur in indoor air. For example, measurements in various contained atmospheres showed between 0.65 ppb and 8.00 ppb[1] in a restaurant and over 20 ppb in an apartment after a cigarette was smoked (Harsch, 1977). Methyl chloride was generally the predominant halomethane found in indoor air and was typically present at between two to ten times the outdoor level (Harsch, 1977; NAS, 1978). It was suggested by NAS (1978) that these elevated indoor levels may be due to cigarette smoking. It is of interest to note that the mean concentrations of methyl chloride over Los Angeles, Phoenix and Oakland in one study were reported to be 3.00 ppb, 2.39 ppb and 1.07 ppb, respectively. Methyl chloride from natural sources is present in the atmosphere at the parts per trillion level (e.g., ocean waters) and at the parts per billion level in urban atmospheres from anthropogenic sources (e.g., cigarette smoke). EPA (1980) has suggested that the local high concentrations of methyl chloride (at parts per million levels) found in occupational settings present the greatest risk to health known to result from exposure to that compound.

Consumer products

The list of consumer products containing halocarbons which may contribute to indoor pollution is quite extensive and includes both aerosol and liquid products; e.g., deodorants and antiperspirants, hair products, mouth products, vaporizers, inhalants containing bronchodilator drugs, women's personal hygiene products, room fresheners, bathroom and window cleaners, cleaning agents and spot removers, furniture polish, spray paints and adhesives, frying-pan sprays, pesticides, nail-polish removers and miscellaneous products for home hobbies and/or crafts. Active ingredients vary from one product to another and complete list of propellant(s) and active ingredients rarely appears on a can.

The propellant gases which were extensively used in the past included vinyl chloride, dichlorodifluoromethane (F-12) and trichlorofluoromethane (F-11). Methylene chloride is now increasingly employed as the propellant in aerosol dispensers, in place of the fluorocarbons.

---

[1] Parts per 10^9

Spray paints often contain methyl chloride, toluene and xylene, while cleaning agents contain tetrachloroethylene, 1,1,1-trichloroethane (methyl chloroform) and petroleum-derived solvents (National Research Council, 1981). Aerosol spot removers used for cleaning and/or protecting clothing, carpets, upholstery and wall-paper may frequently contain tetrachloroethylene (perchloroethylene) or 1,1,1-trichloroethane (Otson et al., 1984) as the active ingredient. The amount of tetrachloroethylene in such products could be similar to the quantity of propellant. Bridbord et al. (1975) have suggested that, under these circumstances, peak concentrations of this agent in excess of 100 ppm might occur during and immediately following spraying. Another source of indoor exposure to halogenated hydrocarbons can arise from solvents such as trichloroethylene (TCE), which has been used as a dry-cleaning and degreasing agent. Under certain situations where air exchange may be low, TCE could build up to appreciable airborne concentrations. Bridbord et al. (1975) tabulated indoor levels for some commonly used solvents by assuming that these attain a concentration of only 0.1% of their saturation level in air (Table 1. It should be noted that carbon tetrachloride, trichloroethylene, tetrachloroethylene and 1,1,1-trichloroethane have been detected in both ambient and indoor air (Capurro, 1973; Hester et al., 1974; Otson et al., 1984).

Table 1. Predicted indoor concentrations of halogenated hydrocarbon solvents in air[a]

Solvent	Temperature (°C)	Indoor concentration in air[a] (ppm)
Trichloroethylene	20	80
Tetrachloroethylene	26	26
1,2-Trichloroethane	21	26
1,1,1-Trichloroethane	20	130
Carbon tetrachloride	23	130
Methylene chloride	24	520

[a] Corresponding to 0.1% of saturation levels (Bridbord et al., 1975)

In recent studies reported by Otson et al. (1984), airborne levels of 1,1,1-trichloroethane (as well as propane and petroleum distillates) generated during the use of aerosol-type fabric protectors were monitored by means of the NIOSH charcoal tube (P & CAM 127) (NIOSH, 1977a,b), a glass bulb grab sampler and the "Gas badge" (Abcor, 1978) passive device techniques. Although 1982 ACGIH threshold limit values - short-term exposure limit (TLV-STEL) (ACGIH, 1982) were readily exceeded in an unventilated test room, the generated vapors quickly dispersed and STEL and 8-hour time-weighted average (TWA)-STEL were not exceeded when fabric was sprayed with 450 g of fabric protector in an unconfined area within a home. Hence, it was suggested that even small quantities of these products should not be used in a confined space. For example, levels of 1,1,1-trichloroethane ranged from 1 430 mg/m³ in an unventilated home to 143 mg/m³ in a ventilated room. In another case, the level of 1,1,1-trichloroethane in an unventilated room was 16 mg/m³, compared to undetectable amounts of the agent in a ventilated room.

The potential hazards due to the use of paint removers containing methylene chloride were reported by Stewart and Hake (1976). The use of these materials in a large interior room results in the absorption of a significant amount of solvent, its prompt metabolism to CO and an elevation of blood carboxy hemoglobin (COHb) level. The greater the minute-respiratory volume or the poorer the room ventilation, the greater the absorption of methylene chloride and the higher the COHb level. Use of the paint remover for a period of 3 hours following the directions on the label can easily produce a COHb level of 5% to 10% of saturation. Exposure for longer periods or under conditions of poorer ventilation would result in even higher COHb elevations.

It is difficult to make health hazard assessments for methylene chloride exposure levels arising from intermittent use of paint removers in a home environment because the necessary limits have not been established. It was found by Otson et al. (1981) that when more than 200 g of the paint removers were applied in an unventilated test room, the methylene chloride 1 and 8 h TWA values at 3 locations usually exceeded the 0.87 g/m³ ACGIH TLV-STEL. Grab sample values obtained after 30, 60, 240 and 480 min usually exceeded this value as well. In this study, both the NIOSH charcoal tube and a glass bulb grab sampling method were employed, with the latter exhibiting a sampling efficiency of 95%, precision of better than ± 5% and a detection limit of 0.01 g/m³, which was useful in determining the variation with time of methylene chloride levels in air, using gas chromatography with flame-ionization detection.

It should be noted that more than 30% of the methylene chloride consumed in the U.S. was used as a paint remover in 1980. The use of methylene chloride as an aerosol propellant, however, has increased rapidly in the past few years and represented 20% of total consumption in 1980 (Anon, 1980). This increased use of methylene chloride in a wider variety of products and processes can result in increased occupational and consumer exposure (Skrabalak & Babish, 1983).

It is also important to cite the comparison of ambient and indoor air concentrations of carbon tetrachloride near a solvent recovery plant in Maryland. The levels indoors were occasionally 3 to 4 times those measured outdoors, concentrations of 10-45 ppm were measured inside a house near the plant when levels outside were 1 ppm, and the highest indoor concentration recorded was 90 ppm (Bridbord et al., 1975).

The earlier use of fluorocarbons as propellants in a large variety of aerosol products has been mentioned above. Given the frequent sequential use of aerosol products containing the most commonly employed fluorocarbons, F-11 (trichlorofluoromethane) and F-12 (dichlorofluoromethane), as hair sprays and deodorants, peak exposures to mixtures of these fluorocarbons well in excess of 100 ppm might result following combined use of such aerosol products in the indoor environment (Bridbord et al., 1975). Levels of F-12 in homes have been reported to exceed 500 ppb and were generally much greater than simultaneous outdoor concentrations, which were usually 1 ppb or less (Hester et al., 1974).

Other fluorocarbons that have been employed as propellants in aerosols widely used in consumer products (hair products, deodorants and antiperspirants, mouth products, inhalants) include F114 (dichlorotetrafluoroethane), F152a (difluoroethane) and F1426 (monochloridifluoroethane), as well as the halocarbons, trichloroethane and methylene chloride (Bridbord et al., 1975).

Industrial and chemical waste disposal sites

In recent years, there has been increasing recognition that numerous sites exist throughout the United States (as well as in most industrialized countries) where waste chemicals have been dumped carelessly and in profusion.

Of particular notoriety is the "Old Love Canal" site in Niagara Falls, New York. This site was used as a toxic chemicals dump for a 25-30 year period up to 1953, and in the late 1950's a housing development was built on adjacent land. Heavier than normal rains over several years have raised the ground water level, causing buried chemicals to begin seeping into adjacent basements (Beck, 1979). In 1978, analyses in some homes in the area revealed chemical compounds accumulating in basements (EPA, 1978) and elevated concentrations of halogenated hydrocarbons were found in the indoor air (Barkley et al., 1980; Pellizzari, 1980). The estimated levels of 10 halogenated compounds found in air in homes in the Old Love Canal area is shown in Table 2 (Barkley et al., 1980). In addition, Barkley et al. (1980) made a quantitative gas chromatographic-mass spectrophotometric determination of the levels of the halogenated compounds in the breath, blood and urine of the exposed population (Old Love Canal area) and the immediate environment (air and water). Levels of halogenated hydrocarbons in air samples from the general area of Buffalo - Niagara Falls were also determined.

The levels of the volatile halogenated compounds found in the air inside the homes were generally lower than those found outside. Many of the site-specific volatile halogenated compounds identified in air were also found in the drinking water. In addition, these volatile halogenated hydrocarbons also occurred as human body burden, as demonstrated by their presence in urine, blood and breath. In the absence of a wider data base, it was not possible to determine whether or not the levels of halogenated compounds found as body burdens in this study were elevated.

Table 2. Estimated levels of halogenated compounds in air inside homes in "Old Love Canal" area of Niagara Falls, New York[a]

Halogenated compound	Household participant number								
	1	2	3	4	5	6	7	8	9
Chloroform	T[b]	3 400	1 400	2 400	15 000	3 600	950	11 000	3 000
1,2-Dichloro-ethane	-[c]	-	-	-	-	-	100	-	130
1,1,1-Trichloro-ethane	-	T	600	1 200	-	360	430	1 000	600
Carbon tetra-chloride	-	T	700	T	-	30	340	870	350
Trichloro-ethylene	-	-	T	T	-	T	T	-	T
Tetrachloro-ethylene	-	-	600	1 300	-	540	480	800	420
Chlorobenzene	-	-	300	600	-	-	60	-	-
Dichlorobenzene isomer(s)	T	T	500	4 000	170	120	230	280	31 000
Chlorotoluene isomer(s)	-	-	T	1 500	-	-	-	-	-
Trichlorotoluene	-	-	T	-	-	-	-	-	-
Sampling period[d]	10:41-16:30	09:29-15:11	11:03-17:37	08:45-17:20	10:22-17:47	15:48-08:58	16:55-09:26	21:13-08:35	18:00-08:07

[a] Barkley *et al.* (1980), values in mg/m³

[b] Trace

[c] Not detected

[d] Time at beginning (-end) of sample collection

Ambient air in basements of 11 homes and 2 elementary schools near the Old Love Canal dump site were measured in another study (Pellizzari & Bunch, 1979; Pellizzari, 1982). Over 200 chemicals (42 halogenated) were detected by capillary gas chromatography-mass spectrometry-computer analysis. Among the halogenated compounds found in ambient air of households were the aliphatic derivatives 1,2-dichloropropane, 1,3-hexachlorobutadiene and pentachlorbutadiene.

MISCELLANEOUS STUDIES OF HALOCARBONS IN INDOOR AIR

In a study of organic contaminants of indoor air and their relation to outdoor contaminants in the Chicago, Illinois area, over 250 organic compounds (including methyl chloride, methylene dichloride and chloroform) were identified, with average concentrations being about 100 ppb. The frequency and variety of compounds with levels > 1 ppb was greater indoors than outdoors. Surprisingly, the location relative to the industrial operations of the community had little effect on the contaminants found in the homes. In summer, averages of 37 and 16 compounds per home were found indoors and outdoors, respectively, compared to 18 and 11 compounds, respectively, in the winter (Jarke et al., 1981).

Office environments, where small quantities of materials containing volatile ingredients are used occasionally have received little attention (Hollowell & Miksch, 1981; Otson et al., 1983). Materials such as glues, adhesives, thinners, cleaning solvents, type correction fluids, coloring pens, copy machine chemicals, cosmetics and aerosol deodorizers are used in many offices and may result in individual exposure to volatile halocarbons (e.g., dichloromethane from aerosols and 1,1,1-trichloroethane from type correction fluids and cleaning solvents). In a recent study by Otson et al. (1983), the charcoal tube (NIOSH, 1977a-c), the PRO-TEK badge (Dupont, 1980) and Gasbadge (Abcor, 1978) passive sampling devices were employed. These were designed for the monitoring of volatile organic compounds in air and have a range of about 1 ppm/h to more than 1000 ppm/h. Of the 51 organic compounds which could be determined (detection limits, ≃ 0.2 ppm), only dichloromethane, tetrachloroethylene and toluene were identified at levels greater than 0.2 ppm in the air of 8 of the 30 offices.

Chlorinated hydrocarbons, predominantly tetrachloroethylene, 1,1,1-trichloroethane, and trichloroethylene were found at levels of 1 to 100 ppb in the air of a number of office buildings surveyed by Hollowell & Miksch (1981).

ANALYTICAL TECHNIQUES

As for the determination of halocarbons in workplace and ambient atmospheres, the measurement of air concentrations of the halogenated hydrocarbons in indoor environments are mainly carried out by gas chromatography with electron-capture detection (EC-GC) and gas chromatography-mass spectrometry. Long-path IR absorption spectroscopy, usually with preconcentration of whole air, followed by separation of the compounds by GC, is employed to a lesser extent.

Most atmospheric values reported for these compounds have come from EC-GC analyses, which are extremely sensitive for most of the halogenated hydrocarbons and are capable of measuring levels of a few parts per trillion (10^{12}) for such compounds as carbon tetrachloride in sample volumes of only 5 or 10mL of air. While compounds such as carbon tetrachloride, the fluorocarbons F-11 and F-12 and other halocarbons can be easily detected down to 1 ppt in the atmosphere, others, such as methyl chloride and dichloromethane, require concentration by 2 or 3 orders of magnitude (NAS, 1978).

Because of the relatively low concentrations of organic contaminants in ambient air, most GC procedures require a preconcentration step. Early methods involved the use of charcoal or silica gel to adsorb volatile organic compounds, followed by elution with carbon disulfide and analysis of the eluate by GC, using flame-ionization detection (NIOSH, 1977a-c; Kring et al., 1984) or mass spectrometry. A major disadvantage of employing carbon disulfide as a solvent is that it cannot be used with EC or photoionization detectors. Additionally, solvent scrubbing and cryogenic concentration were also employed, but because of dilution, most solvent-scrubbing techniques are insufficiently sensitive for analyses in the parts per billion (10^9) range.

More recently, commercially-available polymers (Brown & Purnell, 1979; Russell & Shadoff, 1977), especially Tenax GC (Pellizzari et al., 1976a,b, 1978; Michael et al., 1980; Kebbekus & Bozzelli, 1982; Krost et al., 1982), Porapak N (Van Tassel et al., 1981) and charcoal (Raymond & Guiochon, 1975; Patzelova et al., 1978), are being increasingly used for concentrating low levels of halo-organic compounds in air.

Gas chromatography-mass spectrometry (GC-MS) is a more recently developed method for the analysis of halogenated hydrocarbons in ambient atmospheres (Cronn & Harsch, 1976; Pellizzari et al., 1976a,b; Rasmussen et al., 1977; Barkley et al., 1980; Jonsson & Berg, 1980; Krost et al., 1982; Pellizzari, 1982). One of the advantages of GC-MS over EC-GC is that the sensitivity of GC-MS is the same for all compounds. The maximum sensitivity of this method is about 0.5 picograms of a compound in a 20-mL sample. GC-MS provides certain identification and is usually not affected by other compounds which may be eluted from the chromatograph. The GC-MS (and the more elegant GC-MS-computer) technique has been particularly useful in the analysis of volatile halocarbons in multimedia environments, including indoor air in residences near hazardous sites (Barkley et al., 1980; Krost et al., 1982; Pellizzari, 1982).

It is also important to cite the utility of personal (passive) samplers for halocarbons, as noted above. These include the "Gasbadge" passive device technique (Abcor, 1978) and Pro-Tek (Dupont, 1980), which are alternatives to the charcoal tube. These techniques allow the vapors to enter the sampler by molecular diffusion and thus do not require the use of sampling pumps or electrical power. The collection element (charcoal-impregnated pads) is removed from the sampler and subsequently analysed for organic vapors (e.g., 1,1,1-trichloroethane, trichloroethylene) using GC with flame-ionization detection (Bamberger et al., 1978; Otson et al., 1984).

REFERENCES

Abcor (1978) Gasbadge™ Organic Vapor Dosimeter Use and Analysis Instructions, Wilmington MD, Abcor Development Corp.

Abraham, H.J., Nagel, R. & Seifert, B. (1981) Passivsammler nach dem Diffusionsprinzip als Hilfsmittel zur Bestimmung der indivuellen Schadstoffbelastung in aussen-und innenluft. Schr. Reihe Verein Wasser, Bodenkd. Luft Hygiene, 52, 363-380

ACGIH (1982) TLVs Threshold Limit Values for Chemical Substances and Physical Agents in the Work Environment with Intended changes for 1982, American Conference of Governmental Industrial Hygienists, Cincinnati, OH

Anon. (1980) Key chemicals: Methylene chloride. Chem. Eng. News, 58, 11

Bamburger, R.L., Esposito, G.G., Jacobs, B.W., Podalak, G.E. & Mazur, J.F. (1978) A new personal sampler for organic vapors. Am. Ind. Hyg. Assoc. J., 39, 701-708

Barkley, J., Bunch, J., Bursey, J.T., Castillo, N., Cooper, J.D., Davis, J.M., Erickson, M.D., Harris, B.S.H., III, Kirkpatrick, M., Michael, L.C., Parks, S.P., Pellizzari, E.D., Ray, M., Smith, D., Tomer, K.B., Wagner, R. & Zweidinger, R.A. (1980) Gas chromatography mass spectrometry computer analysis of volatile halogenated hydrocarbons in man and his environment - a multimedia environmental study. Biomed. Mass Spectrom, 7, 139-147

Beck, E.C. (1979) The Love Canal tragedy. Environ. Prot. Agency J., 5,16-19

Bridbord, K., Brubaker, P.E., Gay, B., Jr & French, J.G. (1975) Exposure to halogenated hydrocarbons in the environment. Environ. Health Perspect., 11, 215-220

Brown, R.H. & Purnell, C.J. (1979) Collection and analysis of trace organic vapor pollutants in ambient atmospheres. J. Chromatogr., 178, 79-90

Capurro, P.U. (1973) Effects of exposure to solvents caused by air pollution with special reference to CCl_4 and its distribution in air. Clin. Toxicol., 6, 109-124

Cronn, D.R. & Harsch, D.E. (1976) Rapid determination of methyl chloride in ambient air samples by GC-MS. Anal. Lett., 9, 1015-1023

Dirinck, P., Veys, J., Decloedt, M. & Schamp, N. (1980) Head-space enrichment on Tenax for characterization and flavor evaluation on some tobacco types. Tob. Int., 182, 125-129

Dube, M.F. & Green, C.R. (1982) Recent Adv. Tob. Sci., 8, 42-102

DuPont (1980) Pro-Tek™ Organic Vapor Air Monitoring Badge Instruction Manual G-AA/G-BB, Wilmington DL, E.I. DuPont Nemours & Co.

EPA (1980) Support Document Health Effects Test Rule: Chloromethane, EPA-560/11-80-015, Washington DC, Office of Pesticides and Toxic Substances

GAO (1980) Indoor Air Pollution: An emerging Public Health Problem, Government Accounting Office, Washington DC, Government Printing Office

Greenberg, R.A., Haley, N.J., Etzel, R.A. & Loda, F.A. (1984) Measuring the exposure of infants to tobacco smoke. New Engl. J. Med., 310, 1075-1078

Guthrie, F.F. & Bowery, P.G. (1967) Pesticide residues in tobacco. Residue Rev., 19, 31-56

Harsch, D. (1977) Study of halocarbon concentrations in indoor environments. Final Report, Contract No. WA6-99-2922-J, Washington DC, US Environmental Protection Agency

Hester, N.E., Stephens, E.R. & Taylor, O.C. (1974) Fluorocarbons in the Los Angeles Basin. J. Air Pollut. Control Assoc., 24, 591-595

Hoffmann, D., Patrianakos, C., Brunnemann, K.D. & Gori, G.B. (1976) Chromatographic determination of vinyl chloride in tobacco smoke. Anal. Chem., 48, 47-50

Hoffmann, D., Haley, N.J., Brunnemann, K.D., Adams, J.D. & Wynder, E.L. (1983) Cigarette smoke: formation, analysis and model studies on the uptake by nonsmokers. Presented at the US-Japan Meeting on "New Etiology of Lung Cancer", Honolulu, Hawaii, March 21-23

Hoffmann, D., Haley, N.J., Adams, J.D. & Brunnemann, K.D. (1984) Tobacco sidestream smoke: uptake by nonsmokers. Presented at International Symposium on Medical Perspectives on Passive Smoking, Vienna, Austria, April 9-12

Hollowell, C.D. & Miksch, R.R. (1981) Sources and concentrations of organic compounds in indoor environments. Bull. NY Acad. Med., 57, 962-977

Holzer, G., Oro, J. & Bertsch, W. (1976) Gas chromatographic-mass spectrometric evaluation of exhaled tobacco smoke. J. Chromatogr., 126, 771-785

Jarke, F.H., Dravnieks, A. & Gordon, S.M. (1981) Organic contaminants in indoor air and their relation to outdoor contaminants. Am. Soc. Heat. Refrig. Air-Cond. Trans., 87, 153-166

Johnson, W.R. & Smith, T.E. (1970) Components of mainstream and sidestream smoke. Abstr. 24th Tobacco Chemists Res. Conf., 24

Johnson, W.R., Hale, R.W., Nedlock, J.W., Grubbs, H.J. & Powell, D.H. (1973) Distribution of products between mainstream and sidestream smoke. Tobacco Sci, 17, 141-144

Jonsson, A. & Berg, S. (1980) Determination of 1,2-dibromoethane, 1,2-dichloroethane and benzene in ambient air using porous polymer traps and gas chromatographic-mass spectrometric analysis with selected ion monitoring. J. Chromatogr., 190, 97-106

Kadaba, P.K., Bhagat, P.K. & Goderger, G.N. (1978) Application of microwave spectroscopy for simultaneous detection of toxic constituents in tobacco smoke. Bull. Environ. Contam. Toxicol., 19, 104-112

Kebbekus, B.B. & Bozzelli, J.W. (1982) Determination of selected toxic organic vapors in air by adsorbent trapping and capillary gas chromatography. J. Environ. Sci. Health, Part A, A17, 713-723

Klus, H. & Kuhn, H. (1982) Distribution of various tobacco smoke constituents in main and sidestream smoke (a review). Beitr. Tabakforsch., 11, 229-265

Kring, E.V., Ansul, G.R., Henry, T.J., Morello, J.A., Dixon, S.W., Vasta, J.F. & Hemingway, R.E. (1984) Evaluation of the standard NIOSH type charcoal tube sampling method for organic vapors in air. Am. Ind. Hyg. Assoc. J., 45, 250-259

Krost, K.J., Pellizzari, E.D., Walburn, S.G. & Hubbard, S.A. (1982) Collection and analysis of hazardous organic emissions. Anal. Chem., 54, 810-817

Michael, L.C., Erickson, M.D., Parks, S.P. & Pellizzari, E.D. (1980) Volatile environmental pollutants in biological matrices with a neadspace purge technique. Anal. Chem., 52, 1836-1841

NAS (1978) Non-fluorinated halomethanes in the environment, Washington DC, National Academy of Sciences

National Research Council (1981) Indoor Pollutants, Washington DC, National Academy Press

NIOSH (1977a) Manual of Analytical Methods, 2nd Ed., Vol. 1, (DHEW (NIOSH) Publication No. 77-157-A) Method No. S-127, Cincinnati OH, National Institute of Occupational Safety and Health

NIOSH (1977b) Manual of Sampling Data Sheets, NIOSH Publ. 77-159, Cincinnati OH, National Institute of Occupational Safety and Health

NIOSH (1977c) Documentation of the NIOSH Validation Tests Publication No. 77-185, Washington DC, National Institute of Occupational Safety and Health, US Department of Health, Education and Welfare

Otson, R., Williams, D.T. & Bothwell, P.D. (1981) Dichloromethane levels in air after application of paint removers. Am. Ind. Hyg. Assoc. J., 42, 56-60

Otson, R., Doyle, E.E., Williams, D.T. & Bothwell, P.D. (1983) Survey of selected organics in office air. Bull. Environ. Contam. Toxicol., 31, 222-229

Otson, R., Williams, D.T. & Bothwell, P.D. (1984) Fabric Protectors. II. Propane, 1,1,1-trichloroethane and petroleum distillates level in air after application of fabric protectors. Am. Ind. Hyg. Assoc. J., 45, 28-33

Patzelova, V., Jansta, J. & Dousek, F.P. (1978) Some properties of a new type of active carbon. J. Chromatogr., 148, 53-59

Pellizzari, E.D. (1980) Improvement of Methodologies and Analysis of Carcinogenic Vapors, Contract No. 68-02-2764, Research Triangle Park NC, US environmental Protection Agency

Pellizzari, E.D. (1982) Analysis for organic vapor emissions near industrial and chemical waste disposal sites. Environ. Sci. Technol., 16, 781-785

Pellizzari, E.D. & Bunch, J.E. (1979) Ambient Air Carcinogenic Vapors: Improved Sampling and Analysis Techniques and Field Studies, EPA-600/-2-79-081, Research Traingle Park NC, US Environmental Protection Agency

Pellizzari, E.D., Bunch, J.E., Berkley, R.E. & McRae, J. (1976a) Collection and analysis of trace organic vapor pollutants in ambient atmosphere. The performance of a Tenax GC cartridge samples for hazardous vapors. Anal. Lett., 9, 45-63

Pellizzari, E.D., Bunch, J.E., Berkley, R.E. & McRae, J. (1976b) Determination of trace hazardous organic vapor pollutants in ambient atmospheres by gas chromatography/mass spectrometry/computer. Anal. Chem., 48, 803-806

Pellizzari, E.D., Carpenter, B.H., Bunch, J.E. & Sawicki, E. (1978) Collection and analysis of trace organic vapor pollutants in ambient atmospheres. Environ. Sci. Technol., 9, 556-560

Pellizzari, E.D., Erickson, M.D. & Zweidinger, R.A. (1979) Formulation of a Preliminary Assessment of Halogenated Organic compounds in Man and Environmental Media, EPA-560/13-79-006, Research Triangle Park NC, US Environmental Protection Agency

Rasmussen, R.A., Harsch, D.E., Sweany, P.H., Krasnec, J.P. & Cronn, D.R. (1977) Determination of atmospheric halocarbons by a temperature-programmed gas chromatographic freezeout concentration method. J. Air Pollut. Control, 27, 579-581

Raymond, A. & Guiochon, G. (1975) The use of graphitized carbon black as a trapping material for organic compounds in light gases before a gas chromatographic analyses. J. Chromatogr. Sci., 13, 173-177

Replace, J.L. (1981) The problem of passive smoking. Bull. NY Acad. Med., 57, 936-946

Russell, J.W. & Shadoff, L.A. (1977) The sampling and determination of halocarbons in ambient air using concentration on porous polymer. J. Chromatogr., 134, 375-384

Schmeltz, I., Hoffmann, D. & Wynder, E.L. (1975) Influence of tobacco smoke on indoor atmosphere. Prev. Med., 4, 66-92

Siefert, B. & Abraham, H.J. (1982) Indoor air concentrations of benzene and some other aromatic hydrocarbons. Ecotoxicol. Environ. Saf., 6, 190-192

Skrabalak, D.S. & Babish, J.G. (1983) Safety standards for occupational exposure to dichloromethane. Regul. Toxicol. Pharmacol., 3, 139-143

Sloan, C.H., Lewis, J.S. & Morie, G.P. (1977) Computerization of gas-phase analysis of cigarette smoking. Tob. Sci., 21, 57

Spengler, J.D. & Sexton, K. (1983) Indoor air pollution: A public health perspective. Science, 221, 9-17

SRI (1979) Atmospheric Measurements of Selected Toxic Organic Chemicals, Menlo Park CA, Stanford Research Institute International

Stedman, R.L. (1968) Chemical composition of tobacco and tobacco smoke. Chem. Rev., 68, 153

Stewart, R.D. & Hake, C.L. (1976) Paint-remover hazard. J. Am. Med. Assoc., 235, 398-401

Tso, T.C. (1972) Physiology and Biochemistry of Tobacco of Plants, Stroudsburg PA, Dowden Hutchinson and Ross Publ.

U.S. Surgeon General (1982) The Health Consequences of Smoking and Cancer, DHHS(PHS)82-50179:212-213, Washington DC, Department of Health and Human Services

Van Tassel, S., Amalfitano, N. & Narang, R.S. (1981) Determination of arenes and volatile haloorganic compounds in air at micogram per cubic meter levels by gas chromatography. Anal. Chem., 53, 2130-2135

Weiss, S.T., Tager, I.B., Schenker, M. & speizer, F.E. (1983) The health effects of involuntary smoking. Am. Rev. Respir. Dis., 128, 933-942

WHO (1979) Health Aspects Related to Indoor Air Quality, Geneva, Switzerland, World Health Organization

Wynder, E.L. & Hoffmann, D. (1967) Tobacco and Tobacco Smoke, Studies in Experimental Carcinogenesis, New York NY, Academic Press

CHAPTER 6

WORKPLACE AIR-SAMPLING FOR GASES AND VAPOURS: STRATEGY, EQUIPMENT, PROCEDURE AND EXPOSURE LIMITS

B. Goelzer

Office of Occupational Health
Division of Non Communicable Diseases
World Health Organization
Avenue Appia
CH-1211 Geneva 27, Switzerland

I.K. O'Neill

International Agency for Research on Cancer
Lyon, France

I. INTRODUCTION

Workers are often exposed to environmental factors and stresses which may have a negative impact on their health and well-being. Among these factors are airborne contaminants, which pose health hazards of varying degrees. Airborne contaminants can be gases, vapours or particles (dust, fumes, mists), and may occur in dangerous concentrations without adequate warning, particularly those in the gaseous state. Many chemicals, moreover, are not well-detected by the senses, especially in a work environment where the multiplicity of factors may mask perception of contaminants without striking effects or properties, such as irritation and odour. Air sampling and analysis has been well developed for halogenated alkanes and alkenes and the analytical methods for air described in this volume (derived mainly from the NIOSH Manuals of Analytical Methods) must be considered in a well-developed framework of sampling methods and strategy.

---

* The present chapter, addressed specifically to the needs of this volume, is a condensed and amended version of a more extensive background document prepared by B. Goelzer and which may be obtained directly from the author at the Office of Occupational Health, WHO, Geneva.

To illustrate the concentration ranges of interest, occupational exposure limits are presented in Table 1, which also indicates the appropriate analytical method in this volume. The purpose of this chapter is to set methods 1 through 11 in the context of the considerable practical difficulties of estimating exposures for groups of workers. Recent advances in biological monitoring and in environmental monitoring equipment hold out the promise of better procedures eventually; however, extensive use of personal monitoring by pumped air samples adsorbed on charcoal will continue. Without expert consideration of planning the sampling strategy, environmental measurements described in methods 1 through 11 will not themselves advance an exposure assessment much beyond what could be obtained by epidemiological techniques, coupled with a few measurements of work-place air; it should be noted that human carcinogenicity studies described in Chapter 1 used job categorisation and questionnaire methods for approximating exposures many years before the advent of the approaches and techniques briefly described in this chapter. Space does not allow a full discussion of all industrial hygiene and toxicological considerations, and the chapter is focussed on exposure monitoring relevant to the (possibly) carcinogenic nature of the substances listed in Table 1.

Table 1. Occupational exposure limits for airborne halogenated alkanes and alkenes (mg/m³) - Analytical method number in this volume

Formula or designation	ACGIH TLV (1984-1985)[a]		DFG MAK values 1984	Method number in this volume for determining airborne levels
	TWA	STEL		
CH_3Cl	105	205	105	7
CH_3Br	20	60	20	8
CH_2Cl_2	350	1740	360	3
$CHCl_3$	50,C	225,C	50	1
Bromoform	5	-	-	1
F-11	5600	-	5600	-
F-22	3500	4375	-	-
CCl_4	30,C	125,C	65	1
1,2-Dichloroethane	40	60	80	1
EDB	C	C	-	4, 11
Methylchloroform	1900	2450	1080	1
Hexachloroethane	100	-	10	1
Halothane	400	-	40	-
Vinylidene chloride	20	80	40	-
1,2-Dichloroethylene	790	1000	790	1
Trichloroethylene	270	1080	260	5
Tetrachloroethylene	335	1340	345	5
Allyl chloride	3	6	3	2
DBCP	-	-	-	11
Epichlorohydrin	10	20	-	6
BCME	0.005, C	-	-	10
CMME	C	C	-	9

[a] 'C' indicates that the compound is considered by ACGIH (ACGIH, 1984), to be a known or suspected carcinogen. TWA is time-weighted average for 8 h, and STEL is the short term exposure limit for 15 min.

The mention of specific recommended values does not in any way represent their endorsement or recommendation either by the authors of this chapter, by IARC or WHO.

2. PRELIMINARY CONSIDERATIONS

The evaluation of airborne contaminants involves consideration of many factors and their integration in a well-designed plan. In order to establish whether or not a health hazard exists as a consequence of exposure to airborne contaminants, it is necessary to a) recognize the hazards, and b) evaluate them qualitatively and quantitatively.

In order to recognize potential hazards (Stellman & Baum, 1973; Key et al., 1977; Patty, 1978; Burgess, 1981; Cralley & Cralley, 1982), it is necessary to have a good knowledge of the work process(es) involved, and of potential pollution sources, raw materials and chemicals utilized. These should include possible impurities, products and by-products, and chemicals accidentally formed and released. An example of a possible "hidden" hazard is bis(chloromethyl)ether, an impurity in other types of ethers (e.g., in chloromethyl ether).

Most of the information needed to recognize hazards can be obtained through a preliminary survey, during which a list of all materials and chemicals utilized should be requested. The toxicity and carcinogenicity of materials should be reviewed, so that priorities can be established before planning the evaluation. Information concerning occupations associated with specific chemicals can be found in the specialized literature (Key et al., 1977; NIOSH, 1981; INRS Fiches Toxicologiques). Part of the recognition procedure recommended by NIOSH (1984) is to furnish the analytical laboratory with 'bulk' samples of workplace air and materials being used there, so that a qualitative analysis may be made before the air monitoring samples are analysed.

One should begin by obtaining a list of chemicals purchased by the plant, as well as their rate of consumption, and the number of workers exposed to the hazards should also be ascertained. Determination of the operations which are most hazardous and/or which workers are likely to be at greatest risk will help to set priorities for hazard evaluation. When the actual sampling is planned, priorities have to be set taking into consideration the severity of the potential health hazard to the worker, as well as the number of workers exposed.

Evaluation of the health hazard posed by an airborne contaminant involves not only the determination of its concentration, but also the determination of the exposure time and probable routes of entry. Of primary concern is the uptake of the substance by the workers. In this respect, the concentration of the substance in the air is one of the determining factors. Other factors include entry of material into the body through skin absorption and ingestion, and the physical activity of the workers (which affects their respiration rate), as well as the eventual use of respirators.

The determination of exposure times and possible routes of entry is achieved by careful observation of the work processes. The cyclic nature of some processes, which may involve daily or seasonal variations causing significant changes in the concentration of chemical agents, must be taken into account. Therefore, the observation period must be long enough to identify the most severe situations. Certain occupational hygiene services have been utilizing video equipment to record working procedures over several hours; the tapes can then be watched later in the office, where replays are possible and the observation is not affected by factors such as heat, noise, irritants, lack of time, etc.

The fact that some chemicals have the ability to penetrate intact skin should not be overlooked when conducting surveys. A worker may, for instance, have his hands constantly dipped in a cleaning solvent which, due to factors such as low volatility, temperature, or exhaust ventilation, is not contaminating the air above permissible levels. Examples of halogenated alkanes and alkenes which can penetrate intact skin include allyl chloride, bromoform, carbon tetrachloride, epichlorohydrin, ethylene dibromide, hexachloroethane, methyl bromide, trichloroethylene and 1,1,1-trichloroethane (ACGIH, 1984, 1985; NIOSH, 1981; Key et al., 1977).

When uptake results from routes of entry other than inhalation, the evaluation of exposure cannot be accomplished solely through air sampling and analysis. In such cases, and also when dealing with certain variable processes, biological monitoring (breath, blood or urine analysis, metabolite analysis, etc.) is a very useful tool for the evaluation of the workers' exposure and uptake. This is discussed in Chapter 7 and methods for biological monitoring are presented in this volume for exposure to numerous compounds.

When the objective is to determine whether a health hazard exists or whether an existing standard is being complied with, the concentration of the agent in question should be compared with available toxicological data or with the values specified in the standard. These values are designated by terms such as "maximum permissible levels" (ILO/WHO, 1969), "maximum allowable concentrations" or "threshold limit values" (Table 1, ACGIH, 1984), which can be defined as concentrations to which workers may be exposed for a working week of given duration (e.g., 40 h) over their entire working life without experiencing adverse health effects. Since the concept of an adverse health effect varies from one country to another, the maximum permissible levels adopted may differ; the concept may be very strict, e.g., the permissible level should, in no case, produce any biological or functional changes, or more flexible, e.g., the permissible level makes allowances for reversible clinical changes. However, when a substance has been found to be carcinogenic, other considerations apply; see for example the most recent A.C.G.I.H. compilation of TLV data (ACGIH, 1984).

It is essential to consult with the analytical laboratory before sampling, to assure that the measurement methods available can meet the defined sampling needs. This step should be an early part of survey planning. Certain numbers of blank samples are required by the analytical laboratory to control for contamination of solid sorbent tubes in use in the field or for analytical difficulties.

3. SAMPLING STRATEGY

The fundamental principle in designing a sampling plan is that sampling for the evaluation of workers' exposure be representative. Keeping in mind the factors which determine the representativeness of the evaluations, an adequate sampling strategy must be planned and followed. The necessity of obtaining the same level of accuracy and precision throughout the entire exposure evaluation (recognition of hazards, planning sampling strategy, choice of sampling procedures, analysis of samples and interpretation of results) cannot be over-emphasized. High-level precision and accuracy achieved in the analytical laboratory would be wasted if the samples analysed had been collected with uncalibrated pumps, thereby introducing gross errors in the air volume sampled. Similarly, results obtained through adequate techniques of air sampling and analysis would be meaningless outside the framework of an appropriate sampling strategy; correct results from samples collected at the wrong place or time are of limited or no use for the evaluation of exposures.

A sampling strategy must answer the questions, "where", "for how long", "when to sample" and "how many samples to collect". Sampling strategy is discussed in the specialized literature (Leidel *et al.*, 1977; Clayton & Clayton, 1978; Cralley & Cralley, 1979; Valic, 1984; WHO, 1984). The main factors to be considered are briefly treated here.

Where to sample

If the objective is to determine whether a health hazard exists, or to establish a dose-response relationship, the sample must be representative of the workers' exposure, and sample collection in the breathing zone of the worker is therefore recommended. "Breathing zone" is defined as a hemispherical zone, with a radius of approximately 30 cm in front of the head. If the objective is to design and evaluate control measures, sampling sites may be selected in relation to the source(s) of pollution. Sampling devices suitable for halogenated alkanes and alkenes may be carried by the worker or fixed at points within the workplace.

However, if workers move around areas where the exposure to the chemical agent varies, the use of personal samplers is necessary. Since it is not feasible to monitor everyone, those workers subjected to the most severe exposure conditions should be selected. Direct-reading instruments are particularly useful in determining the sites where exposure is most severe.

Should the most severe exposures for selected workers fall within permissible levels, it might be concluded that this is so for every member of the group; otherwise, further studies are required. However, for studies concerning the response of workers to different levels of exposure, in terms of health effects, situations other than the most severe should also be investigated. When the purpose of sampling is to establish a dose-response relationship, the levels of exposure of each member of the group must be estimated.

Another approach is to group workers in "exposure zones" (Corn & Esmen, 1979; WHO, 1984) not necessarily defined geometrically, but wich should be identifiable, and should comprise workers who have similarity in respect to work practices, to exposure to hazardous agents and to physical environment. In order to identify workers as being in the same exposure zone, their work environment and practices, as well as their individual characteristics, must be carefully observed, since even those performing the same task may have different exposures because of factors such as, for example, anthropometric differences, variations in work practices, differences in the ventilation of the work station, etc. When a sufficient number of workers is sampled in each zone, the information obtained can be used to describe, by a mean value and a standard deviation, the exposure levels for all workers in the same zone, within a range of concentrations and within certain statistical confidence limits.

Duration of sampling

The decision concerning the duration of sampling largely depends on the sensitivity of the analytical method, concentration fluctuations and normally on the physiological action of the airborne contaminant. Brief periods of relatively high concentrations can remain undetected when a sample is collected over a long period, thereby missing physiologically important exposures for some fast-acting contaminants such as narcotics, irritants or acute systemic poisons. Although for some halogenated alkanes and alkenes biochemical and long-term exposure effects in relation to carcinogenicity are known, the relative importance of brief high exposures is not known.

Depending on the sensitivity of the analytical method and the range of concentrations expected, a minimum volume of air is required. If observation of the operations suggests that appreciable concentration fluctuations are to be expected (e.g., due to opening and closing of doors in dry cleaning units, or opening of polymerization reactors, etc.), this has to be taken into account when establishing the duration of sampling. On the other hand, there is a practical upper limit to the volume of the air sample in cases where the collection medium is likely to reach breakthrough, i.e., the kinetic adsorptive capacity of the solid sorbent. In such situations, this maximum volume, together with the recommended (or acceptable) sampling flow rate, determines the maximum feasible duration of sampling.

When to sample

Unless continuous sampling over the whole period of interest (e.g., a full work shift) is planned, it will be necessary to decide "when to sample", particularly when variations in concentration are likely to occur. Whenever peaks in concentration are expected, the strategy of sampling should be so designed as to detect them. If the nature of the work is such that no appreciable changes in concentration with time are expected, it is a convenient practice to collect the required samples at equal time intervals (e.g., at the end of every hour). If this is not the case, however, the most representative times for sampling must be selected. The pattern of air concentration of chemical agents in the workplace, and the pattern of workers' exposure, must both be observed carefully. Changes in air temperature, air movement and ventilation also affect exposure.

Number of samples

For each particular situation, the optimum number of samples can be determined by statistical methods. The statistical treatment of environmental data has been widely discussed (Hounam, 1965; Gale, 1967; Saltzman, 1972; Leidel & Busch, 1975; Leidel et al., 1977; Swinscow, 1977; Patty, 1979) and is beyond the scope of this chapter. For each location, an adequate number of samples must be collected to ensure the determination of concentration at the 95% (at least) confidence level. For practical reasons and for usual occupational hygiene surveys, many authors (NIOSH, 1973; Ulfvarson, 1975; Leidel et al., 1977) accept the following numbers of samples:

for continuous samples: 2 consecutive 4-hour samples (for an 8-hour TWA)
for "grab" samples or direct-readings: 4-7 samples (for each location)

Ulvarson (1977) has described a statistical evaluation of short-period samples for use when attempting to determine whether a given 8-hour TWA has been exceeded.

4. SAMPLING PROCEDURES AND EQUIPMENT

Introduction

Sampling procedures for occupational hygiene have been widely discussed elsewhere (NIOSH, 1973; Linch, 1974; Cralley & Cralley, 1979; ACGIH, 1983).

Three distinct methods have been employed:

(a) Utilisation of direct-reading instruments or colour-reaction tubes, which give immediate values corresponding to an instantaneous concentration; the same type of data is given by the analysis of grab or "spot" samples

(b) Continuous collection by adsorption from air, pumped during an extended period, usually on to charcoal or Tenax adsorbents, which gives the average concentration of the airborne contaminant over the sampling period.

(c) Utilisation of passive monitoring devices (often in the form of lapel badges) which yield the same type of data as adsorption from pumped air. At the time of writing, comparison of pumped and passive samplers is still in progress in various centres.

Manufacturers of equipment required for the above methods are listed at the end of this chapter, although only adsorbent tubes (b) are employed in the methods for air analysis described in this volume.

Direct- reading instruments

Direct-reading instruments are those which perform quantitative analysis for a specific airborne contaminant, or group of airborne contaminants, and directly indicate the result (NIOSH, 1973; Linch, 1974; McCammon, 1976). Since direct-reading instruments involve neither the usually time-consuming and expensive laboratory analysis, nor the problems posed by sample transportation and storage, they are very useful for field investigations. Their limitations, however, must be known and taken into consideration[1]. A non-comprehensive list of manufacturers is presented in Appendix A to this chapter.

(i) Detector tubes

These tubes are based on a colorimetric reaction; a known volume of contaminated air is drawn through a glass tube containing granules impregnated with a reagent which changes colour in the presence of the contaminant. The extent of the colorimetric reaction is proportional to the concentration of the contaminant. The most common detector tubes are those for short-term sampling e.g. as in Figure 1. Tubes for long-term sampling are also available, but require a continuous sampling pump (Fig. 2).

---

[1] The Office of Occupational Health, WHO, Geneva, is preparing a document on Direct-reading Instruments for Airborne Contaminants in the Work Environment, which gives detailed guidelines for the selection and use of these instruments.

FIG. 1 BELLOW PUMP WITH TUBE INSERTED FOR TRICHLOROETHYLENE MEASUREMENT

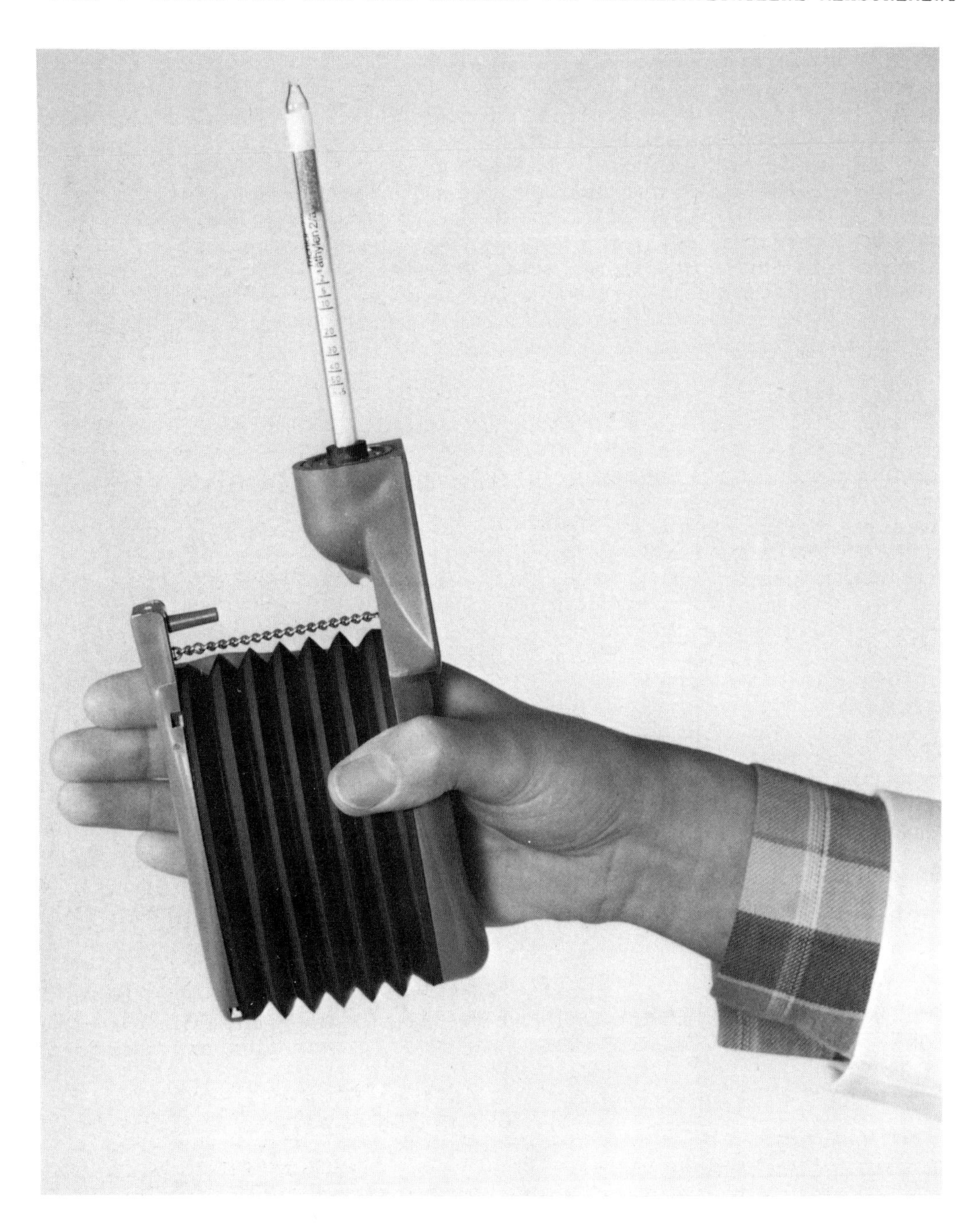

FIG. 2 WORKER WEARING LONG-TERM MEASURING SYSTEM CONSISTING OF PUMP ON SHOULDER STRAP AND DIRECT-READING LONG-TERM DETECTOR TUBE WITHIN BREATHING ZONE

The example shown is of a tube measuring 2.5 to 50 ppm trichloroethylene over sampling period of 4 hours.

The precision and accuracy usually obtained with colorimetric detector tubes is not sufficient for scientific research; for example, in establishing correlations between levels of exposure and health effects. Most detector tubes may not be sensitive enough for evaluating some of the very low concentration limits for carcinogens or suspected carcinogens.

Detector tubes have been widely discussed in the literature (Kitagawa, 1952; Saltzman, 1962; NIOSH, 1973; IUPAC, 1974; Linch, 1975, Roper, 1974; WHO, 1976; Leichnitz, 1977; NIOSH, 1978; Leichnitz, 1979; ACGIH, 1983; Leichnitz, 1983) and are available for the evaluation of a number of halogenated alkanes and alkenes, including the following: allyl chloride, bromoform, carbon tetrachloride, chloroform, 1,2-dichloroethane (ethylene dichloride), 1,1-dichloroethylene, 1,2-dichloroethylene, epichlorohydrin, ethylene dibromide, fluorotrichloromethane (F11), methyl bromide, methyl chloride, methyl chloroform, methylene chloride, tetrachloroethylene (perchloroethylene), trichloroethylene.

Detector tubes should be used with the hand pump for which they were originally designed (ACGIH-AIHA, 1971; NIOSH, 1972; Colen, 1973; Colen, 1974) and which should be routinely calibrated (Kusnetz, 1960; Leichnitz, 1983). If correctly used (AIHA, 1977; Leichnitz, 1983), detector tubes can be a great help in occupational hygiene evaluations, particularly those aimed at establishing the need for control measures.

(ii) Infra-red gas analysers

These instruments can in principle measure any gas or vapour which has absorption lines in the infra-red portion of the electromagnetic spectrum. One example of an infra-red analyser is presented in Figure 3. A large number of halogenated alkanes and alkenes can be measured with commercially available infra-red gas analysers and some examples are presented in Table 2.

FIG. 3a OPTICAL SCHEME OF A MIRAN INFRA-RED ANALYSER

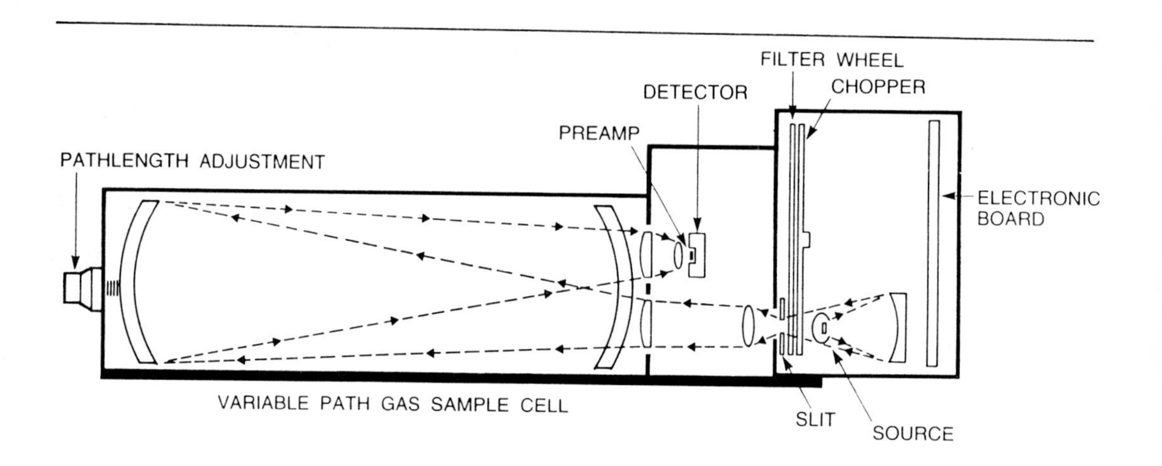

FIG. 3B PORTABLE MIRAN INFRA-RED ANALYSER[a]

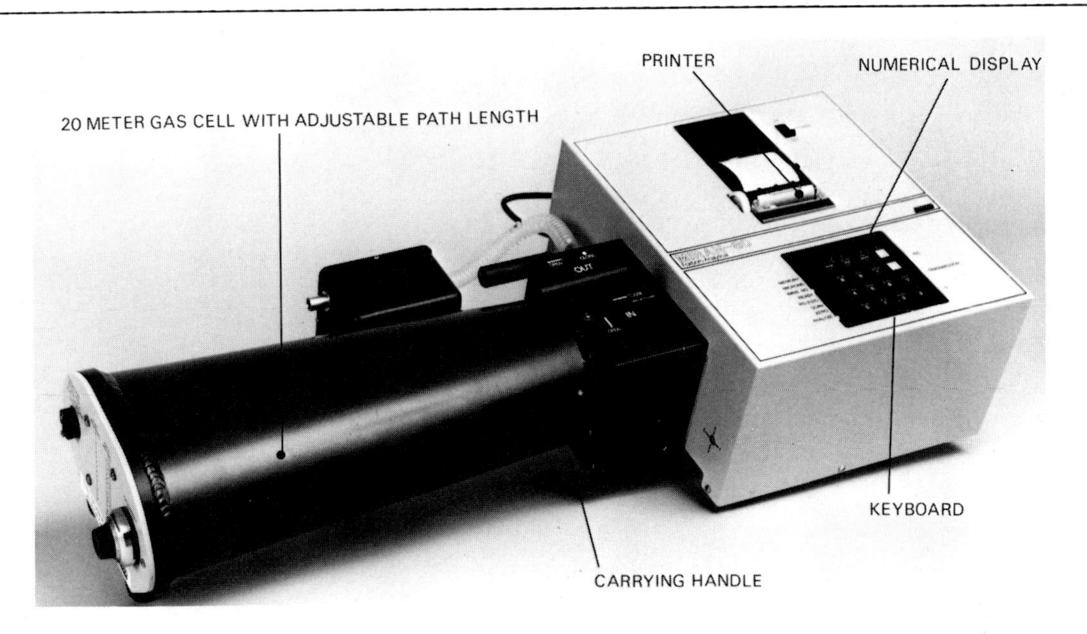

[a] Reproduced with permission of The Foxboro Company, Foxboro, MA 02035, USA

Table 2. Analytical wavelengths and minimum detectable concentrations for some halogenated alkanes and alkenes measurable by a commercial infra-red gas analyser

Chemical	Wavelength (μm)	Minimum detectable amount (ppm) (cell length - 20 m)
Bromoform	8.7	0.05
Carbon tetrachloride	12.6	0.06
Chloroform	13.0	0.06
1,1-Dichloroethane	9.4	0.3
1,2-Dichloroethylene	12.1	0.07
Epichlorohydrin	11.8	0.3
Methyl chloride	13.4	1.5
Methylene chloride	13.3	0.2
Tetrachloroethylene	10.9	0.05
1,1,1-Trichloroethane	9.2	0.06
Trichloroethylene	11.8	0.1

Such analysers, which can be very sensitive, can also be very accurate and precise if properly calibrated and used. Attention must be paid, however, to interference, which may mask the desired spectrum; e.g., from water, carbon dioxide and organic gases and vapours with similar wavelengths. In order to overcome interference, a pre-sampling stage can be utilized (e.g., silica gel for water vapour, a scrubber, etc.) or a secondary wavelength can be used as the analytical wavelength.

(iii) Electrical conductivity analysers

These instruments are based on the principle that air contaminants which form electrolytes in an aqueous solution will change the electrical conductivity of the latter. The method is not specific, but this difficulty may be overcome if concentrations of all other ion-forming contaminants are either constant or insignificant. Instruments in this category are commercially available for methyl chloride, carbon tetrachloride and several other halogenated hydrocarbons, including freons. These instruments may have very good sensitivity, precision and accuracy.

(iv) Flame-ionization detectors

Instruments in this category function by ionizing the gaseous air contaminant and collecting it under the influence of an applied electric field. Maintenance of the flame requires feed gases such as hydrogen and compressed air. There are a number of commercially-available instruments, such as portable organic vapour analysers, portable total hydrocarbon analysers, organic vapour meters, etc. Sensitivities are in the region of the lower ppm's and higher ppb's.

(v) Photoionization detectors

These instruments are similar to flame-ionization detectors in their response to many organic gaseous contaminants, but are safer, since there is no flame. The sensor consists of a sealed UV light source which ionizes the organic air contaminants (but not the major components of air). The ions produced are converted into an electric current which can act on a display unit (or an audio signal). Such instruments can be very sensitive (better than 0.2 ppm) and precise; some are very portable, as is the example shown in Figure 4.

FIG. 4 DIRECT-READING GAS MONITOR WORKING BY PHOTOIONISATION

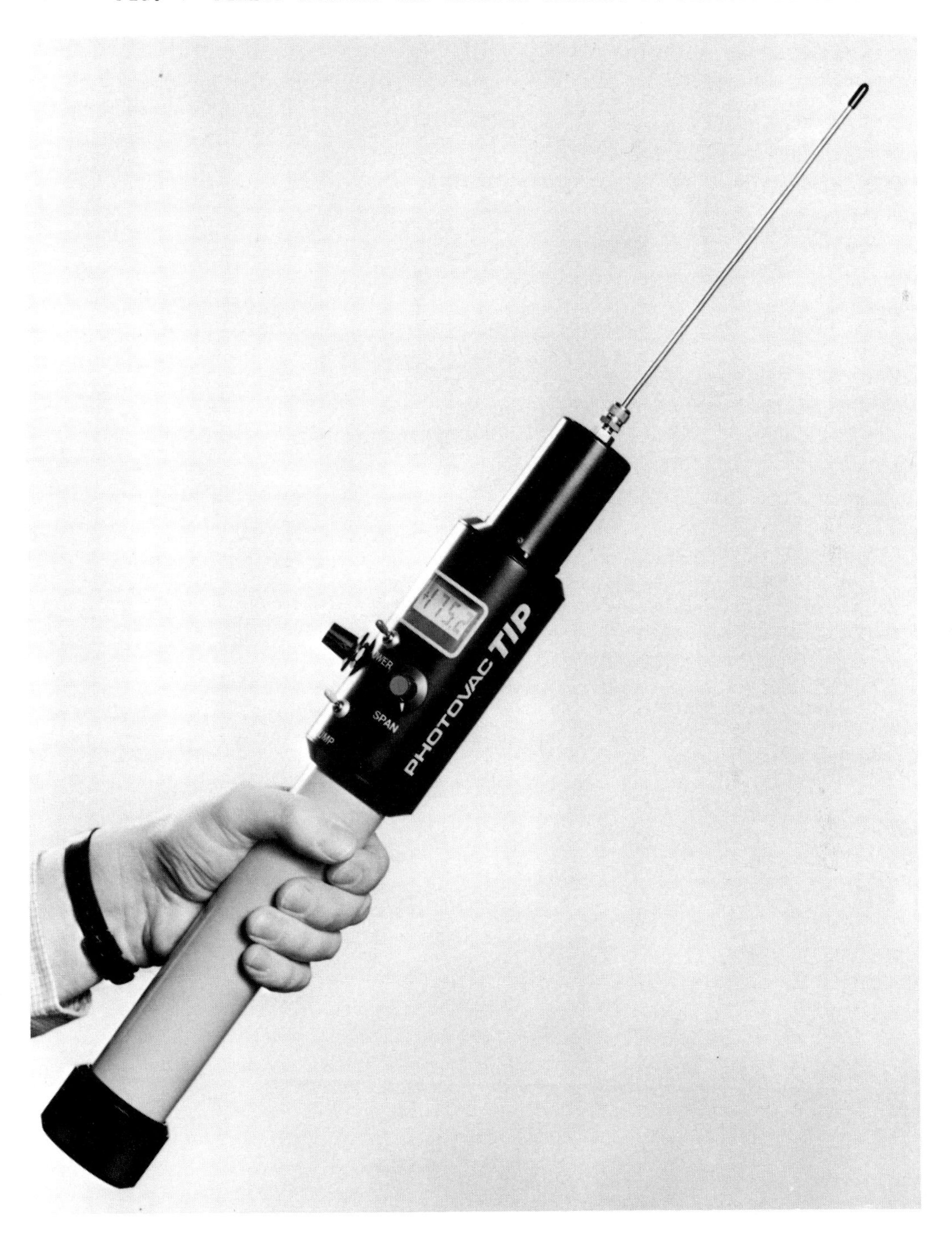

(vi) Portable gas chromatograph

Portable gas chromatographs permit "on-the-spot" field analysis with high sensitivity, precision and accuracy. Columns for a number of specific compounds are commercially available and detailed information can be obtained from the manufacturers. Chromatographs with electron-capture detectors are recommended for halogenated compounds. Although portable gas chromatographs can be very sensitive, not all models are sensitive enough to detect the very low concentrations dealt with when evaluating known and suspect carcinogens.

Spot or grab sampling

Spot or "grab" sampling can be carried out by collection of air samples in flasks, syringes or plastic bags. This type of sampling is not usually suitable when dealing with very low concentrations, since the amounts collected would be insufficient for analysis even with very sensitive methods, and there is a risk of loss by sorption on the walls of the container, or through the walls of plastic bags. This problem has been reviewed (WHO, 1979) and methods of sampling for breath and ambient air described in this volume show that it can be overcome.

Continuous sampling

This type of sampling requires a sampling head, which can be a container such as a flask or a tube holding a collecting medium, and an air mover which can sample air at a constant flow-rate. The air mover is usually a pre-calibrated pump that can operate over a range of pumping speeds.

For the purposes of this volume, the sampling head consists of a glass tube holding an adsorbent or absorbent. The adsorbent most widely used is activated charcoal (Fig. 5) and the adsorbed substance is usually extracted for analysis with carbon disulfide; this procedure is described in this volume for chloroform, carbon tetrachloride, methyl chloroform, hexachloroethane, trichloroethylene, tetrachloroethylene, epichlorohydrin, methyl bromide, bromoform, methylene chloride, 1,1-dichloroethane, and 1,2-dichloroethylene. Adsorption on charcoal is also used, but with another extraction solvent, for ethylene dibromide (benzene : methanol), allyl chloride (benzene) and methyl chloride (methylene chloride). Another adsorbent (Chromosorb 101 porous polymer) is used to trap BCME, and is followed by thermal desorption. An absorbent (potassium 2,4,6-trichlorophenate) on textured glass beads is used for CMME, prior to elution with methanol.

Activated charcoal is an excellent collecting medium since, being non-polar, it has a marked preference for organic gases and vapours relative to atmospheric moisture. However, if the relative humidity is high, of the order of 50-70%, the adsorption capacity decreases markedly. This will present a problem when sampling in very humid climates; in such situations, it may be necessary to decrease the recommended sampling time, since breakthrough (incomplete adsorption) would occur earlier than predicted by studies carried out in drier climates. In order to check for breakthrough, it is usual to have two charcoal sections in series and to analyse them separately, as in the methods described in this volume (NIOSH, 1984).

The most common type of pump is a rotary vane pump, powered by rechargeable batteries, for continuous personal sampling. A great variety exists (ACGIH, 1983), but it is best to use a type that has a variable flow-rate which can be adapted to work against different flow resistances caused by the adsorbent in the sampling head. The precautions to be taken when using these systems are described in NIOSH (1973), ACGIH (1983) and WHO (1984).

Since low flow-rates are used with sorbent tubes (commonly, 0.05-0.2 L/min), pumps with only a low flow-rate range, which are smaller and usually less costly, should be chosen when only sorbent-tube sampling is considered. A great variety of sampling pumps is available (ACGIH, 1983) and a non-comprehensive list of manufacturers is given in Appendix B. Examples of personal pumps are presented in Figure 6.

FIG. 5 CHARCOAL ADSORBENT TUBE, ACCORDING TO NIOSH SPECIFICATIONS

Plastic caps are shown for sealing the tube after use and prior to analysis.

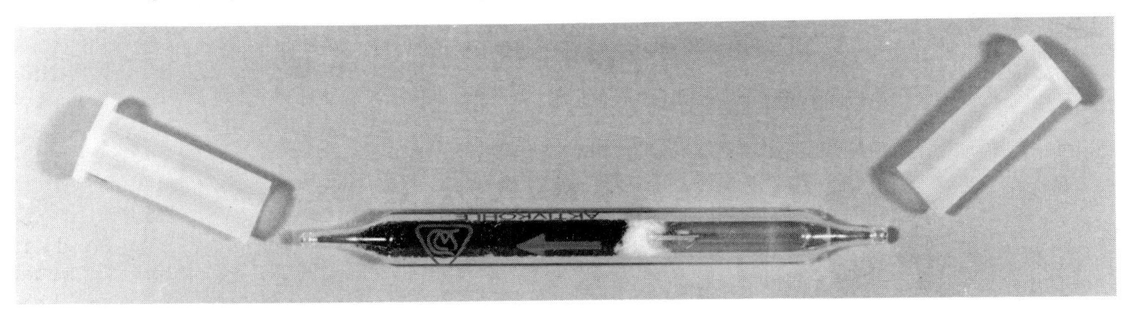

The routine flow-rate calibration of sampling pumps is extremely important, as appreciable errors may be introduced in exposure evaluations by an inaccurate sampling flow-rate. As the flow-rate varies with the resistance offered to air flow, each pump must be individually calibrated with the sampling head in-line.

Passive monitoring devices

Passive monitoring devices, or passive samplers, consist simply of badges with a collecting medium, preceded by either a diffusion zone (a membrane or a quiescent air zone) or a permeation membrane. An example is presented in Figure 7. Such devices rely on molecular diffusion through the membrane or quiescent zone, or on the permeation of the gases and vapours into the collecting medium, which usually consists of activated charcoal, but sometimes of liquid reagents. They do not, therefore, require mechanical air-moving devices and do not present most of the problems associated with a classical

FIG. 6 EXAMPLES OF COMMERCIAL PERSONAL SAMPLING PUMPS

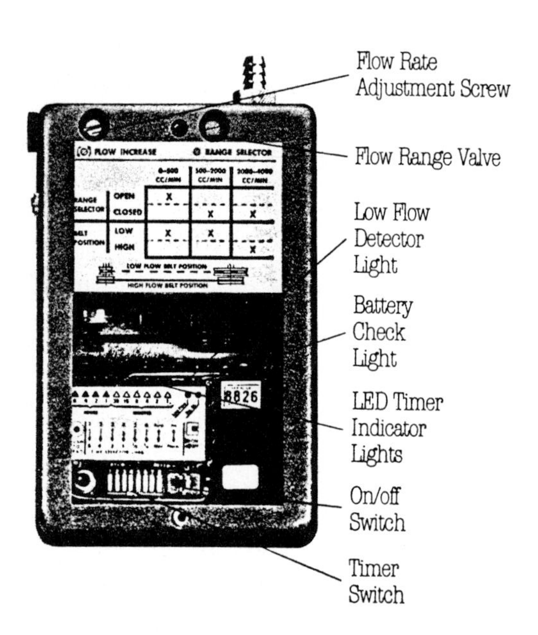

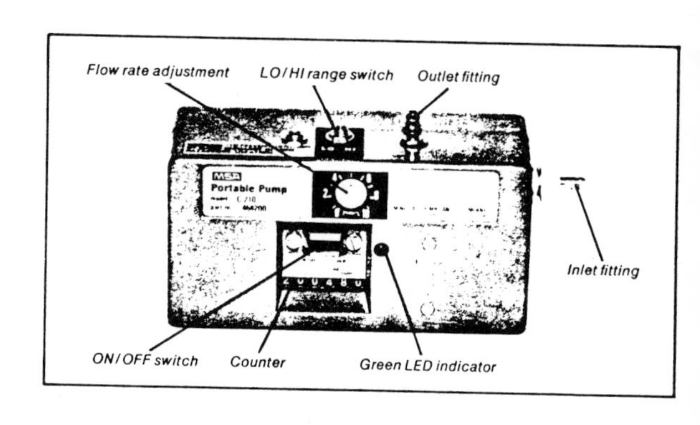

sampling train, such as pump flow-rate calibration, flow-rate checks during sampling, eventual leakages in connections, or interference of the sampling train with workers' movements and the performance of work. In addition to these advantages, passive monitoring devices are small, light-weight and, if properly attached in the worker's breathing zone, do not require periodic checks during the sampling period.

The amount of air contaminant that will diffuse (or permeate) to the collecting medium in a given time depends on its concentration in the sampled air, the atmospheric conditions and the physical characteristics of the monitor. After being exposed to the contaminated air for a known period, the collecting medium is analysed in the laboratory, usually by gas chromatography, although other methods can be used, such as colorimetric methods and infra-red spectroscopy (Purnell et al., 1981; Baker, 1982). Passive monitoring devices are well-discussed in the literature (Bailey & Hollingdale-Smith, 1977; Woebkenberg, 1979; ACGIH, 1980; Lautenberger et al., 1980; Brown et al., 1981; Panwitz, 1981; Feigley & Chastain, 1982; Rose & Perkins, 1982; ACGIH, 1983; Fowler, 1983; Muller & Guenier, 1984; Underhill, 1984).

The influence of air velocity across the face of the passive sampler has been controversial. It is not, however, appreciable over the air velocity ranges usually found in work-places. At very low air velocities, less than 5-6 m/min, the passive samplers are likely to under-sample and therefore introduce a negative bias; at very high face velocities, in the range of

FIG. 7 PASSIVE SAMPLER FILLED WITH 400 mg COCONUT CHARCOAL

Contaminants enter _via_ the two barriers which control the diffusion rate.

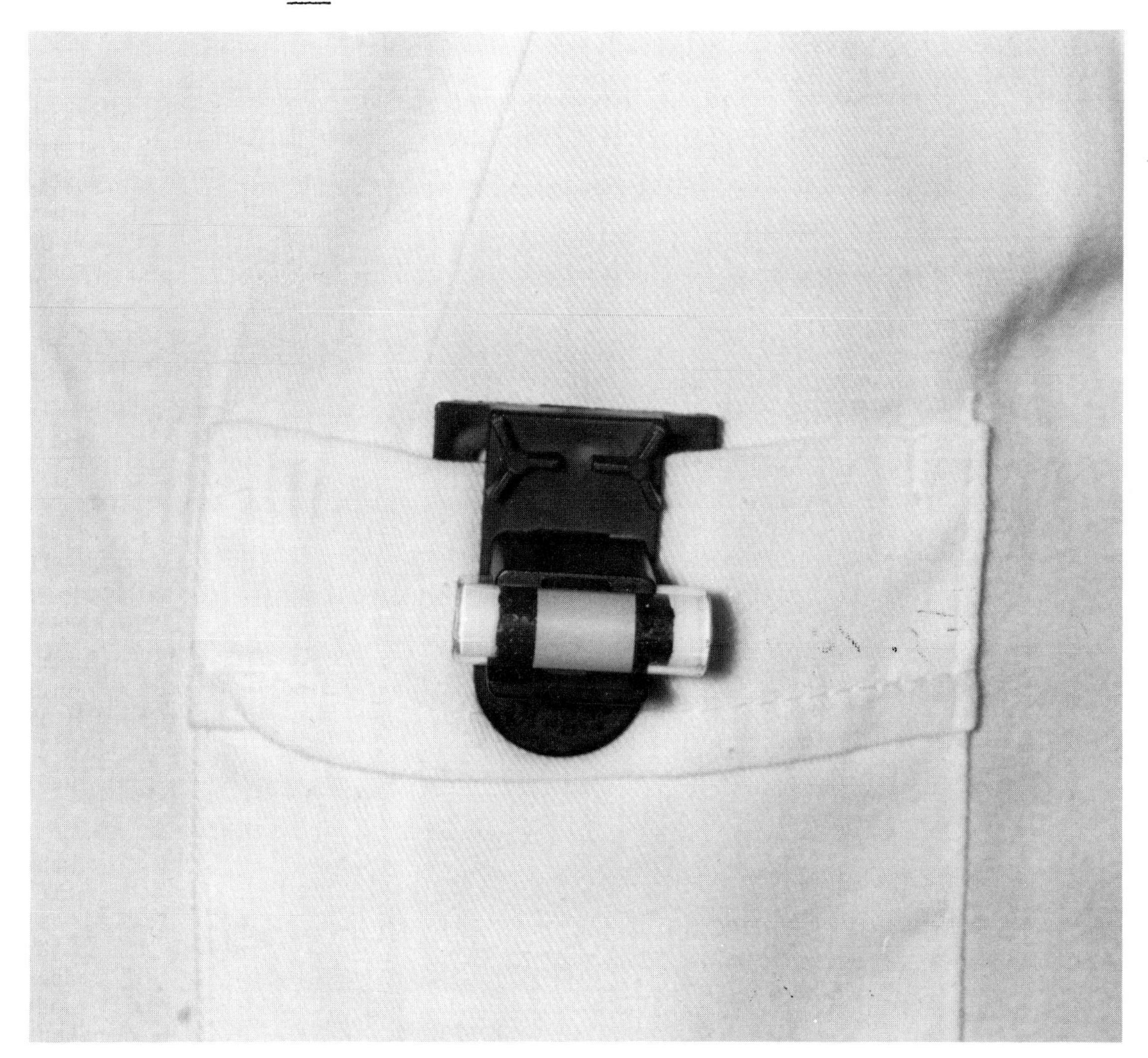

75-90 m/min, they are likely to over-sample because of eddy currents forming inside the sampler. For face velocities between these ranges (which are the most usual in work-places), the sampling rate is not appreciably affected by air movement. Many passive monitoring badges have built-in draught shields.

The first passive sampling unit based on molecular diffusion of the gas or vapour to be measured was designed by Palmes and Gunnison (1973). Since then, passive samplers have been used to sample a number of chemicals and some are commercially available, such as those for methyl chloroform, methyl chloride, carbon tetrachloride, halothane, dichlorodifluormethane (F11), trichloroethylene, methylene chloride, 1,1-dichloroethylene, 1,2-dichloroethylene, 1,2-dichloroethane, allyl chloride, chloroform, epichlorohydrin, ethylene dichloride and 1,1,1-trichloroethane (Bamberger et al., 1978; Halliday & Anderson, 1980; Lautenberger et al., 1980; Mazur et al., 1980; Mazur et al., 1981; Pannwitz, 1981; Reiszner, 1981).

The convenience of passive dosimeters for air sampling in a breathing zone facilitates otherwise problematical or impractical surveys, e.g., for the evaluation of exposures of operating room personnel to waste anaesthetic gases. Dosimeters for halothane and enflurane have been used for this purpose and are commercially available (Halliday & Anderson, 1980; Mazur et al., 1980; Jonas et al., 1981). A disadvantage is that sampling with passive monitoring devices is inadequate for the evaluation of peaks of exposure.

It should be pointed out that, although many studies have shown good correlation between exposure evaluations with passive samplers and charcoal tubes (Lautenberger et al., 1980; Mazur et al., 1980; Guenier & Ferrari, 1981; Hickey & Bishop, 1981; Sesana et al., 1981; Pannwitz, 1984a), there is still need for additional research in order to validate passive monitoring devices for a larger number of air contaminants (Perkins et al., 1981; HSE, 1983). Factors such as sample loading, relative humidity, atmospheric pressure and interferences influence their performance (Gregory & Elia, 1983; Lindenboom & Palmes, 1983). Testing protocols have been developed for this purpose (Brown et al., 1984; Hull & Cassinelli, 1985).

A non-comprehensive list of manufacturers of passive devices is given in Appendix C.

5. CHARCOAL-TUBE SAMPLING PROCEDURE

Prior to the sampling trip, the charcoal tubes must be prepared and individually labelled; charcoal tubes should be capped only with adequate plastic caps and never with caps of rubber. The personal pumps must have their batteries recharged and flow-rates calibrated. Personal pumps operating at flow-rates up to 2 L/min may be conveniently calibrated with the soap-bubble method (Fig. 8).

At the sampling location, the sampling train is attached to the selected worker. The personal pump should be attached to a belt or harness and the charcoal tube should then be opened and fixed so as to remain in a vertical position in the workers breathing zone; e.g., attached to the worker's lapel. There should be no tubing on the inlet side of the charcoal tube.

Once the sampling train is in place, the pump is started, the starting time recorded and the flow-rate checked and noted. The flowmeter of the pump should be watched during sampling and, if necessary, adjustments should be made to keep the flow rate constant. Flow rates are usually in the range of 0.05 to 0.2 L/min when sampling to determine a time-weighted average concentration, and of the order of 1 L/min when sampling to determine a ceiling or short-term concentration.

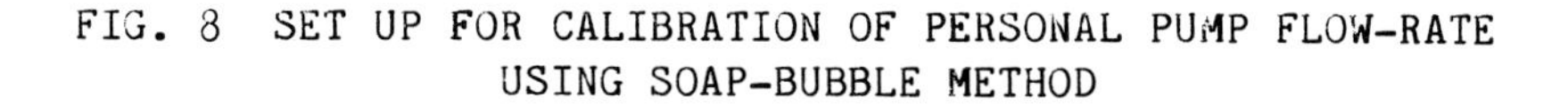

FIG. 8 SET UP FOR CALIBRATION OF PERSONAL PUMP FLOW-RATE USING SOAP-BUBBLE METHOD

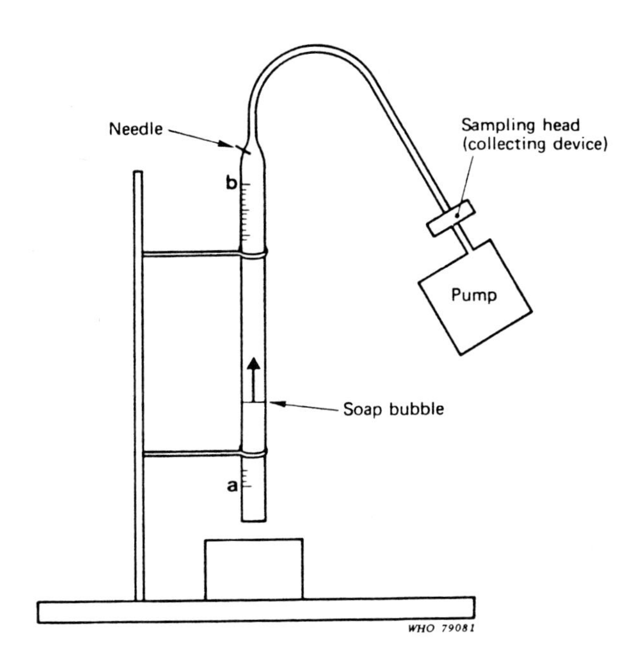

When determining a time-weighted average concentration, 4 hours would be an acceptable sampling time and very low flow rates would be needed to avoid excessive sample volumes and appreciable breakthrough. For short-duration sampling (e.g., 15 min) a higher flow rate (of the order of 1 L/min) can be used. Excessively high flow rates, however, should be avoided when using charcoal tubes, since this may lead to the contaminant being carried through without being adsorbed.

The maximum volume of air to be sampled varies according to the temperature, pressure and humidity of the atmosphere, and the air concentration of the contaminant being collected. The usual recommended volume for a large number of organic gases and vapours is 10 to 15 litres. The NIOSH Manual of Analytical Methods (1984) recommends minimum and maximum air sample volumes, as well as flow rates for specific chemicals. Some examples of maximum sample volumes, taken from various NIOSH methods, are shown in Table 3.

Table 3. Maximum air sample volumes for selected halogenated alkanes and alkenes

Air contaminant	Maximum volume of air sample (litres)[a]
Allyl chloride	100
Bromoform	10
Carbon tetrachloride	17
Chloroform	15
1,2-Dibromoethane	25
1,1-Dichloroethane	10
1,2-Dichloroethylene	3
Epichlorohydrin	30
Ethylene dichloride	10
Hexachloroethane	10
Methyl chloroform	6
Trichloroethylene	15

[a] These values are examples of volumes above which appreciable breakthrough of the sample to the back-up section has been shown to occur.

Breakthrough depends on the air concentration of the contaminant, the amount of charcoal in the tube and on atmospheric conditions (temperature, humidity, etc.). The maximum volume of air to be sampled should be established in the laboratory for the expected sampling conditions and range of concentrations, and may be checked against the maximum volumes which are usually recommended in the analytical procedures. If the sampling rate and duration are well-determined, appreciable breakthrough of the air contaminant to the back-up section of the charcoal tube should not be observed. The back-up section should always be analysed to ensure there is no breakthrough of the adsorbed chemical. If appreciable breakthrough is found (e.g., the back-up section contains 10-25 percent of the amount retained in the front section), the sampling rate and/or time must be decreased.

When the sampling period is over, the stopping time is recorded as the pump is turned off; the charcoal tube is removed from the sampling train, both ends are tightly capped, and it is properly stored and transported to the laboratory where the adsorbed contaminant will be desorbed and analysed by gas chromatography according to the selected analytical procedure.

The adsorbent-tube method has been described above in a general way, with the object of emphasizing the more important aspects to be considered. Further details, which are beyond the scope of this chapter, may be found elsewhere (Guenier & Muller, 1981; NIOSH, 1984; see also the Introduction to Methods 1-8, p. 173).

REFERENCES

ACGIH (1980) Dosimetry for chemical and physical agents, ACGIH Annals Series, Volume 1, American Conference of Governmental Industrial Hygienists, Cincinnati, USA

ACGIH (1983) Air sampling instruments for evaluation of atmospheric contaminants, 6th edition, American Conference of Governmental Industrial Hygienists, Cincinnati, USA

ACGIH (1984) Threshold limit vaues for chemical substances and physical agents in the work environment and biological exposure indices with intended changes for 1984-85, American Conference of Governmental Industrial Hygienists, Cincinnati, USA (produced annually)

ACGIH & AIHA Joint Committee on Direct-Reading Gas Detecting Systems (1971) Direct reading gas detecting tube systems. Am. Ind. Hyg. Assoc. J., 32, 488-489

AIHA (1977) Direct reading colorimetric tubes - a manual of recommended practices, 1st ed., American Industrial Hygiene Association, Akron, USA

Bailey, A. & Hollingdale-Smith, P.A. (1977) A personal diffusion sampler for evaluating time-weighted exposure to organic vapours and gases. Ann. Occup. Hyg., 20, 345-356

Baker, B.B., Jr. (1982) Infrared spectral examination of air monitoring badges. Am. Ind. Hyg. Assoc. J., 43, 98

Bamberger, R.L., Esposito, G.G., Jacobs, B.W. & Podolak, G.E. (1978) A new personal sampler for organic vapors. Am. Ind. Hyg. Assoc. J., 39, 701-708

Brown, R.H., Charlton, J., Saunders, K.J. (1981) The development of an improved diffusive sampler. Am. Ind. Hyg. Assoc. J., 42, 865

Brown, R.H., Harvey, R.P., Purnell, C.J. & Saunders, K.J. (1984) A diffusive sampler evaluation protocol. Am. Ind. Hyg. Assoc. J., 45, 67-71

Burgess, W.A. (1981) Recognition of health hazards in industry, New York, John Wiley & Sons

Clayton, G.D. & Clayton, F.E., eds (1978) Patty's Industrial Hygiene and Toxicology, 3rd Revision, Vol. 1, General Principles, New York, John Wiley & Sons

Colen, F.H. (1973) A study of the interchangeability of gas detector tubes and pumps. Report TR-71. National Institute for Occupational Safety and Health, Morgantown, WV, USA

Colen, F.H. (1974) A study of the interchangeability of gas detector tubes and pumps. Am. Ind. Hyg. Assoc. J., 35, 686-694

Corn, M. & Esmen, N.A. (1979) Workplace exposure zones for classification of employee exposures to physical and chemical agents. Am. Ind. Hyg. Assoc. J., 40, 47

Cralley, L.V. & Cralley, L.J., eds (1979) Patty's Industrial Hygiene and Toxicology, 3rd Revision, Vol. 3, Theory and Rationale of Industrial Hygiene Practice, New York, John Wiley & Sons

Cralley, L.V. & Cralley, L.J. (1982) Industrial Hygiene Aspects of Plant Operations, Vol. 1, Process Flows, New York, MacMillan Publishing Co., Inc.

Feigley, C.E. & Chastain, J.B. 91982) An experimental comparison of three diffusion samplers exposed to concentration profiles of organic vapors. Am. Ind. Hyg. Assoc. J., 43, 227-233

Fowler, K. (1983) Fundamentals of passive vapour sampling. Int. Lab., 13, 40-48

Fraust, C.L. (1975) The use of activated carbon for sampling industrial environs. Am. Ind. Hyg. Assoc. J., 36, 278-284

Gale, H.J. (1967) Some examples of the application of the lognormal distribution in radiation protection. Ann. Occup. Hyg., 10, 39-45

Gregory, E.D. & Elia, V.J. (1983) Sample retentivity properties of passive organic vapor samplers and charcoal tubes under various conditions of sample loading, relative humidity, zero exposure level periods and a competitive solvent. Am. Ind. Hyg. Assoc. J., 44, 88-96

Guenier, J.P. & Ferrari, P. (1981) Echantillonnage des polluants gazeux. Les badges: utilisation et comparaison avec les tubes à charbon actif. Cah. Notes Documentaires, 105, Note No. 1344-105-81

Guenier, J.P. & Muller, J. (1981) Echantillonnage des polluants gazeux - Etude de l'adsorption sur charbon actif. Cah. Notes Documentaires, 103, 197-210

Guenier, J.P. & Muller, J. (1984) Sampling of gaseous pollutants on activated charcoal with 900 mg tubes. Ann. Occup. Hyg., 28, 61-75

Halliday, M.M. & Anderson, J. (1980) Determination of halothane in operating theatre air by using a passive organic vapor dosimeter. Analyst, 105, 289-292

Hickey, J.L.S. & Bishop, C.C. (1981) Field comparison of charcoal tubes and passive vapor monitors with mixed organic vapors. Am. Ind. Hyg. Assoc. J., 42, 264-267

Hill, R. & Fraser, D.A. (1980) Passive dosimetry using detector tubes. Am. Ind. Hyg. Assoc. J., 41, 721-729

Hounam, R.F. (1965) An Application of the Log-normal Distribution to Some Air Sampling Results and Recommendations on the Interpretation of Air Sampling Data. Atomic energy Research Establishment Report AERE-M 1469, London, HMSO

HSE (1983) Methods for the Determination of Hazardous Substances: Protocol for Assessing the Performance of a Diffusive sampler, MDHS 27, Health and Safety Executive, UK

Hull, R.D. & Cassinelli, M.E. (1985) Testing Protocols with Evaluation Criteria and Performance Specifications for Passive Samplers, NIOSH (in press)

ILO (1980) Occupational Exposure Limits for airborne toxic substances, ILO Occupational Safety and Health Series No. 37, International Labour Office, Geneva

ILO/WHO Joint Committee on Occupational Health (1969) Permissible Levels of Occupational Exposure to Airborne Toxic Substances, Technical Report Series No. 415, Geneva, World Health Organization

INRS Cahiers des Notes Documentaires (periodical) Institut National de Recherche et de Securite, Paris, France

INRS Fiches Toxicologiques (series) Institut National de Recherche et de Securite, Paris, France

INRS (1984) Prelevement et analyse de polluants organiques gazeux - Methode utilisee par l'INRS. Cah. Notes Documentaires, 114, 55-61

IUPAC (1974) Analytical methods for use in occupational hygiene, performance standards for detector tubes. Pure Appl. Chem., 40, 39-

Jonas, L.C., Billings, C.E. & Consolacion, L. (1981) Laboratory performance of passive personal samplers for waste anesthetic gas (enflurane) concentrations. Am. Ind. Hyg. Assoc. J., 42, 104-111

Key, M.M., Henschel, A.F., Butler, J., Ligo, R.N. & Tabershaw, I.R., eds (1977) Occupational Diseases: A Guide to their Recognition, US Department of Health, Education and Welfare, NIOSH Publications No. 77-181

Kitagawa, T. (1952) Rapid method of quantitative gas-analysis by means of detector tubes. Kagaku no Ruoiki, 6, 386

Kring, E.V., Ansul, G.R., Henry, T.J., Morello, J.A., Dixon, S.W., Vasta, J.F. & Hemingway, R.E. (1984) Evaluation of the standard NIOSH type charcoal tube sampling method for organic vapors in air. Am. Ind. Hyg. Assoc. J., 45, 250-259

Kusnetz, H.L. (1960) Air flow calibration of direct reading colorimetric gas detecting devices. Am. Ind. Hyg. Assoc. J., 21, 340

Lautenberger, W.J., Kring, E.V. & Morello, J.A. (1980) A new personal badge monitor for organic vapors. Am. Ind. Hyg. Assoc. J., 41, 737-747

Leichnitz, K. (1977) Use of detector tubes under extreme conditions (humidity, pressure, temperature). Am. Ind. Hyg. Assoc. J., 38, 707-711

Leichnitz, K. (1979) How reliable are detector tubes? Drager Rev., 43, 21-26

Leichnitz, K. (1983) Detector Tube Measuring Techniques, Ecomed-Verlagsgesellschaft mbh, FRG

Leidel, N.A. & Busch, K.A. (1975) Statistical methods for the determination of non-compliance with occupational health standards. US Department of Health, Education & Welfare, NIOSH Publication No. 75-159

Leidel, N.A., Busch, K.A. & Lynch, J.R. (1977) Occupational exposure sampling strategy manual. US Department of Health, Education & Welfare, NIOSH Publication No. 77-173

Linch, A.L. (1974) Evaluation of ambient air quality by personnel monitoring. Cleveland OH, USA, CRC Press Inc.

Lindenboom, R.H. & Palmes, E.D. (1983) Effect of reduced atmospheric pressure on a diffusional sampler. Am. Ind. Hyg. Assoc. J., 44, 105-108

Mazur, J.F., Podolak, G.E., Esposito, G.G , Rinchart, D.S. & Glenn, R.E. (1980) Evaluation of a passive dosimeter for collection of 2-bromo-2-chloro-1,1,1-trifluoroethane and 2-chloro-1,1,2-trifluoroethyl difluoromethyl ether in hospital operating rooms. Am. Ind. Hyg. Assoc. J., 41, 317-321

Mazur, J.F., Podolak, G.E., Esposito, G.G , Rinchart, D.S. & Glenn, R.E. (1981) Evaluation of passive monitors for assessing vapor degreaser emissions. Am. Ind. Hyg. Assoc. J. , 42, 752-756

McCammon, C.S. Jr (1976) A summary of the NIOSH evaluation program of portable direct-reading meters. Am. Ind. Hyg. Assoc. J., 37, 289-295

McKee, E.S., McConnaughey, P.W. & Pritts, I.M. Colorimetric personal dosimeters for some inorganic contaminants. Pittsburgh PA, USA, Mine Safety Appliances Co.

Muller, J. & Guenier, J.-P. (1984) Echantillonnage des polluants gazeux: 2. Le point sur les echantillonneurs passifs (badges). Cah. Notes Documentaires, No. 116

NIOSH (1972) A Study of the Interchangeability of Gas Detector Tubes and Pumps. Morgantown WV, USA, National Institute for Occupational Safety and Health (see also, Colen, 1974)

NIOSH (1973) The Industrial Environment - Its Evaluation and Control. (National Institute for Occupational Safety and Health, Cincinnati OH, USA, US Department of Health & Human Services

NIOSH (1978) Guideline for Detector Tube Unit Certification. Morgantown WV, USA, National Institute for Occupational SAfety and Health

NIOSH (1979) BCME Formation and Detection in Selected Work Environments. (National Institute for Occupational Safety and Health, Publication No. 79-118, Cincinnati OH, USA, US Department of Health, Education & Welfare,

NIOSH (1981) Occupational Health Guidelines for chemical Hazards. (National Institute for Occupational Safety and Health) Publication No. 81-123, Cincinnati OH, USA, US Department of Health & Human Services

NIOSH (1984) NIOSH Manual of Analytical Methods, 3rd Ed., Vol. 1, (National Institute for Occupational Safety and Health) Cincinnati OH, USA, US Department of Health & Human Services

Palmes, E.D. & Gunnison, A.F. (1973) Personal monitoring device for gaseous contaminants. Am. Ind. Hyg. Assoc. J., 34, 78-81

Pannwitz, K.-H. (1981) ORSA 5: A new sampling device for vapours of organic solvents. Drager Rev., 48, 8

Pannwitz, K.-H. (1984a) Sampling and analysis of organic solvent vapours in the atmosphere. comparison of active and passive sampling devices. Drager Rev., 52, 19

Pannwitz, K.-H. (1984b) Direct-reading diffusion tubes. Drager Rev., 53, 10

Purnell, C.J., West, N.G., Brown, R.H. (1981) Colorimetric and X-ray analysis of gases collected on passive samplers. Chemistry and Industry, 8594

Reckner, L.R. & Sachder, J. (1975) Charcoal sampling tubes for several organic solvents (National Institute for Occupational Safety and Health) Publication No. 75-184, Cincinnati OH, USA, US Department of Health, Education & Welfare

Reid, F.H. & Halpin, W.R. (1968) Determination of halogenated and aromatic hydrocarbons in air by charcoal tube and gas chromatography. Am. Ind. Hyg. Assoc. J., 29, 390-396

Reizner, K.D. (1981) Permeation Sampling of Methyl Chloride and Dichlorofluoromethane. Preprint extended abstract presented at the 181st National American chemical Society Meeting

Roper, C.P. (1974) The NIOSH detector tube certification program. Am. Ind. Hyg. Assoc. J., 35, 438-442

Rose, V.E. & Perkins, J.L. (1982) Passive dosimetry - state of the art review. Am. Ind. Hyg. Assoc. J., 43, 605-621

Saalwaechter, A.T., McCammon, C.S. Jr, Roper, P. & Carlberg, K.S. (1977) Performance testing of the NIOSH charcoal tube technique for the determination of air concentrations of organic vapors. Am. Ind. Hyg. Assoc. J., 38, 476-486

Saltzman, B.E. (1962) Basic theory of gas indicator tube calibrations. Am. Ind. Hyg. Assoc. J., 23, 112-126

Saltzman, B.E. (1972) Simplified methods for statistical interpretation of monitoring data. J. Air Pollut. Control Assoc., 22, 90

Sansone, E.B. & Jonas, L.A. (1981) Prediction of activated carbon performance for carcinogenic vapors. Am. Ind. Hyg. Assoc. J., 42, 688-690

Sesana, G., Candeloro, F. & Toffoletto, F. (1981) Valutazione dell'esposizione professionale a gas anestetici alogenati confronto fra campionatori personali e dosimetri passivi. Med. Lavoro, 5, 416-419

Severs, L.W. & Skory, L.K. (1975) Monitoring personnel exposure to vinyl chloride, vinylidene chloride and methyl chloride in an industrial work environment. Am. Ind. Hyg. Assoc. J., 36, 669-676

Sherwood, R.J. & Greenbalgh, D.M.S. (1960) A personal air sampler. Ann. Occup. Hyg., 2, 127-132

Stellman, J.M. & Daum, S.M. (1973) Work is dangerous to your health, New York, Random House

Swinscow, T.D.V. (1977) Statistics at square one. Brit. Med. Assoc. (Tavistock Square, London WCIHgJP)

Ulfvarson, U. (1975) Sampling strategy and statistical evaluation of results from sampling and analysis of air contaminants at work places. Arbete och Hälsa, 9

Ulfvarson, U. (1977) Statistical evaluation of the results of measurements of occupational exposure to air contaminants. A suggested method of dealing with short-period samples taken during one whole shift when noncompliance with occupational health standards is being determined. Scand. J. Work Environ. Health, 3, 109-115

Underhill, D.W. (1984) Efficiency of passive sampling by adsorbents. Am. Ind. Hyg. Assoc. J., 45, 306

Valic, F.R. (1984) Detection and analysis of airborne contaminants (field methods). In: Parmeggiani, L., ed., ILO Encyclopaedia of Occupational Health and Safety, Geneva, International Labour Office

Van Mourik, J.H.C. (1965) Experiences with silica gel as absorbent. Am. Ind. Hyg. Assoc. J., 26, 498

White, L.D., Taylor, D.J., Mauer, P.A. & Kupel, R.E. (1970) A convenient optimized method for the analysis of selected solvent vapors in the industrial atmosphere. Am. Ind. Hyg. Assoc. J., 31, 225-232

WHO (1976) Selected Methods of measuring air pollutants. WHO Publication No. 24, pp 54-55 and 77, Geneva, World Health Organization

WHO (1977a) Methods used in Establishing Permissible Levels in Occupational Exposure to Harmful Agents (Technical Report Series, No. 601), Geneva, World Health Organization

WHO (1977b) The SI for the Health Profession, Geneva, World Health Organization

WHO (1978) Use of SI Units in Public Health, Who Chronicle, 32, 99

WHO (1979) Evaluation of Gases and Vapours in the Work Environment, WHO Document No. OCH/79.1

WHO (1984) Evaluation of Airborne Particles in the Work Environment, WHO Offset Publication No. 80, Geneva, World Health Organization

Woebkenberg, M.L. (1979) Current NIOSH research on passive monitors. In: Proceedings of the Symposium on the Development and Usage of Personal Monitors for Exposure and Health Effect Studies, Environmental Protection Agency, EPA-600/9-79-032, pp. 27-33

APPENDIX A

MANUFACTURERS OF DIRECT-READING INSTRUMENTS[a]

1. Manufacturers of Detector Tubes

Bacharach Instrument Company, 301 Alpha Drive, Pittsburgh, PA 15238, USA

Bendix Corporation, Environmental and Process Instruments Division, 12345 Starkey Road, Largo, FL 33543, USA

Dragerwerk AG Lübeck, Moislinger Allee 53/55, D-2400 Lübeck 1, FRG

Gastec Corporation, 3891 Ikebe-cho, Midori-ku, Yokohama 226, Japan

Kitagawa, Komyo Rikagaku Kogyo K.K., 1-8-24 Chuo-cho, Meguro-Ku, Tokyo, Japan

Maimex, 15, D. Nestorov Street, Sofia 1431, Bulgaria

Mine Safety Appliances Company, 600 Penn Center Boulevard, Pittsburgh, PA 15235, USA

2. Manufacturers of other Types of Direct-reading Instruments

Analytical Instrument Development, Route 41 and Newark Road, Avondale, PA 19311, USA

Bacharach Instrument Company, 301 Alpha Drive, Pittsburgh, PA 15238, USA

Beckman Instruments, Inc., Process Instruments Division, 2500 N. Harbor Boulevard, Fullerton, CA 92634, USA

Bendix Corporation, Environmental and Process Instruments Division, 12345 Starkey Road, Largo, FL 33543, USA

C.F. Casella & Co. Ltd., Regent House, Brittania Walk, London, N1 7ND, UK

Devco Engineering, Inc., Control Systems Division, 36 Pier Lane West, Fairfield, NJ 07006, USA

Dragerwerk AG Lübeck, Moislinger Allee 53/55, D-2400 Lübeck 1, FRG

Du Pont Company, Instrument Products Division, Concord Plaza, Clayton Bldg, Wilmington, DE 19898, USA

Foxboro Compan co. Ltd., Regent House, Brittania Walk, London, N1 7ND, UK

Devco Engineering, Inc., Control Systems Division, 36 Pier Lane West, Fairfield, NJ 07006, USA

Dragerwerk AG Lübeck, Moislinger Allee 53/55, D-2400 Lübeck 1, FRG

Du Pont Company, Instrument Products Division, Concord Plaza, Clayton Bldg, Wilmington, DE 19898, USA

Foxboro Company, P.O. Box 5449, South Norwalk, CT 06856, USA

Gas Tech Inc., 331 Fairchild Drive, Mountain View, CA 94043, USA

H-Nu Systems Inc., 30 Ossipee Road, Newton, MA 02164, USA

Icare, 66, Avenue du Prado, B.P. 151, 13253 Marseille Cedex 6, France

Infrared Industries Inc., Western Division, Instrumentation Group, P.O. Box 989, Santa Barbara, CA 93102, USA

Kitagawa, Komyo Rikagaku Kogyo K.K., 1-8-24 Chuo-cho, Meguro-ku, Tokyo 152, Japan

MDA Scientific Inc., 1815 Elmdale Avenue, Glenview, IL 60025, USA

Microsensor Technology Inc., 47747 Warm Springs Blvd, Fremont, CA 94539, USA

Mine Safety Appliances Company, 600 Penn Center Boulevard, Pittsburgh, PA 15235, USA

Societe Française, Oldham, B.P. 962, 62033 Arras Cedex, France

Scott Aviation-Davis Instruments, P.O. Box 751, RT 29 North, Charlottesville, VA 22902, USA

Tracor Inc., Analytical Instruments Division, 6500 Tracor Lane, Austin, TX 78721, USA

XonTech Inc., 6862 Hayvenburst Avenue, Van Nuys, CA 91406, USA

---

[a] The mention of a manufacturer does not in any way represent endorsement or recommendation of their instruments either by the authors of this chapter, by IARC or by WHO.

APPENDIX B

MANUFACTURERS OF PERSOMAL SAMPLING PUMPS[a]

Bendix Corporation, Environmental and Process Instruments Division, 12345 Starkey Road, Largo, FL 33543, USA

C.F. Casella & Co., Ltd., Regent House, Brittania Walk, London, N1 7ND, UK

Compur - elektronic gmnbh, Steiner strasse 15, 8000 München 70, FRG

DuPont Company, Applied Technology Division, Clayton Bldg, Concord Plaza, Wilmington, DE 19898, USA

Gilian Instrument Corp., 1275 Route 23, Wayne, NJ 07470, USA

MDA Scientific Inc., 1815 Elmdale Avenue, Glenview, IL 60025, USA

Mine Safety Appliances Company, 600 Penn Center Boulevard, Pittsburgh, PA 15235, USA

Anatole J. Sipin Company, 505 Eight Avenue New York, N.Y. 10018, USA

SKC Inc., R.D. 1, 395 Valley View Road, Eighty Four, PA 15330, USA

Spectrex Company, 3594 Haven Avenue, Redwood City, CA 94063, USA

---

[a] The mention of a manufacturer does not in any way represent endorsement or recommendation of their instruments either by the authors of this chapter, by IARC or by WHO.

APPENDIX C

MANUFACTURERS OF PASSIVE MONITORING DEVICES[a]

Dragerwerk AG Lübeck, Moislinger Allee 53/55, D-2400 Lübeck 1, FRG

E.I. du Pont de Nemours & Co., Applied Technology Division, Concord Plaza, Clayton Building, Wilmington, DE 19898, USA

GMD Systems, 5345 N. Kedzie Avenue, Chicago, Il 60625, USA

MDA Scientific Company, 808 Busse Highway, Park Ridge, Il 60068, USA

Mine Safety Appliances, 600 Penn Center boulevard, Pittsburgh, PA 15235, USA

Moleculon Research Corporation, 139 Main Street, Cambridge, MA 02142, USA

National Mine Servide Company, Industrial Safety Division, 355 N. Old Steubenville Pike, Oakdale, PA 15701, USA

Reiszner environmental & Analytical Laboratories, Inc., P.O. Box 3341, Baton Rouge, Louisiana 70821, USA

SKC Incorporated, R.D. No. 1, Valley View Road, Eighty Four, PA 15330, USA

3M Company, Occupational Health & Safety, Products Division, 3M Center, 220-7W, St. Paul, MN 55144, USA

---

[a] The mention of a manufacturer does not in any way represent endorsement or recommendation of their instruments either by the authors of this chapter, by IARC or by WHO.

CHAPTER 7

A SURVEY OF THE ANALYSIS OF HALOGENATED ALKANES AND ALKENES IN BIOLOGICAL SAMPLES

L. Fishbein

Department of Health and Human Services
Food and Drug Administration
National Center for Toxicological Research
Jefferson, Arkansas 72079, USA

INTRODUCTION

Halogenated hydrocarbons (particularly chlorinated derivatives) are, quantitatively, amongst the most important products manufactured by the chemical industry. As noted in Chapter 2, numerous members of the C_1-C_4 halogenated alkanes have extensive utility as solvents, fumigants, aerosol propellants, degreasing agents, dry-cleaning fluids, refrigerants, flame-retardants, anaesthetics, cutting fluids and as intermediates in the production of textiles and other chemicals. A wide range of household items, such as degreasing agents, brush cleaners, paints and varnish removers, have contained chloroform, carbon tetrachloride, methylene chloride, trichloroethylene or tetrachloroethylene. Solvents such as trichloroethylene, tetrachloroethylene, 1,1,1-trichloroethane (methyl chloroform) and methylene chloride and the anaesthetic gas halothane have been abused (Garriott & Petty, 1980; Ramsey & Flanagan, 1982). There is increasing concern about the presence of a large number of halogenated hydrocarbons that have been detected in diverse water sources; these include a number of mutagenic and/or carcinogenic derivatives resulting from industrial outfall or the chlorination of water supplies (e.g., trihalomethanes, haloacetonitriles). Potential exposure also arises from the generation of large quantities of waste products during the manufacture of chlorinated solvents or monomers. These can often represent as much as 5% of the main product. Prevalent disposal practices include deep-well disposal, burning the wastes in land-based incinerators and burning them at sea (Miller, 1983). Hence, the halogenated alkanes and alkenes represent one of the most important categories of industrial chemicals, considered from the point of view of volume, variety of uses and environmental and toxicological problems to which they do or may give rise. The populations at risk can include workers, consumers and the general public, with exposure occurring by inhalation, dermal absorption or ingestion. Because of their lipophilic nature, there is additional concern about the extent to which the volatile halogenated alkanes and alkenes will accumulate in the fatty tissues of organisms. the presence of these pollutants in marine organisms has been reported (Pearson & McConnel, 1975; Dickson & Riley, 1976; Ofstad et al., 1981).

While air monitoring is of obvious importance and has been the major means of assessing exposure to the halogenated alkanes and alkenes in the workplace, its limitations are recognized, since this procedure measures only potential exposure by inhalation and does not measure uptake by the exposed individual.

Increasingly, investigators have sought to obviate these shortcomings by biological monitoring. Baselt (1980), Lauwerys (1983) and Monster (1984a,b,c) have reviewed the salient aspects of biological monitoring for industrial exposure to halogenated hydrocarbons, including methyl chloride, methyl bromide, methylene chloride, chloroform, carbon tetrachloride, 1,1,1-trichloroethane (methyl chloroform), trichloroethylene, vinyl chloride and fluorinated anaesthetics (e.g., halothane). The biotransformation of many of these solvents has been reviewed by Lauwerys (1975), Henschler (1977), Toftgard & Gustafsson 91980), WHO (1981) and MacDonald (1983).

Solvents are absorbed through the skin (Stewart & Dodd, 1964). Analyses of solvents in blood specimens drawn during or immediately after the working day may thus lead to erroneous estimations of exposure. Correct estimations may be obtained, however, using specimens drawn in the morning on the day after exposure (Aitio et al., 1984).

ANALYTICAL METHODS

Halogenated hydrocarbons are known to accumulate in various tissues of animals and man (McConnell et al., 1975; Dowty et al., 1976; Davies, 1978; Peoples et al., 1979). Screening biological materials for toxic volatile compounds by chromatographic techniques is acknowledged to be difficult, due to factors such as the diversity of compounds of interest, the resulting complexity of the chromatographic systems and the feasibility of the isolation techniques for handling gaseous and more-or-less volatile compounds.

The halogenated alkanes and alkenes and their metabolites can be determined in biological media by a variety of procedures, the more important of which are gas chromatography (GC) with electron-capture detection and GC-mass spectrometry (MS). These may be combined with solvent extraction, gas-stripping and head-space techniques (Anthony et al., 1978; Balkon & Leary, 1979; Dietz & Singley, 1979; Barkley et al., 1980; Michael et al., 1980; Foerster & Garriott, 1981) for the determination of volatile halogenated pollutants in biological matrices.

Davies (1978) determined blood concentrations of chloroform, trichloroethylene and halothane by GC, using electron-capture detection. The accuracy of the analysis (measured blood concentration as a percentage of the calculated concentration) was as follows: chloroform (5 blood samples in the range

74-636 mg/L), mean 100.4% (SD 1.30%), range 99.3-102.5%; trichloroethylene (5 blood samples in the range 132-206 mg/L), mean 101.6 (SD 1.60), range 99.6-104.0; halothane (11 blood samples in the range 60-379 mg/L), mean 100.7 (SD 2.27), range 97.3-104.0. The anaesthetic agent was initially extracted from the blood with the non-miscible solvent, n-heptane. A Pye Series 104 gas chromatograph was employed, with a column consisting of 15% MS 550 silicone fluid on Universal B support.

Dowty et al. (1976) described the results of GC-MS analysis of low molecular weight, volatile, organic constituents, obtained from cord blood and maternal blood samples collected at birth, as a reflection of transplacentally-acquired compounds. Chloroform and carbon tetrachloride were some of the more than 100 compounds which were found in cord blood in quantities equal to or greater than those in maternal blood. Volatile organic compounds contained in the blood samples were thermally removed by a stream of ultrapure helium, passed over 1 g of Tenax absorbent, thermally desorbed from the polymer, then transferred, by means of a helium stream, from the injection port onto a pre-column (1.5 m × 0.05 cm), coated with 10% GE SF-96 and 1% Igepal CO 880 stationary phase. Separation of the organic components was achieved on this capillary column by programming from ambient temperature to 170°C at 2°/min, after a 10-min, post-injection hold. Components were detected using a flame-ionization detector operated at 300°C. The GC effluent was vented through a heated transfer line (200°C) into the mass spectrometer (DuPont 21-491) by a jet-type separator and the spectra were obtained using an ionization energy of 70 eV.

Dowty et al. (1975) described the analysis of both tetrachloroethylene and carbon tetrachloride in both pooled plasma and New Orleans drinking water. Volatile organic compounds were eluted from water and blood plasma by heating to 95°C under a stream of ultrapure helium. the helium stream was passed through a series of glass condensers to eliminate most of the water vapor. the volatile compounds were trapped on 1 g of poly(p-2,6-diphenylphenylene)-oxide adsorbant (35/60 mesh), attached to the end of a condenser train. After a trapping period of 1 h, the polymer containing the adsorbed compounds was transferred from the collection reservoir to a silylated glass injection port liner (10 mm o.d., 9.2 cm in length, each and plugged with glass wool) which was conditioned at 350°C for 1 h in ultrapure helium prior to use. The liners were placed in the injection port (maintained at 200°C) of a Hewlett-Packard 7620A gas chromatograph and the volatile components transferred by means of a helium stream onto a 1.5 m × 0.5 mm column coated with Emulphor ON-870, held at dry ice-methanol temperature for a 7-min trapping period and finally swept onto a 91.0 m × 0.5 mm stainless-steel capillary column, coated with 10% GE SF-96 and 1% Igepal CO 880. For mass spectral analysis, the carrier gas effluent was passed through a heated transfer line (250°C) into a double-focusing mass spectrometer (DuPont 21-491) by means of a jet-type separator and the spectra were obtained using a 70 eV electron beam. The concentration of carbon tetrachloride in plasma was found to be substantially higher than in drinking water. The lipophilic nature of carbon tetrachloride suggested a bio-accumulation mechanism, if drinking water is the only source of such materials.

A quantitative, GC-MS screening procedure (using a head-space technique) for the determination of a number of volatile compounds of forensic and environmental importance (including 17 chlorinated alkanes and alkenes) in human biofluids was described by Balkon & Leary (1979). A column containing Carbowax 1500 (2%) on Carbopack C was used for the chromatographic analysis. The column oven was programmed from 50°C to 180°C at 8°C/min, with an initial 2 min hold. the final temperature was maintained for 10 min. Helium was used as the carrier gas at a flow rate of 25 mL/min. the injection port temperature was maintained at 250°C and the transfer line into the Hewlett-Packard 5982 A gas chromatograph-mass spectrometer was maintained at 225°C. The column effluent was passed directly into the mass spectrometer <u>via</u> a membrane separator. Quantification approaching the ppm level was achieved by measuring the ion current due to the most intense ion and the corresponding ion current of a simultaneously-run internal standard.

A GC method was developed by Ramsey and Flanagan (1982) for the detection and identification of a number of volatile compounds, including carbon tetrachloride, trichloroethylene, 1,1,1-trichloroethane, trichloroethane, halothane, bromochlorodifluoromethane, dichlorodifluoromethane and trichlorofluoromethane, in whole blood, plasma or serum. After incubation of the blood sample (200 µL) or tissue (200 mg), together with an internal standard (1,1,2-trichloroethane, 1,1,1,2-tetrachloroethane or bromodichloromethane) in a sealed vial, a portion of the head-space was analysed using a 2 m glass column, packed with 0.3% (w/w) Carbowax 20M on Carbopack C, 80/100 mesh. The column oven, after a 2-min isothermal period, was programmed from 35° to 175°C at 5°C/min, then held for 8 min. The effluent was monitored by both flame-ionization and electron-capture detection and peak assignment was effected by means of retention time and relative detector response.

Foerster & Garriott (1981) reviewed procedures for the analysis of 42 volatile substances of toxicological importance in biological samples. The substances examined included 11 halogenated hydrocarbons (e.g., Freon 11 and 12; methylene chloride, chloroform, carbon tetrachloride, 1,1,1-trichloroethane, trichloroethylene, ethylene dibromide, ethylene dichloride and halothane). The extraction methods included solvent extraction and the purge-and-trap techniques (Balkon & Leary, 1979; Krotoszynski <u>et al.</u>, 1979). Solvent extraction is most useful for the analysis of chlorinated compounds, which may be extracted into a hydrocarbon phase and determined by GC with electron-capture detection. The optional procedure for comprehensive volatile screening used direct elution from blood onto Tenax-GC, with GC separation on 60/80 Carbopack B/5% Carbowax 20M.

Garriott and Petty (1980) described a screening procedure for the determination of volatile substances, including methylene chloride, chloroform, ethylene chloride, carbon tetrachloride, 1,1,1-trichloroethane and trichloroethylene, implicated in deaths from inhalant abuse. Following collection of blood samples, one milliliter of head-space vapour was analysed by GC with flame-ionization detection, using a 3 ft (0.91 m) column of Porapak Q maintained at 145°C and an injection port temperature of 135°C.

Michael et al. (1980) evaluated gas stripping and dynamic head-space analysis as methods for purging volatile halogenated organic compounds (e.g., methylene chloride, chloroform, carbon tetrachloride, bromodichloromethane, tetrachloroethylene) from human biological samples, such as urine, blood, milk and adipose tissue. In gas stripping, volatile compounds are purged by bubbling an inert gas through a liquid sample at elevated temperature. The purged organic materials are trapped on the polymeric sorbent, Tenax GC, then thermally desorbed at 270°C for 30 min in a stream of helium for subsequent GC-MS. In dynamic head-space analysis, the gas passes over, rather than through, the solution. (This procedure is particularly useful with samples which foam, such as biological samples). Volatile organic materials have been effectively purged at the 10 ppb level with this technique. GC analysis was performed on a Varian 3700 gas chromatograph with flame-ionization detection, using an 80 m × 0.38 mm i.d. glass SCOT column (1% SE-30), temperature-programmed from 30 to 220°C at 4°/min. Validation experiments involved recovery studies on fortified samples by both GC and mass-balance experiments with ^{14}C-labelled compounds. Recoveries were generally from 60-95%. The techniques described by Michael et al. (1980) have been successfully employed for the analysis of blood and urine collected from individuals residing in the Old Love Canal area of Niagara Falls, NY (Pellizzari et al., 1979; Barkley et al., 1980).

A GC-MS determination of a number of chlorinated aliphatic hydrocarbons (including methylene dichloride, carbon tetrachloride, 1,1,1-trichloroethane, 1,1,2-trichloroethane, trichloroethylene, 1,1-dichloroethane, 1,2-dichloroethane and chloroform) in body fluids and tissues was described by Hara et al. (1980). A tissue specimen (0.1-0.5 g) was weighed and placed in a sample bottle, which was then warmed on a water bath at 40°C for 1 h. The vapor phase (3 mL) was then injected into a GC-MS instrument (a Shimadzu LKD 9000 system, employing a multiple-ion detector and electron-impact ionization). The column was a 1 m × 3 mm i.d. silanized glass tube, packed with Porapak P (80/100 mesh). the temperatures of flash heater, column oven, separator and ion source were 200°C, 155°C, 250°C and 270°C, respectively. The ionization energy was 70 eV.

It was possible to identify all the chlorinated aliphatic hydrocarbons by means of the retention times of the total-ion-current chromatogram and the mass spectral patterns. Quantitative determination was achieved by monitoring the characteristic ion for each compound, where the lower limit of detection was 10 to 20 pg.

Reunanen and Kroneld (1982) determined 10 volatile halocarbons in human serum and urine (as well as in raw and drinking water). The samples were extracted in a single step with petroleum ether and the extracts, without further purification, were anlaysed by high-resolution GC, with electron-capture detection. Since the extraction efficiencies of halocarbons, as well as their electron-capture responses, vary from one compound to another, calibration of the system is essential for reliable quantification (Eklund et al., 1978). Both external and internal calibrations were used to standardize the analytical system. A 20 m × 0.3 mm i.d. SE-52 glass capillary column was coupled to a Varian 2100 gas chromatograph, equipped with a splitter and a ^{63}Ni electron-capture detector. The carrier gas was hydrogen, with a flow rate of 1.7 mL/min(40 cm/s).

The determination of volatile, purgeable halogenated hydrocarbons (chloroform, carbon tetrachloride, 1,2-dichloroethane, trichloroethylene, bromoform, dibromochloromethane and bromodichloromethane) in human adipose tissue and blood serum was achieved by Peoples et al. (1979) using GC-MS. A Tekmar Model LSC-1 liquid sample concentrator was interfaced to a Tracor Model 222 gas chromatograph, equipped with a Hall electrolytic-conductivity detector, operated in the halide-specific mode. The column was a 1.83 m × 6.35 mm i.d. glass U-tube, containing n-octane on 100/120 mesh Porasil-C. The GC was operated with nitrogen carrier gas at a flow-rate of 30 mL/min, an inlet temperature of 140°C and a transfer-line temperature of 210°C. The Hall detector furnace was maintained at 900°C, with a hydrogen flow rate of 40 mL/min and a solvent (n-propanol:distilled water, 1:1, v/v) flow of 0.4 mL/min. A Finnigan Model 4000 GC-MS system, interfaced to a Tekmar liquid sample concentrator, was used to confirm the identities of the compounds quantified by GC.

The procedure for blood serum involved treatment of 1% aqueous antifoam and 0.5 mL of serum in a 5 mL purging device and sample concentrator. After the purge-trap period of 30 min at 115°C, the adsorbed compounds were desorbed and transferred to the analytical column (60°C) by heating the trap at 150°C for 6 min. The GC column was then programmed at 7°/min to 140°C.

In a pilot study, Barkley et al. (1980) used GC-MS computer analysis of selected volatile halogenated hydrocarbons by examining urine, blood and breath of an exposed population (Old Love Canal area, Niagara, NY), as well as air and water samples in the immediate environment. Volatile halogenated organic compounds, including 1,1- and 1,2-dichloroethane, chloroform, 1,1,1-trichloroethane, carbon tetrachloride, trichloroethylene and tetrachloroethylene, were recovered from blood, urine and water samples, using a modification (Michael et al., 1980) of reported methods (Zlatkis et al., 1973a,b, 1974; Bellar & Lichtenberg, 1974; Dowty et al., 1975). Volatile samples were recovered from biological matrices by heating at 50°C and passing a stream of ultra-pure helium. The volatile compounds were trapped in a glass cartridge containing 35/60 mesh Tenax GC sorbent, attached to the end of a condenser train. Cartridge samples were analysed by a thermal desorption technique, using an inlet manifold system (Pellizzari et al., 1978). Volatile halogenated organic compounds which are best quantified by this method are those whose boiling point is less than about 150°C and which possess a moderate-to-low solubility in water. The method has been validated for blood, urine and water.

The volatile halogenated hydrocarbons purged from water, blood and urine were analysed on an LKB 2091 GC-MS, equipped with a 100 m, SE-30 SCOT capillary column and a LKB 2031 data system.

Methyl chloride and methyl bromide

Methyl chloride is a gas at normal temperatures and is used as a methylating agent and as a blowing agent in industrial processes. Inhaled methyl chloride is easily absorbed and rapidly metabolized. In man, approximately 30% of inhaled methyl chloride is exhaled during the first hour after exposure (Morgan et al., 1970). The remainder is metabolized mainly by conjugation with glutathione, yielding s-methyl cysteine, which is excreted in the urine (Van Doorn et al., 1980). Methyl chloride has also been shown to react with human plasma and erythrocytes. In erythrocytes, approximately 40% of the uptake was bound by glutathione, forming S-methyl glutathione (Redford-Ellis & Gowenlock, 1971). S-methyl glutathione is also a metabolite of several other methylating agents (Johnson, 1966; Barnsley, 1968).

Preliminary results obtained on workers exposed to methyl chloride suggest that the biological half-life of S-methyl cysteine in urine is greater than 16 hours. In addition, some individuals are relatively less able to produce S-methyl cysteine following exposure to methyl chloride, suggesting an enhanced susceptibility to the toxic effects of this agent (Van Doorn et al., 1980).

Van Doorn et al. (1980) reported an assay procedure for the detection of methyl thio compounds in urine, based on alkaline hydrolysis and subsequent GC determination of methyl mercaptan in the head-space (vapour phase) over acidified hydrolysates. An F & M Model 402 dual-column gas chromatograph was employed, equipped with a flame-ionization detector and a 1.22 m x 3.2 mm glass column, packed with 10% OV-101 on 100/120 mesh Chromosorb Q. The temperatures of injection port, column and detector were 40°C, room temperature and 100°C, respectively. Helium was employed as carrier gas, with a flow rate of 15-20 mL/min, and hydrogen and air were supplied to the detector at 40 mL/min and 550 mL/min, respectively, for maximum sensitivity.

Methyl bromide is a gaseous agent which is frequently employed as a fumigant for large, enclosed industrial and agricultural areas. It has also been employed as a refrigerant and fire extinguisher. A large fraction of inhaled methyl bromide is rapidly exhaled unchanged, whereas the inorganic bromide metabolite has a serum half-life of about 15 days and is slowly excreted in the urine (Maynert, 1965). A fraction of the absorbed methyl bromide (2-3%) may also be eliminated in the urine as a mercapturic acid conjugate (Drawneek et al., 1964). While blood concentrations of intact methyl bromide after low-level exposure have not been established, inorganic bromide blood concentrations have been measured in 12 asymptomatic methyl bromide workers and found to average 15 mg/L, with a range of 4-36 mg/L (Rathus & Landy, 1961).

The determination of inorganic bromide in serum may be used as an index of exposure to methyl bromide, although the long serum half-life of bromide would appear to limit the usefulness of the measurement (Baselt, 1980). Inorganic bromide can be determined by GC, after derivatization, employing the procedure of Corina et al. (1979).

Methylene chloride

Methylene chloride is rapidly absorbed by the lung and probably also by direct skin contact with the liquid form and is metabolized to carbon monoxide by humans (Stewart et al., 1972a,b; Ratney et al., 1974; Astrand et al., 1975; Divincenzo & Kaplan, 1981a,b). Measurement of methylene chloride in blood or expired air, carbon monoxide in expired air and carboxyhemoglobin in blood can be used to monitor exposure (Stewart et al., 1974a, 1976; Peterson, 1978; Baselt, 1980; Divincenzo & Kaplan, 1981a,b; Lauwerys, 1983; Ott et al., 1983).

Methylene chloride is conveniently assayed in blood and breath by GC with flame-ionization detection (Divincenzo et al., 1971; Stewart et al., 1974a, 1976; Laham & Potvin, 1976; Divincenzo & Kaplan, 1981a,b). In the procedure of Divincenzo et al. (1971) for the determination of methylene chloride in breath, blood and urine, a Varian model 2100 gas chromatograph was employed, with a flame-ionization detector and a 1.83 m, U-shaped glass column, packed with 20% Carbowax 20M on 80/100 mesh Gas Chrom Q. The GC was operated with injection port, column and detector temperatures of 90°C, 25°C and 230°C, respectively, and nitrogen carrier gas. Hydrogen and air flow rates to the detector were 70 and 435 mL/min, respectively.

Stewart et al. (1976) used breath analysis to monitor methylene chloride exposure. A Varian model 2700 gas chromatograph was employed, equipped with a flame-ionization detector and a stainless-steel column (35.6 × 0.3 cm), packed with Poropak Q, 60/80 mesh. The nitrogen carrier gas flow rate was 35 mL/min, with a column temperature of 150°C, injector 230°C and detector 200°C.

The determination of methylene chloride in whole blood plasma or urine by head-space GC for forensic analysis was reported by Anthony et al. (1978). A Perkin-Elmer F-42 automated head-space analyser was used, fitted with a 1.83 m × 3.2 mm o.d. stainless-steel column, containing 0.2% Carbowax 1500 and Carbopack C. The oven was held at 80°C, the detector at 150°C and the nitrogen carrier gas flow rate was 15 mL/min.

Laham and Potvin (1976) described the microdetermination of methylene chloride in blood (1 µl) using a syringeless GC injection system. Analyses were performed on a Perkin-Elmer Model 900 gas chromatograph, equipped with a flame-ionization detector and 1.83 m × 3.2 mm stainless steel columns. Column A was packed with 30% Carbowax 20M on chromosorb W and operated at an oven temperature of 100°C, with a helium flow rate of 30 mL/min. Column B was packed with Porapak Q (100/120 mesh) and operated at an oven temperature of 150°C, with a helium flow rate of 30 mL/min. Since this procedure requires a relatively small sample of blood, it could be applied to the analysis of children's blood obtained from the earlobe or the finger (after accidental ingestion or inhalation of household solvents, brush cleaners, varnish removers, etc.). Additionally, physiological effects can be investigated as a function of time, along with biochemical monitoring.

The transplacental passage of methylene chloride in pregnant rats was described by Anders and Sunram (1982). Blood methylene chloride concentrations were determined following heptane extraction, using a Hewlett-Packard 5730A gas chromatograph, equipped with a ^{63}Ni electron-capture detector and a 1.83 m x 2 mm i.d. glass column, packed with 0.2% Carbowax 1500 on Carbopack C (80/100 mesh) and operated at 90°C with argon-methane (19:1, v/v) carrier gas.

Chloroform and carbon tetrachloride

Although both chloroform and carbon tetrachloride can be detected in blood and expired air from exposed individuals (Stewart et al., 1961; Stewart & Dodd, 1964; Morgan et al., 1970), it is generally considered that the paucity of data precludes the evaluation of the relationship between human exposure to these solvents and their levels in biological media (Baselt, 1980; Lauwerys, 1983).

Whereas chloroform and carbon tetrachloride and most other related halogenated hydrocarbons are eliminated to some extent via the lungs without biochemical transformation, some part of the absorbed hydrocarbons is metabolized in the liver to toxic radicals, which in turn may initiate intracellular lipid peroxidation with subsequent structural as well as functional disturbances of hepatic cell organalles (Pohl, 1979; Reynolds & Moslen, 1980; Toftgard & Gustafsson, 1980; Vierke et al., 1982; MacDonald, 1983).

Chloroform undergoes extensive biotransformation in man, with the formation of carbon dioxide and hydrochloric acid. Less than 0.01% of the absorbed dose was found in an 8 h urine (Fry et al., 1972). The hepatotoxicity of chloroform has been suggested to result from the conversion of chloroform to phosgene ($COCl_2$) (Pohl, 1979).

The metabolism of carbon tetrachloride in man has not been elucidated definitively. A probable metabolic route based on in-vitro and in-vivo studies in monkeys suggests homolytic and heterolytic cleavage, reductive dechlorination and hydrolytic dechlorination, with formation of chloroform and phosgene (Toftgard & Gustafsson, 1980; MacDonald, 1983).

As noted previously, GC procedures involving both flame-ionization and electron-capture detection have been employed for the determination of chloroform and carbon tetrachloride in biological media (Stewart et al., 1961; Stewart & Dodd, 1964; Morgan et al., 1970; Dowty et al., 1975, 1976; Davies, 1978; Balkon & Leary, 1979; Barkley et al., 1980; Garriott & Petty, 1980; Hara et al., 1980; Michael et al., 1980; Foerster & Garriott, 1981; Ramsey & Flanagan, 1982; Goldermann et al., 1983).

Trichloroethylene

Trichloroethylene vapours are readily absorbed by the lungs (Fernandez et al., 1975; Monster et al., 1979a) and liquid trichloroethylene can also be readily absorbed through the skin (Stewart & Dodd, 1964; Sata & Nakajima, 1978). Relatively small amounts (e.g., less than 10%) of absorbed trichloroethylene are eliminated unchanged in expired air (Fernandez et al., 1975) and the major portion is metabolized via the sequence trichloroethylene oxide → trichloroacetaldehyde → chloralhydrate, thence to trichloroethanol (largely as a glucuronide conjugate, urochloralic acid) and trichloroacetic acid, which are excreted in the urine. Small amounts of trichloroethanol are excreted by the lung (Ertle et al., 1972; Müller et al., 1972, 1974; Kimmerle & Eben, 1973a,b; Lauwerys, 1975, 1983; Monster et al., 1979a; Monster, 1984a). The biological half-lives of trichloroethanol and trichloroacetic acid differ, both in urine and blood. That of trichloroacetic acid (70-100 h), because of binding to plasma proteins, is greater than that of trichloroethanol (10-15 h) (Müller et al., 1972). The half-life of trichloroethylene in exhaled air and blood depends upon the length of exposure and the time elapsed before sampling (Monster, 1984a).

Numerous biological monitoring procedures have been proposed for evaluating exposure to trichloroethylene by measuring the solvent per se, or its metabolites (Tanaka & Ikeda, 1968; Stewart et al., 1970a, 1974b; Nomiyama, 1971; Ogata et al., 1971; Ertle et al., 1972; Ikeda et al., 1972; Imamura & Ikeda, 1973; Kimmerle & Eben, 1973a,b; Müller et al., 1974, 1975; Lauwerys, 1975, 1983; Monster et al., 1979a. Baselt, 1980; Ziglio et al., 1983; Monster, 1984a).

The limitations of biological monitoring of trichloroethylene or its metabolites as a measure of industrial exposures have been cited (Baselt, 1980; Lauwerys, 1983; Monster, 1984a). According to Baselt (1980), for example, the determination of urinary metabolites has not been fully successful as an index of exposure, due to factors including inter-individual variation in metabolism, accumulation of metabolites during chronic exposure, delayed urinary excretion (especially for trichloroacetic acid) and the similarity of metabolic products produced in man by other halogenated agents, such as chloral hydrate, tetrachloroethylene and trichloroethane.

Although blood trichloroacetic acid concentrations have been found to exhibit the least inter-individual variation due to the factors mentioned above, so that this measurement appeared to be the most promising for monitoring purposes (Müller et al., 1972; Monster et al., 1979a), it was noted that trichloroacetic acid tends to accumulate with repeated exposure, and thereafter leaves the body slowly ($t_{1/2}$ = 70-90 h) (Baselt, 1980).

Müller et al. (1972) suggested that simultaneous determination of trichloroacetic acid and trichloroethanol in blood (or plasma provides reliable information concerning previous exposure to trichloroethylene.

Concentrations of trichloroethylene in alveolar air have been reported to correlate well with the actual environmental trichloroethylene level, or with the time-weighted average exposure, depending on when the specimen is obtained (Stewart et al., 1970a, 1974b; Kimmerle & Eben, 1973b; Fernandez et al., 1975; Monster et al., 1979a; Baselt, 1980; Lauwerys, 1983). Analogously to the trichloroethylene concentration in expired air, the concentration of the solvent in blood reflects either the most recent exposure, when blood is sampled during exposure, or the time-weighted average exposure when blood is collected 16 hours later (Lauwerys, 1983).

Although good correlations have been found between trichloroethanol in blood and exhaled air (Müller et al., 1972; Monster et al., 1979a), it should be noted that the concentration in exhaled air during exposure is about 20 000 times lower than in blood.

Monster (1984c) recently reviewed the most salient procedures for the determination of trichloroethylene, trichloroethanol and trichloroacetic acid in biological media, with a primary focus on their utility for monitoring industrial exposures.

A variety of GC procedures have been described for the determination of trichloroethylene, trichloroethanol and trichloroacetic acid in blood. Most of these determinations are based on separate analysis of these constituents, using extraction or head-space analysis for trichloroethylene and for free and total trichloroethanol. Determinations of trichloroacetic acid are based on decarboxylation to chloroform, or esterification to trichloroacetic acid methyl ester. Methods have been published for trichloroethylene, free trichloroethanol and chloral hydrate (Kimmerle & Eben, 1973); for trichloroacetic acid in plasma (Müller et al., 1972; Ziglio, 1979); for total trichloroethanol in plasma (Müller et al., 1975); for trichloroethylene, trichloroacetic acid, free trichloroethanol and ethanol (Herbolsheimer & Funk, 1974); for chloral hydrate, free trichloroethanol, total trichloroethanol, trichloroacetic acid and monochloroacetic acid (Olgata & Saeki, 1974); for free trichloroethanol, total trichloroethanol, chloral hydrate and trichloroacetic acid (Breimer et al., 1974); for trichloroethylene, total trichloroethanol and trichloroacetic acid (Monster & Boersma, 1975); for trichloroethanol, trichloroacetic acid and chloral hydrate (Garrett & Lambert, 1966) and for free trichloroethanol, total trichloroethanol and trichloroacetic acid (Vesterburg et al., 1975). While most of the above determinations are based on separate determinations of trichloroethylene, trichloroethanol and trichloroacetic acid and require extractions and hydrolysis, the procedure of Monster and Boersma (1975) allows the simultaneous determination of trichloroethylene, total trichloroethanol and trichloroacetic acid.

The determination of trichloroethanol and trichloroacetic acid in urine has been accomplished by very sensitive and specific GC techniques (Ertle et al., 1972; Breimer et al., 1974; Buchet et al., 1974; Ogata & Saeki, 1974; Stewart et al., 1974b; Fernandez, 1976; Nomiyama et al., 1978). In addition, a number of spectrophotometric methods have been described for the determination of trichloroethanol and trichloroacetic acid in urine (Tanaka & Ikeda, 1968; Weichardt & Bardodej, 1970; Ogata et al., 1970, 1974). These methods involve the hydrolysis of trichloroethanol glucuronide in urine, followed by its oxidation to trichloroacetic acid, which is then determined by the Fujiwara reaction (alkaline pyridine) (Fujiwara, 1976). (This reaction is not specific and the natural presence of chromophore substances in biological fluids may interfere.)

A variety of methods for the determination of trichloroethylene in exhaled air have been reported. Most of these require collecting exhaled air (alveolar or mixed exhaled air) in glass tubes or bags (Saran, Tedlar), followed by determination via infrared spectroscopy (Stewart et al., 1970a; Schäcke et al., 1973) or gas chromatography (Kimmerle & Eben, 1973; Müller et al., 1974; Stewart et al., 1974b; Monster & Boersma, 1975).

The procedure of Breimer et al. (1974) for the determination of chloral hydrate, trichloroethanol and trichloroacetic acid in blood and urine employs head-space GC, with electron-capture detection. Detection limits are 0.5 mg/L for chloral hydrate and trichloroethanol in blood or urine and 0.1 mg/L for trichloroacetic acid. A Hewlett-Packard gas chromatograph, model 402, with a ^{63}Ni electron-capture detector was employed, using a 1.8 m × 3 mm i.d. glass column, packed with 10% OV-17 on Gas Chrom Q, 80/100 mesh. The injector block was maintained at 150°C, the column at 125°C and the detector at 200°C. Flow rates were 20 mL/min for the nitrogen carrier gas and 60 mL/min for the purge gas (argon:methane, 19:1, v/v).

Nomiyama et al. (1978) determined the trichloroethylene metabolites, chloral hydrate, trichloroethanol, monochloroacetic acid and trichloroacetic acid, at levels of 0.1-5 ng in urine, employing a single column and GC with electron-capture detection. A Hitachi gas chromatograph (model K53) was employed, with a 2 m × 3 mm i.d. glass column, filled with Chromosorb W, 60/80 mesh, coated with 10% Reoplex 400. Temperatures of the injection block and oven were 180°C and 130°C, respectively. Nitrogen was used as carrier and scavenger gas at 20 and 80 mL/min, respectively. The recovery figures for the metabolites ranged from 94 to 112%.

In the procedure of Ertle et al. (1972) for the determination of trichloroethanol, a 0.2 mL blood sample (taken from the ear lobe) was shaken with 1.6 mL n-hexane and applied directly to a 1.83 m × 3.2 mm column containing 8% SF 1265 (QF-1) on Chromosorb P (AW DMCS) (100/120 mesh). The column and injection port temperatures were 130°C and 150°C, respectively. A Hewlett-Packard model 5750 gas chromatograph with a ^{63}Ni electron-capture detector was used, with helium carrier gas (76 mL/min) and 10% methane in argon as detector gas (74 mL/min). The detector temperature was 240°C. For the determination of trichloroethanol in urine, 1 mL of urine and 1 mL of beta-glucuronidase solution in an acetate buffer (pH 4.5) was introduced into an ampoule which was incubated for 18 h at 37°C, then diluted with acetone to 50 mL; 1 µL was analysed by GC, using a 1.22 m × 3.2 mm column containing 10% Carbowax on 60/80 mesh Chromosorb W (AW DMCS). The column and injection port temperatures were 150°C and 190°C, respectively (Ertle et al., 1972).

Tetrachloroethylene (perchloroethylene)

Tetrachloroethylene is absorbed mainly through the lungs and, to a lesser extent, the liquid form may also be absorbed through the skin (Stewart & Dodd, 1964). The capacity of humans to metabolize tetrachloroethylene is limited and the compound is largely excreted unchanged (80-100%) in exhaled air (Ogata et al., 1971; Ikeda et al., 1972). Less than 3% is excreted in urine as trichloroacetic acid (Fernandez et al., 1976; Monster et al., 1979b). The biological half-life of tetrachloroethylene has been estimated from pulmonary excretion data to be 72 h (Guberan & Fernandez, 1974). A small amount of tetrachloroethylene absorbed in the body is probably oxidized to perchloroethylene oxide and subsequently rearranged to trichloroacetyl chloride, which is then hydrolyzed to trichloroacetic acid (Yllner, 1961; Daniel & Gage, 1963).

Aspects of the biological monitoring of tetrachloroethylene have been reviewed by Baselt (1980), Lauwerys (1980, 1983) and Monster (1984b).

Although the main biological tests for evaluating exposure to tetrachloroethylene are the determination of tetrachloroethylene in expired air, the determination of trichloroacetic acid in urine and the determination of tetrachloroethylene and trichloroacetic acid in blood, a number of limitations should be cited (Baselt, 1980; Lauwerys, 1983; Monster, 1984b). Blood concentrations have not been sufficiently employed in measuring worker exposure. Measured blood levels of tetrachloroethylene in volunteers exposed at rest are still too scarce to permit the proposal of a biological threshold limit value for this parameter (Stewart et al., 1970b; Monster et al., 1979b). Urine concentrations of tetrachloroethylene metabolites (trichloroacetic acid) appear to be of limited usefulness in monitoring exposure, since the metabolic pathways are saturated at very low levels (Ikeda et al., 1972). Breath concentrations of tetrachloroethylene, measured subsequent to exposure, are reported to relate quite well to the amount of chemical absorbed and, possibly, to the blood level of the compound (Piotrowski, 1977; Baselt, 1980).

Gas chromatographic methods for measuring tetrachloroethylene in air and blood have been described (Stewart et al., 1970b; Essing et al., 1972; Weichart & Lindner, 1975; Caplan et al., 1976; Riihimaki & Pfäffli, 1978; Ziglio, 1979). Methods for the determination of trichloroacetic acid in blood and urine have been noted above in the section concerning trichloroethylene.

1,1,1-Trichloroethane

Trichloroethane (methyl chloroform) is absorbed primarily by inhalation and, to a small extent, by skin contact with the liquid form. It is mainly eliminated unchanged in expired air, with 60-80% of the absorbed dose appearing within 1 week, and traces appearing in the post-exposure expired breath over periods as long as 1 month (Stewart et al., 1969; Monster et al., 1979c). Trichloroethane is slowly metabolized in man by oxidation to trichloroethanol, which is excreted as a glucuronide conjugate (urochloralic acid) in urine over a period of 5-12 days and which accounts for about 2% of the absorbed dose. Trichloroethanol can also be detected in expired air (Monster et al., 1979c). Trichloroacetic acid is formed from trichloroethanol as a further oxidation product and is excreted in the urine to the extent of about 0.5% of the absorbed dose (Baselt, 1980; Lauwerys, 1983).

The use of alveolar air specimens has been suggested for monitoring exposure to trichloroethane (Stewart, 1971; Astrand et al., 1973). Caperos et al. (1982) proposed that the most suitable method for estimating exposure is the determination of urinary trichloroethanol, both before and after a work-shift. However, it should be noted that while the concentrations of urinary metabolites of trichloroethane appeared to correlate with both intensity and duration of exposure, the data available were considered insufficient to allow a biological threshold limit value to be proposed for these substances (Stewart et al., 1969; Seki et al., 1975; NIOSH, 1976; Baselt, 1980). Methods for the determination of trichloroethane in blood are given by Pellizzari et al. and by Pekari and Aitio in Methods 25 and 26, respectively, in the present volume.

Halothane

Halothane (2-bromo-2-chloro-1,1,1-trichloroethane), since its introduction in 1956 for clinical use, has become one of the most commonly employed anaesthetic agents and is generally administered at levels of 0.2-2% for maintenance of anaesthesia. As much as 25% of the halothane absorbed during the course of anaesthesia is metabolized. Isolated cases of hepatotoxicity have been associated with halothane-induced anaesthesia. Chronic exposure to trace amounts of anaesthetics in the operating room may also constitute a potential hazard for operating room personnel, to be associated with an increased risk of cancer, spontaneous miscarriages and births involving congenital anomalies and possibly mutagenic effects (IARC, 1976). The mutagenicity of halothane has been reviewed by Baden et al. (1976), Garro & Phillips (1977); McCoy et al. (1977), Waskell (1978) and Edmunds et al. (1979). The absorption, biotransformation and storage of halothane has been reviewed by Cohen (1975), Holaday (1977) and Marier (1982) and aspects of biological monitoring by Lauwerys (1983).

The major urinary metabolites of halothane in man are trifluoroacetic acid (TFA), N-trifluoroacetyl-2-aminoethanol and N-acetyl-S-(2-bromo-2-chloro-1,1-difluoroethyl)-L-cysteine. A dehydrofluorination intermediate, 2-bromo-2-chloro-1,1-difluoroethylene, has also been proposed (Cohen et al., 1975). Patients metabolize one-fourth to one-half of absorbed halothane. One fraction is excreted in the urine, principally as TFA, bromide and chloride. Under normal conditions, a second fraction undergoes irreversible binding to microsomal proteins and lipids, without interfering with cell functions. A third pathway of biotransformation is activated under anaerobic conditions, resulting in release of fluoride and increased binding to phospholipids, and may be associated with hepatotoxicity (Holaday, 1977).

The production of metabolites of halothane by humans has been measured mainly in urine and blood, while sweat and faeces have been shown to account for less than 1% of the absorbed dose (Cascorbi et al., 1970). Trifluoroacetic acid (TFA) is the principal fluorine-containing urinary metabolite of halothane (as well as of the other volatile halogenated anaesthetics, isoflurane and fluoroxene). TFA can also be detected in blood, where its biological half-life is between 50 to 70 hours (Dallmeier & Henschler, 1981). At the end of a one-week exposure, a steady-state concentration of TFA is found in blood and urine and there is a linear relationship between the blood or urine concentration of the metabolite and the airborne halothane concentration. The monitoring of TFA concentration in blood at the end of a work week has been suggested by Dallmeier & Henschler (1981) to be preferable to the monitoring of TFA in urine. It was estimated that an exposure to 10 or 5 ppm halothane for 5 days (8 h/day) will lead to blood concentrations of TFA of 5 or 2.5 mg/L, respectively.

A variety of gas chromatographic procedures have been reported for the determination of TFA in biological media (principally blood) (Douglas et al., 1970; Brachet-Liermain et al., 1971; Atallah & Geddes, 1972; Cole et al., 1975; Urich et al., 1977; Witte et al., 1977; Davies, 1978; Jones, 1978; Maiorino et al., 1980; Ramsey & Flanagan, 1982; Reunanen & Kroneld, 1982). These techniques involve primarily either solvent extraction or head-space analysis, with detection by flame-ionization (Cole et al., 1975; Jones, 1978) or electron-capture (Douglas et al., 1970; Atallah & Geddes, 1972; Davies, 1978; Reunanen & Kroneld, 1982).

Bromide ion is an end product of halothane metabolism and its determination has been accomplished by a variety of procedures (Archer, 1972; Cornelis et al., 1975; Johnstone et al., 1975; Kainz et al., 1976), including gas chromatography (Archer, 1972; Wells & Cimbura, 1973; Corina et al., 1979; Maiorino et al., 1980), which is the method of choice because of enhanced sensitivity and simplicity.

Corina et al. (1979) reported a GC procedure which permitted the routine measurement of bromide at the 10-100 mg/L level in both serum and urine from subjects undergoing halothane anaesthesia. Bromide ion was converted to a 2,4-dimethylphenol derivative and assayed using a Pye 104 chromatograph, fitted with a 2.1 m x 2 mm i.d. glass column, packed with 3% OV-17 on Diatomite CQ, 100/120 mesh, and a flame-ionization detector. The column temperature was 145°C and the injector temperature 200°C; the carrier gas was nitrogen, at a flow rate of 15 mL/min. The available GC methods for bromide (Archer, 1972; Wells & Cimbura, 1973) have lower detection limits of about 100 mg Br^-/L. By using charcoal absorbent and an effective internal standard (2,3-dibromocyclohexanol), Corina et al. (1979) increased the sensitivity to a lower limit of 10 mg Br^-/L.

Maiorino et al. (1980) reported a procedure for the simultaneous quantitative determination of TFA and bromide in urine, plasma or serum, for use in in-vivo pharmacological studies. The method does not require extraction, separation, sample transfer, lyophilization or other isolation procedures which would result in losses of the two metabolites. The biological fluid is treated with dimethylsulfate in an acidic medium, resulting in the formation of the methylester of trifluoroacetic acid and methyl bromide. The volatile derivatives are then isolated from the samples via a head-space technique and separated by GC, with flame-ionization detection. The method was applicable to the determination of the two metabolites in the 0.1-10 mmol/L concentration range. A Varian 3700 gas chromatograph was employed, equipped with a 1.8 m x 2 mm i.d. nickel column containing Porapak Q (100/120 mesh). The injection port was maintained at 130°C, the column at 110°C and the detector at 150°C.

Flow rates of nitrogen (carrier), hydrogen and air were 30, 30 and 300 mL/min, respectively.

Miscellaneous halogenated derivatives

Epichlorohydrin (3-chloro-1,2-epoxypropane) is a monomer widely used in the manufacture of epoxy resins (IARC, 1976). It has been determined by a variety of different techniques, such as spectrophotometry (Daniel & Gage, 1963; Eminger & Vlacil, 1978), infrared spectroscopy (Patterson, 1954; Adamek & Peterka, 1971); gas chromatography (Ullman & Houston, 1981) and mass spectrometry (van Lierop, 1978). Epichlorohydrin has been recently determined in blood by GC and selected ion monitoring (DePetrocellis et al., 1982). A Carlo Erba model G1 gas chromatograph was employed, with flame-ionization detection and a 2 m × 4 mm i.d. glass column, packed with 100/120 mesh Gas Chrom Q, coated with 3% OV-17. The column oven was held at 60°C, the injection port at 150°C and the detector at 250°C. The nitrogen carrier gas flow rate was 35 mL/min and the air and hydrogen flow rates were adjusted to give maximal detector response. An LKB 2091 mass spectrometer was used, with a Model 2130 computer system for data acquisition and calculation. The gas chromatographic conditions were the same as above, except that helium was used as the carrier gas. Selected-ion monitoring was performed with 70 eV ionizing electrons and ion currents were measured at m/e 62 and 64 for epichlorohydrin, m/e 106 and 108 for the internal standard (epibromohydrin) and at m/e 57, which is common to both compounds. The detector responses (flame-ionization and selected-ion monitoring) were linear over a range of epichlorohydrin concentrations from 50 µg/L to 100 mg/L in blood.

Ethylene dichloride (1,2-dichloroethane) is widely employed as an intermediate in the synthesis of vinyl chloride, anti-knock substances, pesticides and fumigants. Ethylene dichloride has been determined in biological samples by a variety of GC procedures (Balkon & Leary, 1979; Peoples et al., 1979; Barkley et al., 1980; Garriott & Petty, 1980; Hara et al., 1980; Zuccato et al., 1980; Foerster & Garriott, 1981; Reunanen & Kroneld, 1982). Zuccato et al. (1980) developed a simple head-space GC method for the determination of ethylene dichloride in biological samples from exposed animals, including blood, liver, lung, spleen, brain, kidney and epididymal adipose tissue. A Carlo Erba model 2150 gas chromatograph was employed, with flame-ionization detection and a 4 m × 6 mm o.d. (4 mm i.d.) glass column, packed with Tenax GC and operated at 150°C. The injector port and detector temperatures were 225°C and the nitrogen carrier gas flow rate was 43 mL/min. Methylene chloride was used as an internal standard.

REFERENCES

Adamek, P. & Peterka, V. (1971) The determination of epichlorohydrin in aqueous solutions in the presence of glycerin, monochlorohydrin and glycidol. Analyst, 96, 806-809

Aitio, A., Pekari, K. & Järvisalo, J. (1984) Skin absorption as a source of error in biological monitoring. Scand. J. Work Environ. Health, 10, 317-320

Anders, M.W. & Sunram, J.M. (1982) Transplacental passage of dichloromethane and carbon monoxide. Toxicol. Lett., 12, 231-234

Anthony, R.M., Bost, R.O., Thompson, W.L. & Sunshine, I. (1978) Paraldehyde, toluene and methylene chloride analysis by headspace gas chromatography. J. Anal. Toxicol., 2, 262-264

Archer, A.W. (1972) A gas chromatographic method for the determination of increased bromide concentrations in blood. Analyst, 97, 428-432

Astrand, I., Kilborn, A., Wahlberg, I. & Ovrum, P. (1973) Methylchloroform exposure. Work Environ. Health, 10, 69-81

Astrand, I., Ovrum, P. & Carlsson, A. (1975) Exposure to methylene chloride. I. Its concentration in alveolar air and blood during rest and exercise and its metabolism. Scand. J. Work Environ. Health, 1, 78-94

Atallah, M.M. & Geddes, I.C. (1972) The gas chromatographic estimation of haloethane in blood using electron capture detector unit. Br. J. Anaesth., 44, 1035-1039

Baden, J.M., Brinkenhoff, M., Wharton, R.S., Hitt, B.A., Simmon, V.F. & Mazze, R.I. (1976) Mutagenicity of volatile anaesthetics. Anaesthesiology, 45, 311-318

Balkon, J. & Leary, J.A. (1979) An initial report on a comprehensive, quantitative, screening procedure for volatile compounds of forensic and environmental interest in human biofluids by GC/MS. J. Anal. Toxicol., 3, 213-215

Barkley, J., Bunch, J., Bussey, J.T., Castillo, N., Cooper, S.D., Davis, J.M., Erickson, M.D., Harris, B.S.H., III, Kirkpatrick, M., Michael, L.L., Parks, S.P., Pellizzari, E.D., Ray, M., Smith, D., Tomer, K.B., Wagner, R. & Zweidinger, R.A. (1980) Gas chromatography mass spectrometry computer analysis of volatile halogenated hydrocarbons in man and his environment - a multimedia environmental study. Biomed. Mass Spectrom., 7, 139-147

Barnsley, E.A. (1968) The metabolism of methyl methanesulphonate in the rat. Biochem. J., 106, 18P-19P

Baselt, R.C. (1980) Biological Monitoring Methods for Industrial chemicals, Davis, CA, Biomedical Publications

Bellar, T.A., Lichtenberg, J.J. & Kroner, R.C. (1974) Occurrence of organohalides in chlorinated drinking waters. J. Am. Water Works Assoc., 66, 703-706

Brachet-Liermain, A., Ferrus, L. & Caroff, J. (1971) Electron capture detection and measurement of Fluothane (halothane) in blood. J. Chromatogr. Sci., 9, 49-53

Breimer, D.D., Ketelaars, H.C.J. & Van Rossumm, J.M. (1974) Gas chromatographic determination of chloral hydrate, trichloroethanol and trichloroacetic acid in blood and urine employing head-space analysis. J. Chromatogr., 88, 55-63

Buchet, J.P., Lauwerys, R. & Roels, H. (1974) Le dosage par chromatographie en phase gazeuse de métabolites urinaires du trichloroéthylène: l'acide trichloroacetique et le trichloroethanol. Arch. Mal. Prof., 35, 395-402

Caperos, J.R., Droz, P.O., Hake, C.L., Humbert, B.E. & Jacot-Guillarmod, A. (1982) 1,1,1-Trichloroethane. Clin. Toxicol., 9, 69-74

Cascorbi, H.F., Blake, D.A. & Helrich, M. (1970) Differences in the biotransformation of halothane in man. Anaesthesiology, 32, 119-123

Chang, L.W. (1977) Pathologic changes following chronic exposures to halothane: A Review. Environ. Health. Perspect., 21, 195-210

Cohen, E.N., Trudell, J.R., Edmunds, H.N. & Watson, E. (1975) Urinary metabolites of halothane in man. Anesthesiology, 43, 392-401

Cole, W.J., Salamonsen, R.F. & Fish, K.J. (1975) A method for the gas chromatographic analysis of inhalation anaesthetics in whole blood by direct injection into a simple precolumn device. Br. J. Anesth., 47, 1043-1047

Corina, D.L., Ballard, K.E., Grice, K., Eade, O.E. & Lucas, K. (1979) Bromide measurement in serum and urine by an improved gas chromatograpic method. J. Chromatogr., 162, 382-387

Cornelis, R., speecke, A. & Hoste, J. (1975) Multielement serum standard for neutron activation analysis. Anal. Chim. Acta, 78, 317-320

Dallmeier, E. & Henschler, D. (1981) Halothan-Belastung am Arbeitsplatz im Operationssaal. Dtsch. Med. Wochenschr., 106, 324-328

Daniel, J.W. & Gage, J.C. (1963) Metabolism of ^{36}Cl-labeled trichloroethylene and tetrachloroethylene in the rat. Biochem. Pharmacol., 12, 795-802

Davies, D.D. (1978) A method of gas chromatography using electron capture detection for the determination of blood concentrations of haloethane, chloroform and trichloroethylene. Br. J. Anesthesiol., 50, 147-155

De Petrocellis, L., Tirtoreto, M., Paglialunga, S., Paesani, R., Airoldi, L., Ramoscastaneda, E. & Pantarotto, C. (1982) Determination of epichlorohydrin in blood by gas chromatography and selected ion monitoring. J. Chromatogr., 240, 218-223

Dickson, A.G. & Riley, J.P. (1976) The distribution of short-chain halogenated aliphatic hydrocarbons in some marine organisms. Mar. Pollut. Bull., 7, 167-169

Dietz, E.A. & Singley, K.F. (1979) Determination of chlorinated hydrocarbons in water by headspace gas chromatography. Anal. Chem., 11, 1809-1814

Divincenzo, G.D. & Kaplan, C.J. (1981a) Uptake, metabolism and elimination of methylene chloride vapor by humans. Toxicol. Appl. Pharmacol., 59, 130-140

Divincenzo, G.D. & Kaplan, C.J. (1981b) Effect of exercise or smoking on the uptake, metabolism and excretion of methylene chloride vapor. Toxicol. Appl. Pharmacol., 59, 141-148

Divincenzo, G.D., Yanno, F.J. & Astill, B.D. (1971) The gas chromatographic analysis of methylene chloride in breath, blood and urine. Am. Ind. Hyg. Assoc. J., 32, 387-391

Douglas, R., Hill, D.W. & Wood, D.G.L. (1970) Methods for the estimation of blood halothane concentrations by gas chromatography. Br. J. Anaesth., 42, 119-123

Dowty, B., Carlisle, D., Laseter, J.L. & Storer, J. (1975) Halogenated hydrocarbons in New Orleans driniing water and blood plasma. Science, 187, 75-77

Dowty, B., Laseter, J.L. & Storer, J. (1976) The transplacental migration and accumulation in blood of volatile organic constituents. Pediatr. Res., 10, 696-701

Drawneek, W., O'Brien, M.J., Goldsmith, H.J. & Bourdillion, R.E. (1964) Industrial methylbromide poisoning in fumigators. Lancet, 2, 855-856

Edmunds, H.N., Baden, J.M. & Simmon, V.F. (1979) Mutagenicity studies with volatile metabolites of halothane. Anesthesiology, 51, 424-429

Eklund, G., Josefsson, B. & Roos, C. (1978) Determination of volatile halogenated hydrocarbons in tap water. High Resolut. Chromatogr. Column Chromatogr., 1, 34-40

Eminger, S. & Vlacil, F. (1978) Photometric determination of alpha-epichlorohydrin in air. Chem. Prum., 28, 525-528

Ertle, T., Henschler, D., Müller, G. & Spassowski, M. (1972) Metabolism of trichloroethylene in man. I. Significance of trichloroethanol in long-term exposure conditions. Arch. Toxikol., 29, 171-188

Essing, H.G., Schäcke, G. & Schaller, K.H. (1972) Arbeitsmedizinische Untersuchungen zur dynamik der perchloräthylens im organiismus. In: Medichern, 1st Int. Symp. Werksärtze Chemischen Industrie; Ludwigshaven, pp. 375-385

Fernandez, J.G., Humbert, B.E., Droz, P.O. & Caperos, J.R. (1975) Exposition au trichloroethylene. Bilan de l'absorption, de l'excretion et du metabolisme chez des sujets humains. Arch. Mal. Prof., 36, 397-407

Fernandez, J., Suberan, E. & Caperos, J. (1976) Experimental human exposures to tetrachloroethylene vapor and elimination in breath after inhalation. Am. Ind. Hyg. Assoc. J., 37, 143-150

Flint, R.W., lawson, C.D. & Standit, S. (1975) The application of trace element analysis by X-ray fluorescence to human blood serum. J. Lab. Clin. Med., 85, 155-160

Foerster, E.H. & Garriott, J.C. (1981) Analysis for volatile compounds in biological samples. J. Anal. Toxicol., 5, 241-244

Fry, B.J., Taylor, T. & Hathway, D.E. (1972) Pulmonary elimination of chloroform and its metabolite in man. Int. Pharm. ther., 196, 98-111

Fujiwara, K. (1976) New reaction for the detection of chloroform. Sitz. Nat. Ges. Rostock., 6, 33-40

Garrett, E.R. & Lambert, H.J. (1966) Gas chromatographic analysis of trichloroethanol, chloral hydrate, trichloroacetic acid and trichloroethanol glucuronide. J. Pharm. Sci., 55, 812-817

Garriott, J. & Petty, C.S. (1980) Death from inhalant abuse: toxicological and pathological evaluation of 34 cases. Clin. Toxicol., 16, 305-315

Garro, A.J. & Phillips, R.A. (1977) Mutagenicity of the halogenated olefin, 2-bromo-2-chloro-1,1-difluoroethylene, a presumed metabolite of the inhalation anesthetic halothane. Environ. Health Perspect., 21, 65-69

Goldermann, L., Gellert, J. & Teschke, R. (1983) Quantitative assessment of carbon tetrachloride levels in human blood by head-space gas chromatography: application in a case of suicidal carbon tetrachloride intoxication. Intensive Care Med., 9, 131-135

Guberan, E. & Fernandez, J. (1974) Control of industrial exposure to tetrachloroethylene by measuring alveolar concentrations; theoretical approach using a mathematical model. Br. J. Ind. Med., 31, 159-167

Hara, K., Nagata, T., Kishimoto, E. & Kojima, T. (1980) Gas chromatographic/-mass spectrometric demonstration of chlorinated aliphated hydrocarbons in biological materials. Jpn J. Legal Med., 34, 507-513

Henschler, D. (1977) Metabolism of chlorinated alkenes and alkanes as related to toxicity. J. Environ. Pathol. Toxicol., 1, 125-133

Herbolsheimer, R. & Funk, L. (1974) Gas chromatographische Bestimmung von Trichloräthylen, Trichloroëthanol, Trichloressigsäure und Athanol in einen Analysegang aus einer Probe. Arch. Toxikol., 32, 209-215

Holaday, D.A. (1977) Absorption, biotransformation and storage of halothane. Environ. Health Perspect., 21, 165-169

Humbert, B.E. & Fernandez, J.A. (1976) Simultaneous determination of trichloroacetic acid and trichloroethanol by gas chromatography. Int. Arch. Occup. Environ. Health, 36, 235-241

IARC (1976) IARC Monographs on the Evaluation of Carcinogenic Risk of Chemicals to Man, Vol. 11, Cadmium, Nickel, some Epoxides, Miscellaneous Industrial chemicals and General Considerations on Volatile Anaesthetics, Lyon, International Agency for Research on Cancer, pp. 131, 285-293

Ikeda, M., Ohtsuji, H., Imamura, T. & Komoike, Y. (1972) Urinary excretion of total trichloro-compounds, trichloroethanol, and trichloroacetic acid as a measure of exposure to trichloroethylene and tetrachloroethylene. Br. J. Ind. Med., 29, 328-333

Imamura, T. & Ikeda, M. (1973) Lower fiducial limit of urinary metabolite level as an index of excessive exposure to industrial chemicals. Br. J. Ind. Med., 30, 289-296

Johnson, M.K. (1966) Metabolism of iodomethane in the rat. Biochem. J., 98, 38-43

Johnstone, R.E., Kennell, E.M., Behar, M.G., Brummond, W., Ebersole, R.C. & Shaw, L.M. (1975) Increased serum bromide concentration after halothane anesthesia in man. Anaesthesiology, 42, 598-601

Jones, D.J. (1978) Rapid gas chromatographic assay for volatile anesthetics in blood. J. Pharmacol. Methods, 1, 155-160

Kainz, G., Boehm, G. & Sontag, G. (1976) Determination of preservatives containing bromine in wine. Mikrochim. Acta, 2, 135-139

Karashima, D., Shigematsu, A., furukawa, T., Nagayoshi, T. & Matsumoto, I. (1977) Esterification of trifluoroacetic acid with phenyldiazomethane for quantitative gas chromatographic analysis. Methods involving separation from biological materials. J. Chromatogr., 130, 77-86

Kimmerle, G. & Eben, A. (1973a) Metabolism, excretion and toxicology of trichloroethylene after inhalation. I. Experimental exposure on rats. Arch. Toxicol., 30, 115-126

Kimmerle, G. & Eben, A. (1973b) Metabolism, excretion and toxicology of trichloroethylene after inhalation. II. Experimental human exposure. Arch. Toxikol., 30, 127-138

Krotoszynski, B.K., Bruneau, G.M. & O'Neill, H.J. (1979) Measurement of chemical inhalation exposure in urban population in the presence of endogenous effluents. J. Anal. Toxicol., 3, 225-234

Laham, S. & Potvin, M. (1976) Microdetermination of dichloromethane in blood with a syringeless gas chromatographic injection system. Chemosphere, 6, 403-411

Lauwerys, R. (1975) Biological criteria for selected industrial toxic chemicals: A review. Scand. J. Work Environ. Health, 1, 139-172

Lauwerys, R. (1983) Industrial Chemical Exposure: Guidelines for Biological Monitoring, Davis CA, Biomedical Publications, pp. 81-96

Lee, R.C., Sipes, I.G., Gandolfi, A.J. & Brown, B.R., Jr (1980) Factors influencing halothane hepatoxicity in the rat hyposic model. Toxicol. Appl. Pharmacol., 52, 267-277

MacDonald, T.L. (1983) Chemical mechanisms of halocarbon metabolism. CRC Crit. Rev. Toxicol., 11, 85-120

Maiorino, R.H., Gandolfi, A.J. & Sipes, I.G. (1980) Gas chromatographic method for the halothane metabolites, trifluoroacetic acid and bromide in biological fluids. J. Anal. Toxicol., 4, 250-254

Marier, J.R. (1982) Halogenated hydrocarbon environmental pollution: the special case of halogenated anaesthetics. Environ. Res., 28, 212-239

Maynert, E.W. (1965) Sedatives and hyponotics. I. Nonbarbiturates. In: DiPalma, J.R., ed., Drills Pharmacology in Medicine, 3rd ed., New York, McGraw-Hill, p. 184

McConnell, G., Ferguson, O.M. & Pearson, C.R. (1975) Chlorinated hydrocarbons and the environment. Endeavor, 34, 13-18

McCoy, E.C., Hankel, R., rosenkranz, H.S., Guiffrida, J.G. & Bizzari, D.V. (1977) Detection of mutagenic activity in the urines of anaesthesiologists: A preliminary report. Environ. Health Perspect., 21, 221-223

Michael, L.C., Erickson, M.D., Parks, S.P. & Pellizzari, E.D. (1980) Volatile environmental pollutants in biological matrices with a headspace purge technique. Anal. Chem., 52, 1836-1841

Miller, S. (1983) Chlorinated hydrocarbon wastes. Environ. Sci. Technol., 17, 290A-291A

Monster, A.C. (1984a) Trichloroethylene. In: Aitio, A., Riihimaki, V. & Vainio, H., eds, Biological Monitoring and surveillance of workers Exposed to Chemicals, Washington, Hemosphere Publ. Co., pp. 111-129

Monster, A.C. (1984b) Tetrachloroethylene. In: Aitio, A., Riihimaki, V. & Vainio, H., eds, Biological Monitoring and surveillance of workers Exposed to Chemicals, Washington, Hemosphere Publ. Co., pp. 131-139

Monster, A.C. (1984c) 1,1,1-Trichloroethane. In: Aitio, A., Riihimaki, V. & Vainio, H., eds, Biological Monitoring and surveillance of workers Exposed to Chemicals, Washington, Hemosphere Publ. Co., pp. 141-150

Monster, A.C. & Boersma, G. (1975) Simultaneous determination of trichloroethylene and metabolites in blood and exhaled air by gas chromatography. Int. Arch. Occup. Environ. Health, 35, 155-163

Monster, A.C., Boersma, G. & Duba, W.C. (1979a) Kinetics of trichloroethylene in repeated exposure of volunteers. Int. Arch. Occup. Environ. Health, 42, 283-292

Monster, A.C., Boersma, G. & Steenweg, H. (1979b) Kinetics of tetrachloroethylene in volunteers, influence of exposure concentrations and work load. Int. Arch. Occup. Environ. Health, 42, 303-309

Monster, A.C., Boersma, G. & Steenweg, H. (1979c) Kinetics of 1,1,1-trichloroethane in volunteers; influence of exposure concentration and workload. Int. Arch. Occup. Environ. Health, 42, 293-301

Morgan, A., Black, A. & Belcher, D.R. (1970) The excretion in breath of some aliphatic halogenated hydrocarbons following administration by inhalation. Ann. Occup. Hyg., 13, 219-233

Müller, G., Spassovski, M. & Henschler, D. (1972) Trichloroethylene exposure and trichloroethylene metabolites in urine and blood. Arch. Toxicol., 29, 335-340

Müller, G., spassovski, M. & Henschler, D. (1974) Metabolism of trichloroethylene in man. II. Pharmacokinetics of metabolites. Arch. Toxicol., 32, 283-295

Müller, G., Spassovski, M. & Henschler, D. (1975) Metabolism of trichloroethylene in man. III. Interaction of trichloroethylene and ethanol. Arch. Toxicol., 33, 173-189

NIOSH (1976) Occupational Exposure to 1,1,1-Trichloroethane (Methylchloroform), Cincinnati, OH, National Institute for Occupational Safety and Health

Nomiyama, K. (1971) Estimation of trichloroethylene exposure by biological materials. Int. Arch. Arbeitsmed., 27, 281-292

Nomiyama, H., Nomiyama, K. & Uchiki, H. (1978) Gas-liquid chromatographic determination of trichloroethylene metabolites in urine. Am. Ind. Hyg. Assoc. J., 39, 506-510

Ogata, M., Takatsuka, Y. & Tomokuni, K. (1970) A simple method for quantitative analysis of urinary trichloroethanol and trichloroacetic acid as an index of trichloroethylene exposure. Br. J. Ind. Med., 27, 378-381

Ogata, M., Takatsuka, Y. & Tomokuni, K. (1971) Excretion of organic chlorine compounds in the urine of persons exposed to vapors of trichloroethylene and tetrachloroethylene. Br. J. Ind. Med., 28, 386-391

Ogata, M. & Saeki, T. (1974) Measurement of chloral hydrate, trichloroethanol, trichloroacetic acid and monochloracetic acid in the serum and urine by gas chromatography. Int. Arch. Arbeitsmed., 33, 49-58

Ogata, M., Tomokuni, K. & Asahara, H. (1974) Simple microdetermination of trichloroethanol glucuronide and trichloroacetic acid in urine. Int. Arch. Arbeitsmed., 32, 203-215

Ofstad, E.B., Drangsholt, H. & Carlberg, G.E. (1981) Analysis of volatile halogenated organic compounds in fish. Sci. Total Environ., 20, 205-215

Ott, M.G., Skory, L,K., Holder, B.B., Bronson, J.M. & Williams, P.R. (1983) Health evaluation of employees occupationally exposed to methylene chloride. Scand. J. Work Environ. Health, Suppl. 1, 9, 1-7

Patterson, W.A. (1954) Infrared absorption bands characteristic of the Oxirane ring. Anal. Chem., 26, 823-835

Pearson, C.R. & McConnel, G. (1975) Chlorinated C_1 and C_2 hydrocarbons in the marine environment. Proc. R. Soc. Lond., B, 189, 305-322

Pellizzari, E.D., Carpenter, B., Bunch, J. & Sawicki, E. (1978) Collections and analyses of trace organic vapor pollutants in ambient atmospheres. Environ. Sci. Technol., 9, 556-560

Pellizzari, E.O., Erickson, M.D. & Zweidinger, R.A. (1979) Formulation of a Preliminary Assessment of Halogenated Organic Compounds in Man and Environmental Media. EPA-560/13-79-116, Washington DC, U.S. Environmental Protection Agency, 469 pp.

Peoples, A.J., Pfaffenberger, c.D., Shafik, T.M. & Enos, H.F. (1979) Determination of volatile purgeable halogenated hydrocarbons in human adipose tissue and blood serum. Bull. Environ. Contam. Toxicol., 23, 244-249

Peterson, J.E. (1978) Modeling the uptake, metabolism and excretion of dichloromethane in man. Am. Ind. Hyg. Assoc. J., 39, 41-47

Piotrowski, J.K. (1977) Exposure Tests for Organic compounds in Industrial Toxicology, Washington DC, U.S. Government Printing Office, pp. 98-101

Pohl, L.R. (1979) Biochemical toxicology in chloroform. In: Hodgson, E., Bend, J.R. & Philpot, R.M., eds, Reviews in Biochemical Toxicology, Vol. 1, Amsterdam, Elsevier, North-Holland, pp. 79-107

Ramsey, J.D. & flanagan, R.J. (1982) Detection and identification of volatile organic compounds in blood by headspace gas chromatography as an aid to the diagnosis of solvent abuse. J. Chromatogr., 240, 423-444

Rathus, E.M. & Landy, P.J. (1961) Methyl bromide poisoning. Br. J. Ind. Med., 18, 53-57

Ratney, R.S., Wegman, D.H. & Elkins, H.B. (1974) The in vivo conversion of methylene chloride to carbon monoxide. Arch. Environ. Health, 28, 223-226

Redford-Ellis, M. & Gowenlock, A.H. (1971) Studies on the reaction of chloromethane with human blood. Acta Pharmacol., 30, 36-48

Reunanen, M. & Kroneld, R. (1982) Determination of volatile hydrocarbons in raw and drinking water, human serum and urine by electron capture GC. J. Chromatogr. Sci., 20, 449-454

Reynolds, E.S. & Moslen, M.T. (1980) Environmental liver injury: halogenated hydrocarbons. In: Farber, E. & Fisher, M.M., eds, Toxic Injury of the Liver, Part B, New York, Basel, Marcel Dekker, pp. 541-596

Riihimaki, v. & Pfäffli, P. (1978) Percutaneous absorption of solvent vapors in man. Scand. J. Work Environ. Health, 4, 73-85

Sata, A. & Nakajima, T. (1978) Differences following skin or inhalation exposure in the absorption and excretion kinetics of trichloroethylene and toluene. Br. J. Ind. Med., 35, 48

Schäcke, G., Essing, H.G., Schaller, K.H., Lehrl, S. & Nikolay, J. (1973) Untersuchungen zur kinetik des Trichloroäthylens und seiner Metaboliten beim Menschen. In: 12 Johrestagung Deutschen Gesellschaft Arbeitsmedizin, Stuttgart, Gentoner, pp. 335-342

Seki, Y., Urashima, Y., Aikawa, H. et al. (1975) Trichloro compounds in the urine of humans exposed to methyl chloroform at sub-threshold levels. Int. Arch. Arbeitsmed., 34, 39-49

Stewart, R.D. (1971) Methyl chloroform intoxication. J. Am. Med. Assoc., 215, 1789-1792

Stewart, R.D. & Dodd, H.C. (1964) Absorption of carbon tetrachloride, trichloroethylene, tetrachloroethylene, methylene chloride and 1,1,1-trichloroethane through the human skin. Am. Ind. Hyg. Assoc. J., 25, 439-446

Stewart, R.D., Arbor, A., Erley, D.S. (1961) Human exposure to carbon tetrachloride vapor. J. Occup. Med., 3, 586-590

Stewart, R.D., Gay, H.H., Schaffer, A.W. et al. (1969) Experimental human exposure to methylchloroform vapor. Arch. Environ. Health, 19, 467-472

Stewart, R.D., Dodd, H.C., Gay, H.H., Schaffer, A.W. & Erley, D.S. (1970a) Experimental human exposure to trichloroethylene. Arch. Environ. Health, 20, 64-71

Stewart, R.D., Baretta, E.D., Dodd, H. & Torkelson, T.R. (1970b) Experimental exposure to tetrachloroethylene. Arch. Environ. Health, 20, 224-229

Stewart, R.D., Fisher, T.N., Hosko, M.J., Peterson, J.E., Baretta, E.D. & Dodd, H.C. (1972a) Carboxyhemoglobin elevation after exposure to dichloromethane. Science, 176, 295-296

Stewart, R.D., Fisher, T.N., Hosko, M.J., Peterson, J.E., Baretta, E.D. & Dodd, H.C. (1972b) Experimental human exposure to methylene chloride. Arch. Environ. Health, 25, 342-348

Stewart, R.D., Hake, C.L., Forster, H.V., Lebron, A.J., Peterson, J.E. & Wu, A. (1974a) Methylene Chloride: Development of a Biologic Standard for the Industrial Worker by Breath Analysis. Report No. NIOSH-MCOW-ENVM-MC-74-9, cincinnati, OH, National Institute of Occupational Safety and Health

Stewart, R.D., Hake, C.L., Lebrun, A.J., Peterson, J.E., forster, H.V., Hosko, M.J., Newton, P.E. & Dodd, H.C. (1974b) Biological Standards for the Industrial Worker by Breath Analysis: Trichloroethylene, HEW Publication (NIOSH) No. 74-133, Washington DC, U.S. Department of Health, Education and Welfare

Stewart, R.D., Hake, C.L. & Wu, A. (1976) Use of breath analysis to monitor methylene chloride exposure. Scand. J. Work Environ. Health, 2, 57-70

Tanaka, S. & Ikeda, M. (1968) A method for determination of trichloroethanol and trichloroacetic acid in urine. Br. J. Ind. Med., 25, 214-219

Toftgard, R. & Gustafsson, J.A. (1980) Biotransformation of organic solvents. Scand. J. work Environ. Health, 6, 1-8

Ullman, A.H. & Houston, R. (1981) Determination of residual epichlorohydrin in middle cut alkyl glycidyl ethers by headspace gas chromatography. J. Chromatogr., 211, 398-402

Urich, R.W., Bowerman, D.L., Wittenberg, P.H., Pierce, A.F., Schisler, D.K., Levisky, J.A. & Pflug, J.L. (1977) Head space mass spectrometric analysis for volatiles in biological specimens. J. Anal. Toxicol., 1, 195-199

Van Doorn, R., Borm, P.J.A., Leijdekkers, C.M., Henderson, P.T., Reuvers, J. & VanBergen, T.J. (1980) Detection and identification of s-methyl cysteine in urine of workers exposed to methyl chloride. Int. Arch. Occup. Environ. Health, 46, 99-109

Van Lierop, J.B.H. (1978) Simple and rapid determination of epichlorohydrin at the lower parts per billion level by gas-chromatography-mass fragmentography. J. Chromatogr., 166, 609-610

Vesterberg, O., Gorczak, J. & Krasts, M. (1975) Methods for measuring trichloroethanol and trichloroacetic acid in blood and urine after exposure to trichloroethylene. Scand. J. work Environ. Health, 1, 243-249

Vierke, W., Gellert, J. & Teschke, R. (1982) Head-space gas chromatographic analysis for rapid quantitative determination of carbon tetrachloride in blood and liver of rats. Arch. Toxicol., 551, 91-99

Waskell, L. (1978) A study of the mutagenicity of anesthetics and their metabolites. Mutat. Res., 57, 141-153

Weichardt, H. & Bardodej, Z. (1970) Die Bestimmung von Trichloräthanol im Urin von Tri-Arbeitern. Zentralbl. Arbeitsmed., 20, 219-221

Weichardt, H. & Lindner, J. (1975) Gesundheitsgefahren durch Perchloräthylen im Chemisch - Reinigungsbetrieben aus arbeitsmedizinisch-taxikologischen Sicht. Staub. Reinhalt. Luft., 35, 416-420

Wells, J. & Cimbura, G. (1973) Determination of elevated bromide levels in blood by gas chromatography. J. Forensic Sci., 18, 437-442

WHO (1981) Recommended Health-Based Limits in Occupational Exposure to Selected Organic Solvents. Technical Report Series No. 664, Geneva, World Health Organization

Witte, L., Narr, H., Fuhrhop, J.H., Doenicke, A. & Crote, B. (1977) Quantitative analysis of trifluoroacetic acid in body fluids in patients treated with halothane. J. Chromatogr., 143, 329-334

Yllner, S. (1961) Urinary metabolites of ^{14}C-tetrachloroethylene in mice. Nature, 191, 820-821

Ziglio, G. (1979) Determinazione gas chromatografica del livello ematico di acido trichloroacetico in soggetti non professionalmente esposti a trichloro-e tetrachloroetilene. Ig. Mod., 72, 876-900

Ziglio, G., Fara, G.M., Beltramelli, G. & Pregliasco, F. (1983) Human environmental exposure to trichloro- and tetrachloroethylene from water and air in Milan, Italy. Arch. Environ. Contam. Toxicol., 12, 57-64

Zlatkis, A., Bertsch, W., Lichtenstein, H.A., Tishbee, A., Shumbo, F., Liebich, H.M., Coscil, A.M. & Fleischer, N. (1973a) Profile of volatile metabolites in urine by gas chromatography-mass spectrometry. Anal. Chem., 45, 763-767

Zlatkis, A., Betsch, W., Lichtenstein, H.A., Tishbee, A., Shumbo, F., Liebich, H.M., Coscil, A.M. & Fleischer, N. (1973b) Concentration and analysis of volatile urinary metabolites. J. Chromatogr. Sci., 11, 299-302

Zlatkis, A., Bertsch, W., Bafus, D. & Liebich, H.M. (1974) Analysis of trace volatile metabolites in serum and plasma. J. Chromatogr., 91, 379-383

Zuccato, E., Marcucci, F. & Mussini, E. (1980) GLC determination of ethylene chloride in biological samples. Anal. Lett., 13(B5), 363-370

II. METHODS OF SAMPLING AND ANALYSIS

THE DETERMINATION OF HALOGENATED ALKANES AND ALKENES IN AIR

THE DETERMINATION OF HALOGENATED ALKANES AND ALKENES IN AIR

INTRODUCTION

A. MacKenzie Peers

Methods 1 to 8, which follow, are based on the use of activated-charcoal sampling tubes and personal sampling pumps, and employ similar, if not identical, sampling and other procedures. To avoid needless repetition, some of these are described below. They have been condensed from the introductory part of the NIOSH Manual of Analytical Methods, 3rd Ed., Vol. 1, U.S. Department of Health and Human Services (1984).

Apparatus

The activated-charcoal sampling tubes specified in the methods may be obtained from the following sources:

Dragerwerk AG
Postfach 1339
Moislinger Allee 53/55
2400 Lubeck 1, FRG

SKC, Inc.
R.D.1
395 Valley View Road
Eighty Four, PA, 15330
USA

Supelco, Inc.
Supelco Park
Bellefonte, PA 16823-0048
USA

Personal sampling pumps must be capable of maintaining a constant (±5%) flow rate over the desired time interval, with a representative sampler in line. In general, all high-flow pumps can maintain at least 3 kPa pressure drop across the sampling tube at 1 L/min for 8 h. Some pumps can maintain a 7.5 kPa pressure drop at flow rates of 2 to 3 L/min. Most low flow-rate pumps (0.01 to 0.2 L/min) can be used with the available sampling tubes without problems.

All pumps need to be calibrated prior to use, with a representative sampler in line. As a minimum, calibration should be carried out before and after each sampling programme, if not each day. Calibration may employ wet-test or dry-gas meters, but the use of a primary standard, such as a soap-bubble meter, is recommended. Set up the calibration system so that the pump draws air through the rate-meter, then the sampler tube. Place a water manometer between the sampler tube and the pump. The pressure drop across the sampler should not exceed 33 cm of water. If the pump is fitted with a rotameter (flow-rate indicator), record the air temperature and pressure during calibration.

Sampling procedure

1. Immediately before sampling, break both ends of the glass sorbent tube to provide openings of at least one-half of the internal diameter of the tube.

2. Connect the sorbent tube to the pump with flexible tubing, with the smaller (back-up) section nearest the pump. Position the tube vertically to avoid channeling.

3. Prepare the field blanks (see below) at about the same time that sampling is begun. At least two field blanks are required per ten samples.

4. Check the flow rate during sampling. If it is not constant (± 5%) discard the sample.

5. With each set of field samples, take two to four replicate samples for control of reproducibility.

6. If sampling pump is fitted with rotameter only, record initial and final air temperatures and pressures during sampling. The volume of air sampled, V, is given by

$$V = Ft(P_c T_s / P_s T_c)^{0.5}$$

where,

F = indicated flow rate (L/min)
t = sampling time (min)
P_c = pressure during calibration
P_s = pressure during sampling (same units as P_c)
T_c = temperature during calibration (°K)
T_s = temperature during sampling (°K)

7. Immediately after sampling, seal the ends of the tubes with plastic (not rubber) caps and pack securely for shipment.

Blanks and spikes

Reagent blanks: These give the analyte contribution from the reagents employed in preparing samples for analysis.

Media blanks: These give the analyte contribution from the collection medium (e.g., charcoal tubes), plus the reagents. Media blanks are measured using new sampling tubes which are not opened until the start of the analytical procedure.

Field blanks: These give the analyte contribution due to the reagents, the collection medium and any contamination which may occur during handling, shipping and storing before analysis.

Analyst spikes: These are prepared by adding known amounts of the analytes of interest to unused sampling tubes, and serve to measure the recovery (desorption efficiency) of the analyte from the collection medium.

Blind spikes: These are prepared in the same way as analyst spikes, but the amounts of analyte added are unknown to the analyst. Blind spikes are used for quality control.

NOTE: All the sampling tubes employed for blanks and spikes which are to be analysed with a given set of samples must be taken from the same lot as those employed for the collection of that set of samples.

METHOD 1

MULTISUBSTANCE METHOD FOR THE DETERMINATION OF INDIVIDUAL HALOGENATED COMPOUNDS IN INDUSTRIAL AIR

A. MacKenzie Peers

Adapted from Method 1003, NIOSH Manual of Analytical Methods, 3rd Ed., Vol. 1 (1984)

1. SCOPE AND FIELD OF APPLICATION

This personal sampling pump and sorption-tube method is suitable for the simultaneous determination of two or more of the substances listed in Table 1 (section 11) by appropriate modification of the gas chromatographic conditions. Useful ranges vary with the analyte (see Table 2, section 12). the estimated limit of detection is 10 μg/sample. High humidity during sampling decreases the trapping efficiency and the breakthrough volume.

2. REFERENCES

NIOSH (1977) Documentation of the NIOSH Validation Tests, U.S. Department of Health, Education and Welfare, Publ. (NIOSH) 77-185

Eller, P.M., ed. (1984) NIOSH Manual of Analytical Methods, 3rd Ed., Vol. 1, U.S. Department of Health and Human Services

3. DEFINITIONS

Not applicable

4. PRINCIPLE

A known volume of air is drawn through a tube with two sections containing activated coconut charcoal which adsorbs the gaseous organic componds. These are subsequently desorbed in carbon disulfide (containing an internal standard for quantification purposes) and are determined by gas chromatography (GC), with flame-ionization detection (FID). A calibration curve is employed for quantification and a correction is applied for desorption efficiency.

5. HAZARDS

Carbon disulfide is toxic and an acute fire and explosion hazard (flash point, -30°C) and should be handled only in a hood. The analytes (Table 1) are toxic as well and should also be handled with care in a ventilated hood.

6. REAGENTS[1]

Carbon disulfide	Chromatographic quality
Toluene	Reagent grade
Analytes (Table 1)	Reagent grade
n-Hexane	GC purity
Nitrogen or Helium	Purified, for GC carrier gas
Hydrogen	Pre-purified, for FID
Air	Filtered, for FID
Extraction solvent	Carbon disulfide containing 0.1% (v/v) toluene, or other suitable internal standard (see Table 2)
Stock standard solution of bromoform	10 g/L in *n*-hexane
Working standard solution of analytes	Add known amounts of stock standard solution (in the case of bromoform) or of pure analyte to the extraction solvent in a 10-mL volumetric flask and dilute to the mark. Prepare at least five concentrations of each analyte over the appropriate ranges shown in Table 2.

---

[1] Reference to a company and/or product is for the purpose of information and identification only and does not imply approval or recommendation of the company and/or product by the International Agency for Research on Cancer, to the exclusion of others which may also be suitable.

7. APPARATUS[1] (see INTRODUCTION, p. 173)

Air sampler tubes	Glass tubes, 7 cm x 6 mm o.d., 4 mm i.d., flame-sealed ends, containing two sections of 20/40 mesh activated (600°C) coconut-shell charcoal (front, 100 mg; back, 50 mg), separated by a 2-mm urethane foam plug. A sylilated glass-wool plug precedes the front section and a 3-mm urethane foam plug follows the back section. The pressure drop across the tube must be < 3.4 kPa at 1 L/min air flow rate.
Personal sampling pumps	0.01 to 0.2 L/min, with flexible connecting tubing.
Gas chromatograph	With FID and peak-area integrator. See Table 2 for column specifications.
Vials	2-mL, glass, with PTFE-lined crimp caps
Volumetric flasks	10-mL
Syringe	10-μL, readable to 0.1 μL
Pipette	1-mL, with bulb

8. SAMPLING

8.1 Follow sampling instruction in INTRODUCTION, p. 173

8.2 Sample at an accurately known rate between 0.01 and 0.2 L/min to obtain the total sample volume indicated in Table 2. Samples (in capped samplers) are stable at least 1 week at 25°C.

---

[1] Reference to a company and/or product is for the purpose of information and identification only and does not imply approval or recommendation of the company and/or product by the International Agency for Research on Cancer, to the exclusion of others which may also be suitable.

9. PROCEDURE

9.1 Blank tests (see INTRODUCTION, p. 173)

9.1.1 Field blanks: analyse with air samples, as in 9.4-9.6.

9.1.2 Media blanks: set aside three sampler tubes from each new batch of charcoal used for sampling. Extract and analyse with air samples, as in 9.4-9.6.

9.2 Check tests

Analyse three blind spikes and three analyst spikes to check the calibration and desorption efficiency curves (see INTRODUCTION, p. 173).

9.3 Test portion

Not applicable

9.4 Sample extraction

9.4.1 Place front and back sorbent sections in separate vials (discard glass-wool and foam plugs).

9.4.2 Add 1.0 mL extraction solvent to each vial and close with crimp cap.

9.4.3 Allow to stand 30 min, with occasional agitation. Retain for GC analysis.

9.5 GC operating conditions

Column specifications and temperatures	see Table 2 (section 12)
Carrier gas, nitrogen or helium	30 mL/min

9.6 Analyte determination

9.6.1 Inject 5 μL of sample extract (9.4.3) manually onto the GC using solvent-flush technique (see Notes on Procedure, section 12), or with autosampler.

9.6.2 Measure peak areas of analyte(s) and internal standard on the same chromatogram. Divide peak area of analyte by that of internal standard to obtain relative peak area.

9.7 Calibration curve

9.7.1 Calibrate daily by injecting 5-μL aliquots of each of the five working standard solutions of the analyte onto the GC.

9.7.2 Measure peak areas of analyte and internal standard on the same chromatogram and calculate relative peak area (9.6.2) for each concentration.

9.7.3 Prepare calibration curve showing relative peak area *versus* amount (mg) of analyte per mL.

9.8 Desorption efficiency (DE)

9.8.1 For each batch of charcoal used for sampling, prepare three spiked tubes at each of five levels in the calibration range (e.g., spike with the amounts of analyte in 1 mL of each of the five working standard solutions). Proceed as follows:

9.8.2 Remove and discard back (50 mg) sorbent section of a fresh sampler.

9.8.3 Inject a known amount of analyte (see 9.8.1) directly onto front sorbent section by means of a microlitre syringe. Cap tube and allow to stand overnight.

9.8.4 Extract and analyse as in 9.4-9.6

9.8.5 Using the calibration curve, calculate mg analyte recovered at each spiking level and prepare a graph of desorption efficiency (DE) *versus* mg analyte recovered (DE = mass of analyte recovered/mass of spike).

10. METHOD OF CALCULATION

Using the relative peak area (9.6.2) and the calibration curve (9.7.3), calculate the mass concentration, ρ, of the analyte in the air samples from the equation,

$$\rho = 10^3(m_s - m_b)/V \quad (mg/m^3)$$

where, m_s = sum of analyte masses (corrected for DE) found in front and back sections of sample tube (mg)

m_b = average sum of analyte masses (corrected for DE) found in front and back sections of media blanks (mg)

V = volume of air sample (L)

NOTE: If $10\ m_b > m_s$ report breakthrough and possible sample loss.

11. REPEATABILITY AND REPRODUCIBILITY

The method has been tested using spiked samples and generated atmosphere (see NIOSH, 1977, section 2), with the results shown in Table 1.

Table 1. Results of laboratory validation tests

Compound	Range studied (mg/m^3)	Sample size (L)	Coefficient of variation (%)	
			GC determination	Overall
Bromoform	3 - 10	10	4.7	7.1
Carbon tetrachloride	65 - 299	15	3.7	9.2
Chloroform	100 - 416	15	4.7	5.7
1,1-Dichloroethane	218 - 838	10	1.1	5.7
1,2-Dichloroethylene	475 -1 915	3	1.7	5.2
Ethylene dichloride	195 - 819	3	1.2	7.9
Hexachloroethane	5 - 25	10	1.4	12.1
Methyl chloroform	904 -3 790	3	1.8	5.4

12. NOTES ON PROCEDURE

The solvent-flush injection technique is carried out as follows. The 10-μL syringe is first flushed with solvent several times to wet the barrel and plunger. Three μL of solvent are drawn into the syringe. The needle is then removed from the solvent and the plunger is pulled back about 0.2 μL to separate the solvent from the sample with a pocket of air to be used as a marker. The needle is immersed in the sample and a suitable aliquot is withdrawn, taking into account the volume of the needle, since the sample in the needle will be completely injected. After the needle is removed from the sample, and prior to injection, the plunger is pulled back 1.2 μL to minimize evaporation of the sample from the tip of the needle. Duplicate injections should show no more than 3% difference in peak area.

Table 2. Sampling and GC conditions

Compound	Air sample volume (L)		Breakthrough in dry air	Working range (mg/m^3)	Column[a]	Temp. (°C) column injector detector	Internal standard	Range (mg per sample)
	Min	Max						
Bromoform	1 @ 0.5 ppm	10	48 L @ 10 mg/m^3	0.5 to 15 (10 L)	A	130 170 210	n-pentadecane	0.005 to 0.15
Carbon tetrachloride	2 @ 10 ppm	17	45 L @ 338 mg/m^3	16 to 480 (15 L)	B	60 155 200	decane	0.2 to 7
Chloroform	1.5 @ 50 ppm	15	46 L @ 473 mg/m^3	25 to 750 (15 L)	B	75 155 200	n-undecane	0.4 to 10
1,1-Dichloroethane	1 @ 100 ppm	10	18.3 L @ 838 mg/m^3	40 to 1 215 (10 L)	A	50 100 175	-	0.4 to 12
1,2-Dichloroethylene	0.5 @ 200 ppm	3	5.4 L @ 1 909 mg/m^3	80 to 2 370 (3 L)	A	60 170 210	-	0.2 to 7
Ethylene dichloride	1 @ 50 ppm	10	29 L @ 821 mg/m^3	40 to 1 215 (3 L)	A	70 225 250	octane	0.1 to 4
Hexachloroethane	1 @ 1 ppm	10	48 L @ 23 mg/m^3	1 to 30 (10 L)	C	110 170 210	n-tridecane	0.01 to 0.3
Methylchloroform	0.5 @ 350 ppm	6	9.5 L @ 3 788 mg/m^3	190 to 5 700 (3 L)	A	70 225 250	octane	0.6 to 17

[a] A = 3 m × 3 mm stainless steel, 10% SP-1000 on 80/100 mesh Supelcoport;
B = 6 m × 3 mm, otherwise same as A;
C = 3 m × 6 mm glass, 3% SP-2250 on 80/100 mesh Supelcoport

13. SCHEMATIC REPRESENTATION OF PROCEDURE

Draw air sample through charcoal sampler tube
(front section, 100 mg; back section, 50 mg)

↓

Place front and back solvent sections in separate vials,
add 1.0 mL carbon disulfide (+ standard) and cap

↓

Stand 30 min with occasional agitation

↓

Inject 5-μL aliquot of extract onto GC

↓

Inject 5 μL aliquots of standard solutions of analytes onto GC
and prepare calibration curve

↓

Determine desorption efficiency in the calibration range,
using three spiked tubes at each of five levels.
Extract and analyse by GC, as above.

↓

Calculate mass concentration (mg/m³) of analyte in air sample,
correcting for desorption efficiency and media blanks

14. ORIGIN OF THE METHOD

Method 1003, NIOSH Manual of Analytical Methods, 3rd Ed., Vol. 1, Cincinnati OH, National Institute for Occupational Safety and Health (1984)

Contact point: Dr P.M. Eller
Division of Physical Sciences and Engineering
National Institute for Occupational Safety
and Health
4676 Columbia Parkway
Cincinnati, OH 45236, USA

METHOD 2

THE DETERMINATION OF ALLYL CHLORIDE IN AIR

A. MacKenzie Peers

Adapted from method 1000, NIOSH Manual of Analytical Methods, 3rd Ed., Vol. 1 (1984)

1. SCOPE AND FIELD OF APPLICATION

This personal sampling pump and sorption tube method is suitable for the determination of allyl chloride (3-chloro-1-propene) in air over the range 0.5 to 10 mg/m^3 (100 L air sample). The method may be applied to 15-min samples, taken at 1 L/min. Breakthrough occurred between 210 and 240 min when sampling air containing 7.56 mg allyl chloride/m^3 at 0.94 L/min and 0% relative humidity.

2. REFERENCE

Eller, P.M., ed. (1984) NIOSH Manual of Analytical Methods, 3rd Ed., Vol. 1, U.S. Department of Health and Human Services

3. DEFINITION

Breakthrough: $\geq$ 5% of analyte found on back-up sorbent section.

4. PRINCIPLE

A known volume of air is drawn through a glass tube with two sections (front and back-up) containing activated charcoal which adsorbs gaseous organic compounds. The latter are desorbed in benzene and determined by gas chromatography (GC) with flame-ionization detection (FID). Standard solutions and a calibration curve are employed for quantification.

5. HAZARDS

Allyl chloride is toxic by inhalation, strongly irritant to tissue and a fire and explosion risk. Benzene is also toxic and a fire and explosion risk and is a suspected human carcinogen. Hexane is highly flammable and moderately toxic. All these substances should be handled only in a ventilated hood.

6. REAGENTS[1]

Allyl chloride	Reagent grade
Benzene	Chromatographic quality
n-Hexane	Chromatographic quality
Nitrogen	Purified, for GC
Hydrogen	Pre-Purified, for FID
Air	Filtered, for FID
Allyl chloride stock standard solution	7.5 g/L. dilute 7.5 mg allyl chloride (80.0 μL at 20°C) to 10 mL with hexane
Allyl chloride working standard solutions	Prepare at least five working standard solutions over the range 0.01 to 1.5 g/L. Add known amounts (13.3 μL-2.0 mL) of stock standard solution to 10-mL volumetric flasks and dilute to the mark with benzene.

7. APPARATUS[1] (see INTRODUCTION, p. 173)

Air sampler tubes	Glass tubes, 7 cm x 6 mm o.d., 4 mm i.d., flame-sealed ends, containing two sections of 20/40 mesh activated (600°C) coconut-shell charcoal (front, 100 mg; back, 50 mg), separated by a 2-mm urethane foam plug. A sylilated glass-wool plug precedes the front section and a 3-mm urethane foam plug follows the back section. The pressure drop across the tube must be < 3.4 kPa at 1 L/min air flow rate.
Personal sampling pump	0.01 to 1 L/min, with flexible connecting tubing

---

[1] Reference to a company and/or product is for the purpose of information and identification only and does not imply approval or recommendation of the company and/or product by the International Agency for Research on Cancer, to the exclusion of others which may also be suitable.

Gas chromatograph	With FID and peak-area integrator. column, 1.2 m x 6 mm o.d., stainless-steel, packed with 50/80 mesh Porapak Q.
Vials	2-mL, with PTFE-lined crimp caps
Syringe	10-μL, readable to 0.1 μL
Volumetric flasks	10-mL
Pipette	1-mL, with bulb.

8. SAMPLING

8.1 Follow sampling instructions in INTRODUCTION, p. 173.

8.2 Sample at an accurately known rate between 0.01 and 1 L/min to obtain sample volume of 15 to 100 L. Samples are stable at least six weeks at 25°C.

9. PROCEDURE

9.1 Blank tests (see INTRODUCTION, p. 173)

9.1.1 Field blanks: analyse with air samples, as in 9.4-9.6.

9.1.2 Media blanks: set aside three sampler tubes from each new batch of charcoal used for sampling. Extract and analyse with air samples, as in 9.4-9.6.

9.2 Check tests

Analyse three blind spikes and three analyst spikes to check the calibration and desorption efficiency curves (see INTRODUCTION, p. 173).

9.3 Test portion

Not applicable

9.4 Sample extraction

9.4.1 Place front and back sections of each sampler in separate 2-mL vials (discard glass-wool and foam plugs).

9.4.2 Add 1.0 mL benzene to each vial and close with a crimp cap.

9.4.3 Allow to stand 30 min, with occasional agitation. Retain for GC analysis.

9.5 GC operating conditions

Column temperature	160 to 200°C
Injection temperature	185 to 200°C
Detector temperature	250°C
Carrier (nitrogen) flow rate	30 mL/min
Injection volume	5 µL

9.6 Analyte determination

Inject 5 µL of sample extract (9.4.3) onto GC, using solvent-flush technique (see Method 1, section 12), or autosampler, and record peak area.

9.7 Calibration curve

9.7.1 Calibrate daily by injecting 5 µL aliquots of each of the working standard solutions onto the GC.

9.7.2 Record peak areas and prepare calibration curve of peak area *versus* mg allyl chloride per mL.

9.8 Desorption efficiency (DE)

9.8.1 For each batch of charcoal used for sampling, prepare three spiked tubes at each of five levels in the calibration range (e.g., spike with the amounts of analyte in 1 mL of each of the five working standard solutions). Proceed as follows:

9.8.2 Remove and discard back (50 mg) sorbent section of a fresh sampler.

9.8.3 Inject a known amount of analyte (see 9.8.1) directly onto front sorbent section by means of a microlitre syringe. Cap tube and allow to stand overnight.

9.8.4 Extract and analyse as in 9.4-9.6

9.8.5 Using the calibration curve, calculate mg analyte recovered at each spiking level and prepare a graph of desorption efficiency (DE) *versus* mg analyte recovered (DE = mass of analyte recovered/mass of spike).

10. METHOD OF CALCULATION

Using the peak area (9.6) and the calibration curve (9.7.2), calculate the mass concentration, ρ, in the air sample from the equation given in Method 1, section 10.

11. REPEATABILITY AND REPRODUCIBILITY

The method has been validated over the range 1.8-7.2 mg/m^3 at 25°C and 101 kPa, using a 100 L sample. the overall coefficient of variation was 7.1%. Desorption efficiency was 0.943 in the range 0.15 mg - 0.6 mg allyl chloride per sample.

12. NOTES ON PROCEDURE

Not applicable

13. SCHEMATIC REPRESENTATION OF PROCEDURE

Draw air sample through charcoal sampler tube
(front section, 100 mg; back section, 50 mg)

↓

Place front and back solvent sections in separate vials,
add 1.0 mL benzene and cap

↓

Stand 30 min with occasional agitation

↓

Inject 5-µL aliquot of extract onto GC

↓

Inject 5 µL aliquots of standard solutions of analytes onto GC
and prepare calibration curve

↓

Determine desorption efficiency in the calibration range,
using three spiked tubes at each of five levels.
Extract and analyse by GC, as above.

↓

Calculate mass concentration (mg/m^3) of analyte in air sample,
correcting for desorption efficiency and media blanks

14. ORIGIN OF THE METHOD

Method 1000, NIOSH Manual of Analytical Methods, 3rd Ed., Vol. 1, Cincinnati OH, National Institute for Occupational Safety and Health (1984)

Contact point: Dr P.M. Eller
Division of Physical Sciences and Engineering
National Institute for Occupational Safety and Health
4676 Columbia Parkway
Cincinnati, OH 45236, USA

METHOD 3

THE DETERMINATION OF METHYLENE CHLORIDE IN AIR

A MacKenzie Peers

Adapted from Method 1005, NIOSH Manual of Analytical Methods, 3rd Ed., vol. 1 (1984)

I. SCOPE AND FIELD OF APPLICATION

This personal sampling pump and sorption tube method is suitable for the determination of methylene chloride (dichloromethane) in industrial atmospheres. The estimated limit of detection is 0.35 to 10.4 g/m^3 for a 1-L air sample. No interferences have been identified. Breakthrough occurred at 18.5 min when sampling air containing 6.7 g/m^3 methylene chloride at 0.187 L/min and 0% relative humidity. High air humidity decreases the breakthrough volume.

2. REFERENCES

NIOSH (1977) Documentation of the NIOSH Validation Tests, U.S. Department of Health, Education and Welfare, Publ. (NIOSH) 77-185

Eller, P.M., ed. (1984) NIOSH Manual of Analytical Methods, 3rd Ed., vol. 1, U.S. Department of Health and Human Services

3. DEFINITION

Breakthrough: $\geq$ 5% of analyte found on back-up sorbent section.

4. PRINCIPLE

A known volume of air is drawn through two glass sampler tubes in series, containing activated charcoal (100 mg, front tube; 50 mg, back-up tube) which adsorbs the methylene chloride. The latter is subsequently desorbed in carbon disulfide containing an internal standard and is determined by gas chromatography (GC) with flame-ionization detection (FID). A calibration curve is used for quantification and a correction is applied for desorption efficiency.

5. HAZARDS

Carbon disulfide is toxic and an acute fire and explosion hazard. Methylene chloride is also moderately toxic. Both compounds should be handled under a ventilated hood.

6. REAGENTS[1]

Carbon disulfide	Chromatographic quality
Methylene chloride	
Nitrogen or helium	Purified, for GC
Hydrogen	Pre-purified, for FID
Air	Filtered, compressed, for FID
Extraction solvent	Carbon disulfide containing 0.1% (v/v) decane, benzene, or other internal standard
Methylene chloride working standard solutions	Prepare five standard solutions of methylene chloride over the range 0.1-100 mg/10 mL (add known amounts of methylene chloride to the extraction solvent in 10-mL volumetric flasks and dilute to the mark).

7. APPARATUS[1] (see INTRODUCTION, p. 173)

Air sampler tubes	Separate front and back-up glass tubes, each 7 cm × 6 mm o.d., 4 mm i.d. flame-sealed ends, containing activated (600°C) coconut-shell charcoal (front tube, 100 mg; back-up tube, 50 mg). A sylilated glass-wool plug is placed at each end of each tube. The pressures drop across the tubes at an air flow-rate of 1 L/min must be ≤ 3.4 kPa. (Two commercially-available tubes, each containing 150 mg charcoal in two beds, may be used.)

---

[1] Reference to a company and/or product is for the purpose of information and identification only and does not imply approval or recommendation of the company and/or product by the International Agency for Research on Cancer, to the exclusion of others which may also be suitable.

Personal sampling pump	0.01 to 0.2 L/min, with flexible connecting tubing.
Gas chromatograph	3 m x 3 mm stainless-steel column, packed with 10% SP-1000 on 80/100 mesh Supelcoport, with FID and peak-area integrator. (Alternative columns: 10% TCEP on 80/100 Chromosorb PAW; SP-2100, SP-2100 with 0.1% carbowax 1500 or DB-1 fused silica capillary)
Vials	2-mL, with PTFE-lined caps
Syringe	10-µL, readable to 0.1 µL
Volumetric flasks	10-mL
Pipette	1-mL, with bulb.

8. SAMPLING

8.1 Break the ends of the tubes just before sampling and connect the two tubes with a short length of flexible tubing.

8.2 Follow the sampling instructions in the INTRODUCTION, p. 173, and sample at an accurately-known rate between 0.01 and 0.2 L/min for a total sample volume of 0.5 to 2.5 L.

9. PROCEDURE

9.1 Blank tests (see INTRODUCTION, p. 173)

9.1.1 Field blanks: analyse with air samples, as in 9.4-9.6

9.1.2 Media blanks: set aside three complete (front and back) sampler tubes from each new batch of charcoal used for sampling. Analyse with air samples, as in 9.4-9.6.

9.2 Check tests

Analyse three blind spikes and three analyst spikes to check the calibration and desorption efficiency curves (see INTRODUCTION, p. 173).

9.3 Test portion

Not applicable

9.4 Sample extraction

9.4.1 Place sorbent sections from front and back-up tubes in separate 2-mL vials (discard glass-wool plugs).

9.4.2 Add 1.0 mL extraction solvent to each vial and close with PTFE-lined cap.

9.4.3 Allow to stand 30 min, with occasional agitation. Retain for GC analysis.

9.5 GC operating conditions

Column temperature	60 to 90°C
Injection temperature	200 to 225°C
Detector temperature	250°C
Carrier gas flow rate	30 mL/min (nitrogen or helium)

9.6 Analyte determination

9.6.1 Inject 5 µL of sample extract (9.4.3) manually onto the GC using solvent-flush technique (see Method 1, section 12), or with auto-sampler.

9.6.2 Measure peak areas of analyte(s) and internal standard on the same chromatogram. Divide peak area of analyte by that of internal standard to obtain relative peak area.

9.7 Calibration curve

9.7.1 Calibrate daily by injecting 5-µL aliquots of each of the five working standard solutions of the analyte onto the GC.

9.7.2 Measure peak areas of analyte and internal standard on the same chromatogram and calculate relative peak area (9.6.2) for each concentration.

9.7.3 Prepare calibration curve showing relative peak area *versus* amount (mg) of analyte per mL.

9.8 Desorption efficiency (DE)

9.8.1 For each batch of charcoal used for sampling, prepare three spiked tubes at each of five levels in the calibration range (e.g., spike with the amounts of analyte in 1 mL of each of the five working standard solutions). Proceed as follows:

9.8.2 Remove and discard back (50 mg) sorbent section of a fresh sampler.

9.8.3 Inject a known amount of analyte (see 9.8.1) directly onto front sorbent section by means of a microlitre syringe. Cap tube and allow to stand overnight.

9.8.4 Extract and analyse as in 9.4-9.6

9.8.5 Using the calibration curve, calculate mg analyte recovered at each spiking level and prepare a graph of desorption efficiency (DE) versus mg analyte recovered (DE = mass of analyte recovered/mass of spike).

10. METHOD OF CALCULATION

Using the relative peak area (9.6.2) and the calibration curve (9.7.3), calculate the mass concentration, ρ, of the analyte in the air samples from the equation,

$$\rho = 10^3(m_s - m_b)/V \quad (mg/m^3)$$

where, m_s = sum of analyte masses (corrected for DE) found in front and back sections of sampler (mg)

m_b = average sum of analyte masses (corrected for DE) found in front and back sections of media blanks (mg)

V = volume of air sample (L)

NOTE: If $10\ m_b > m_s$ report breakthrough and possible sample loss.

11. REPEATABILITY AND REPRODUCIBILITY

The method has been validated over the range 1.7 to 7.1 g/m³, using a 1-L sample (25°C, 108 kPa). The overall coefficient of variation was 7.3%, with an average desorption efficiency of 95.3%.

12. NOTES ON PROCEDURE

Not applicable

13. SCHEMATIC REPRESENTATION OF PROCEDURE

Draw air sample through charcoal sampler tubes
(front section, 100 mg; back section, 50 mg)

↓

Place front and back solvent sections in separate vials,
add 1.0 mL carbon disulfide (+ standard) and cap

↓

Stand 30 min with occasional agitation

↓

Inject 5-μL aliquot of extract onto GC

↓

Inject 5 μL aliquots of standard solutions of analytes onto GC
and prepare calibration curve

↓

Determine desorption efficiency in the calibration range,
using three spiked tubes at each of five levels.
Extract and analyse by GC, as above.

↓

Calculate mass concentration (mg/m^3) of analyte in air sample,
correcting for desorption efficiency and media blanks

14. ORIGIN OF THE METHOD

Method 1005, NIOSH Manual of Analytical Methods, 3rd Ed., Vol. 1, Cincinnati, OH, National Institute for Occupational Safety and Health (1984)

Contact point: Dr P.M. Eller
Division of Physical Sciences and Engineering
National Institute for Occupational Safety and Health
4676 Columbia Parkway
Cincinnati, OH 45236, USA

METHOD 4

THE DETERMINATION OF ETHYLENE DIBROMIDE IN AIR

A. MacKenzie Peers

Adapted from Method 1008, NIOSH Manual of Analytical Methods, 3rd Ed., vol. 1 (1984)

1. SCOPE AND FIELD OF APPLICATION

This personal sampling pump and sorption tube method is suitable for the determination of ethylene dibromide (1,2-dibromoethane) in air over the range 0.04-200 µg per sample (0.002-8 mg/m³ for a 25-L air sample). Breakthrough was determined with one lot of charcoal in dry air containing 446 mg ethylene dibromide/m³. After 4 h sampling (48-L air), the effluent concentration was 2% of the influent, so that the working capacity was at least 21 mg. The working range can thus be considerably increased if the sample extract is diluted before analysis. No interference has been observed with the GC column employed.

2. REFERENCE

Eller, P.M., ed. (1984) NIOSH Manual of Analytical Methods, 3rd ed., Vol. 1, U.S. Department of Health and Human Services

3. DEFINITIONS

Not applicable

4. PRINCIPLE

A known volume of air is drawn through a glass tube with two sections containing activated charcoal which adsorbs organic vapours. The latter are desorbed in benzene:methanol containing an internal standard for quantification purposes. Ethylene dibromide is determined by gas chromatography (GC), with an electron-capture detector (ECD). A correction is applied for desorption efficiency.

5. HAZARDS

Benzene and ethylene dibromide are carcinogens and can be absorbed through the skin. Benzene and methanol are flammable. These compounds should be handled in a ventilated hood and suitable gloves should be worn.

6. REAGENTS[1]

Nitrogen	Purified, for GC
Benzene	Pesticide quality
Methanol	Pesticide quality
Ethylene dibromide	High-purity (density = 2.169 g/mL at 25°C)
1,1,2,2-Tetrachloroethane (1.587 g/mL at 25°C), or 1,2-Dibromopropane (1.923 g/mL at 25°C)	For internal standard use
Extraction solvent	Benzene:methanol (99:1, v/v) containing an internal standard (see above) at a concentration of 80 mg/L
Stock standard solution of ethylene dibromide	10 mg/L. Dissolve 50 mg of ethylene dibromide in benzene in a 25-mL volumetric flask and dilute to the mark with benzene. Prepare stock standard solution by diluting 50 µL of above solution to 10 mL with benzene (10 µg/mL).
Working standard solutions of ethylene dibromide	Prepare at least five working standard solutions over the range 0.01-0.8 µg/10 mL (add known amounts of stock standard solution to 10-mL volumetric flasks and dilute to the mark with the extraction solvent).

---

[1] Reference to a company and/or product is for the purpose of information and identification only and does not imply approval or recommendation of the company and/or product by the International Agency for Research on Cancer, to the exclusion of others which may also be suitable.

7. APPARATUS[1] (see INTRODUCTION, p. 173)

Air sampler tubes	Glass tubes, 7 cm × 6 mm o.d., 4 mm i.d., flame-sealed ends, containing two sections of 20/40 mesh activated (600°C) coconut-shell charcoal (front, 100 mg; back, 50 mg), separated by a 2-mm urethane foam plug. A sylilated glass-wool plug precedes the front section and a 3-mm urethane foam plug follows the back section. The pressure drop across the tube must be < 3.4 kPa at 1 L/min air flow rate.
Personal sampling pump	0.02-0.2 L/min, with flexible connecting tubing.
Gas chromatograph	With ^{63}Ni electron-capture detector and peak-area integrator. Column, 1.8 m × 4 mm i.d., packed with 3% OV-210 on 80/100 Gas Chrom Q (alternative packing, GP 20% SP-2100/0.1% Carbowax 1500 on 100/120 Chromosorb WHP).
Syringes	10-µL, readable to 0.1 µL, and 50 µL, readable to 1 µL
Volumetric flasks	10-mL, 25-mL and 500-mL
Pipette	10-mL, with bulb.

8. SAMPLING

8.1 Follow sampling instructions in INTRODUCTION, p. 173.

8.2 Sample at an accurately-known rate between 0.02 and 0.2 L/min, to obtain total volume of 0.1 to 25 L.

8.3 Pack in an insulated container with solid carbon dioxide (dry ice) for shipping. Store at -25°C or below (samples are thus stable for two weeks).

---

[1] Reference to a company and/or product is for the purpose of information and identification only and does not imply approval or recommendation of the company and/or product by the International Agency for Research on Cancer, to the exclusion of others which may also be suitable.

9. PROCEDURE

9.1 Blank tests (see INTRODUCTION, p. 173)

9.1.1 Field blanks: analyse with air samples, as in 9.4-9.6.

9.1.2 Media blanks: set aside three sampler tubes from each new batch of charcoal used for sampling. Extract and analyse with air samples, as in 9.4-9.6.

9.2 Check tests

Analyse three blind spikes and three analyst spikes to check the calibration and desorption efficiency curves (see INTRODUCTION, p. 173).

9.3 Test portion

Not applicable

9.4 Sample extraction

9.4.1 Place front and back sorbent sections in separate 10-mL volumetric flasks (discard glass-wool and foam plugs).

9.4.2 Add 10 mL of extraction solvent (6) to each flask and close with stopper.

9.4.3 Allow to stand 60 min, with occasional agitation. Retain for GC.

9.5 GC operating conditions

Column temperature	50°C
Injector temperature	175°C
Detector temperature	315°C
Carrier flow-rate	35 mL/min
Injection volume	5 µL

Retention times under these conditions are: ethylene dibromide, 2.2 min; 1,2-dibromopropane, 2.9 min; 1,1,2,2-tetrachloroethane, 4.1 min.

NOTE: The ECD should be optimized for the smallest possible amount of analyte.

9.6 Analyte determination

9.6.1 Inject 5 µL of sample extract (9.4.3) manually onto the GC using solvent-flush technique (see Method 1, section 12), or with autosampler.

9.6.2 Measure peak areas of analyte(s) and internal standard on the same chromatogram. Divide peak area of analyte by that of internal standard to obtain relative peak area.

9.7 Calibration curve

9.7.1 Calibrate daily by injecting 5-µL aliquots of each of the five working standard solutions of the analyte onto the GC.

9.7.2 Measure peak areas of analyte and internal standard on the same chromatogram and calculate relative peak area (9.6.2) for each concentration.

9.7.3 Prepare calibration curve showing relative peak area *versus* amount (mg) of analyte per mL.

9.8 Desorption efficiency (DE)

9.8.1 For each batch of charcoal used for sampling, prepare three spiked tubes at each of five levels in the calibration range. Proceed as follows:

9.8.2 Remove and discard back (50 mg) sorbent section of a fresh sampler.

9.8.3 Inject a known amount of stock standard solution (6) directly onto front sorbent section by means of a microlitre syringe. Cap and allow to stand overnight.

9.8.4 Extract and analyse, as in 9.4-9.6.

9.8.5 Using the calibration curve, calculate µg analyte recovered at each spiking level and prepare a graph of desorption efficiency (DE) *versus* µg analyte recovered.

10. METHOD OF CALCULATION

Using the relative peak area (9.6.2) and the calibration curve (9.7.3), calculate the mass concentration, ρ, of the analyte in the air samples from the equation,

$$\rho = 10^3(m_s - m_b)/V \quad (\mu g/m^3)$$

where, m_s = sum of analyte masses (corrected for DE) found in front and back sections of sample tube (µg)

m_b = average sum of analyte masses (corrected for DE) found in front and back sections of media blanks (µg)

V = volume of air sample (L)

NOTE: If 10 $m_b > m_s$ report breakthrough and possible sample loss.

11. REPEATABILITY AND REPRODUCIBILITY

Only the recovery from spiked charcoal has been systematically evaluated. Using two lots of charcoal, average recoveries (DE) ranged from 0.85 to 0.93. The coefficient of variation for the analytical method (31 samples, pooled) was 4.4%.

12. NOTES ON PROCEDURE

Not applicable

13. SCHEMATIC REPRESENTATION OF PROCEDURE

Draw air sample through charcoal sampler tube

Place front and back sorbent sections in separate 10-mL flasks, add 10 mL benzene:methanol (99:1, v/v) + internal standard and stopper

↓

Let stand 60 min with occasional agitation

↓

Inject aliquot of extract onto GC

↓

Inject aliquots of standard solutions of analyte onto GC and prepare calibration curve

↓

Determine desorption efficiency in calibration range using 3 spiked front sections at each of 5 levels

↓

Calculate mass concentration (µg/m³) of ethylene dibromide in air sample

14. ORIGIN OF THE METHOD

Method 1008, NIOSH Manual of Analytical Methods, 3rd Ed., Vol. 1, Cincinnati, OH, National Institute for Occupational Safety and Health (1984)

Contact point: Dr P.M. Eller
Division of Physical Sciences and Engineering
National Institute for Occupational Safety and Health
4676 Columbia Parkway
Cincinnati, OH 45236, USA

METHOD 5

THE DETERMINATION OF TRICHLOROETHYLENE AND TETRACHLOROETHYLENE IN AIR

A. MacKenzie Peers

Adapted from Methods S 335 and S 336, NIOSH Manual of analytical Methods, 2nd Ed., Vol. 3 (1977)

1. SCOPE AND FIELD OF APPLICATION

This personal sampling pump and sorption tube method is suitable for the individual determination of the title compounds. These can probably be determined simultaneously as well by GC temperature programming between 70° and 90°C. The limits of detection (useful ranges) depend on the analyte and are shown below for typical sample volumes.

Analyte	Sample volume	Probable useful range	Breakthrough volume[a]
Trichloroethylene	3 L	0.108-3.22 g/m³	18.5 L at 2.27 g/m³
Tetrachloroethylene	3 L	0.136-4.06 g/m³	21 L at 2.75 g/m³

[a] Corresponding to effluent concentration > 5% of that in influent. Breakthrough volumes are decreased by high relative humidity

2. REFERENCES

Taylor, D.G., ed. (1977) NIOSH Manual of Analytical Methods, 2nd ed., Vol. 3, U.S. Department of Health, Education and Welfare

Eller, P.M., ed., (1984) NIOSH Manual of Analytical Methods, 3rd ed., Vol. 1, U.S. Department of Health and Human Services

3. DEFINITIONS

Not applicable

4. PRINCIPLE

A known volume of air is drawn through a tube with two sections containing activated coconut charcoal which adsorbs the gaseous organic componds. These are subsequently desorbed in carbon disulfide (containing an internal standard for quantification purposes) and are determined by gas chromatography (GC), with flame-ionization detection (FID). A calibration curve is employed for quantification and a correction is applied for desorption efficiency.

5. HAZARDS

Carbon disulfide is toxic and an acute fire and explosion hazard. It should be handled (like the title compounds) only in a ventilated hood.

6. REAGENTS[1]

Carbon disulfide	Chromatographic quality
Trichloroethylene	Reagent grade
Tetrachloroethylene	Reagent grade
Octane	Reagent grade
Nonane	Reagent grade
Nitrogen	Purified, for GC
Hydrogen	Pre-purified, for FID
Air	Filtered, compressed, for FID
Extraction solvent A (for trichloroethylene)	Carbon disulfide containing a known amount (e.g., 2 mg/L) of octane as internal standard
Extraction solvent B (for tetrachloroethylene)	Carbon disulfide containing a known amount of nonane (e.g., 2 mg/L) as internal standard

[1] Reference to a company and/or product is for the purpose of information and identification only and does not imply approval or recommendation of the company and/or product by the International Agency for Research on Cancer, to the exclusion of others which may also be suitable.

Trichloroethylene working standard solutions	Add known amounts of trichloroethylene to extraction solvent A in a 10-mL volumetric flask and dilute to the mark with extraction solvent A. Prepare at least five concentrations over the appropriate range
Tetrachloroethylene working standard solutions	Add known amounts of tetrachloroethylene to extraction solvent B in a 10-mL volumetric flask and dilute to the mark with solvent B. Prepare at least five concentrations over the appropriate range.

7. APPARATUS[1] (see INTRODUCTION, p. 173)

Air sampler tubes	Glass tubes, 7 cm × 6 mm o.d., 4 mm i.d., flame-sealed ends, containing two sections of 20/40 mesh activated (600°C) coconut-shell charcoal (front, 100 mg; back, 50 mg), separated by a 2-mm urethane foam plug. A sylilated glass-wool plug precedes the front section and a 3-mm urethane foam plug follows the back section. The pressure drop across the tube must be < 3.4 kPa at 1 L/min air flow rate.
Personal sampling pumps	0.1 to 0.2 L/min, with flexible connecting tubing
Gas chromatograph	With FID and peak-area integrator. column, 3 m × 3.2 mm stainless-steel, packed with 10% OV-101 on 100/120 mesh Supelcoport
Vials	2-mL, glass, PTFE-lined caps
Volumetric flasks	10-mL
Syringes	10-mL, readable to 0.1 μL, and other sizes for preparation of standards
Pipette	1-mL, with bulb

[1] Reference to a company and/or product is for the purpose of information and identification only and does not imply approval or recommendation of the company and/or product by the International Agency for Research on Cancer, to the exclusion of others which may also be suitable.

8. SAMPLING

8.1 Follow sampling procedure in INTRODUCTION, p. 173.

8.2 At concentrations of 1 g/m³ or more, a sample volume of 3 L is recommended (sample for 15 min at 0.2 L/min). At lower concentrations, a volume of 10 L is recommended. Sample at 0.2 mL/min, or less.

9. PROCEDURE

9.1 Blank tests (see INTRODUCTION, p. 173)

9.1.1 Field blanks: analyse with air samples, as in 9.4-9.6.

9.1.2 Media blanks: set aside three sampler tubes from each new batch of charcoal used for sampling. Extract and analyse with air samples, as in 9.4-9.6.

9.2 Check tests

Analyse three blind spikes and three analyst spikes to check the calibration and desorption efficiency curves (see INTRODUCTION, p. 173).

9.3 Test portion

Not applicable

9.4 Sample extraction

9.4.1 Place front and back sorbent sections in separate 2-mL vials (discard glass-wool and foam plugs)

9.4.2 Add 1.0 mL of appropriate extraction solvent (A or B) to each vial and close with PTFE-lined cap.

9.4.3 Allow to stand 30 min, with occasional agitation. Retain for GC analysis.

9.5 GC operating conditions

Column temperature:	
trichloroethylene	70°C
tetrachloroethylene	90°C
Injection temperature	225°C
Detector temperature	250°C
Nitrogen flow-rate	30 mL/min
Hydrogen (FID) flow-rate	35 mL/min
Air (FID) flow-rate	400 mL/min
Injection volume	5 μL

9.6 Analyte determination

9.6.1 Inject 5 μL of sample extract (9.4.3) manually onto the GC using solvent-flush technique (see Method 1, section 12), or with auto-sampler.

9.6.2 Measure peak areas of analyte(s) and internal standard on the same chromatogram. Divide peak area of analyte by that of internal standard to obtain relative peak area.

9.7 Calibration curve

9.7.1 Calibrate daily by injecting 5-μL aliquots of each of the five working standard solutions of the analyte onto the GC.

9.7.2 Measure peak areas of analyte and internal standard on the same chromatogram and calculate relative peak area (9.6.2) for each concentration.

9.7.3 Prepare calibration curve showing relative peak area *versus* amount (mg) of analyte per mL.

9.8 Desorption efficiency (DE)

9.8.1 For each batch of charcoal used for sampling, prepare three spiked tubes at each of five levels in the calibration range (e.g., spike with the amounts of analyte in 1 mL of each of the five working standard solutions). Proceed as follows:

9.8.2 Remove and discard back (50 mg) sorbent section of a fresh sampler.

9.8.3 Inject a known amount of analyte (see 9.8.1) directly onto front sorbent section by means of a microlitre syringe. Cap tube and allow to stand overnight.

9.8.4 Extract and analyse as in 9.4-9.6

9.8.5 Using the calibration curve, calculate mg analyte recovered at each spiking level and prepare a graph of desorption efficiency (DE) *versus* mg analyte recovered (DE = mass of analyte recovered/mass of spike).

10. METHOD OF CALCULATION

Using the relative peak area (9.6.2) and the calibration curve (9.7.3), calculate the mass concentration, ρ, of the analyte in the air samples from the equation,

$$\rho = 10^3(m_s - m_b)/V \quad (mg/m^3)$$

where, m_s = sum of analyte masses (corrected for DE) found in front and back sections of sample tube (mg)

m_b = average sum of analyte masses (corrected for DE) found in front and back sections of media blanks (mg)

V = volume of air sample (L)

NOTE: If $10\ m_b > m_s$ report breakthrough and possible sample loss.

11. REPEATABILITY AND REPRODUCIBILITY

11.1 Trichloroethylene

This method was validated over the ange 0.52-2.18 g/m³ (24.5°C, 101 kPa), using a 3-L sample. The coefficient of variation for the total sampling and analytical method over the above range was 8.2%.

11.2 Tetrachloroethylene

This method was validated over the range 0.655-2.75 g/m³ (25°C, 102 kPa), using a 3-L sample. the coefficient of variation for the total method over the above range was 5.2%.

12. NOTES ON PROCEDURE

Not applicable

13. SCHEMATIC REPRESENTATION OF PROCEDURE

Draw air sample through charcoal sampler tube
(front section, 100 mg; back section, 50 mg)

↓

Place front and back solvent sections in separate vials,
add 1.0 mL carbon disulfide (+ standard) and cap

↓

Stand 30 min with occasional agitation

↓

Inject 5-μL aliquot of extract onto GC

↓

Inject 5 μL aliquots of standard solutions of analytes onto GC
and prepare calibration curve

↓

Determine desorption efficiency in the calibration range,
using three spiked tubes at each of five levels.
Extract and analyse by GC, as above.

↓

Calculate mass concentration (mg/m^3) of analyte in air sample,
correcting for desorption efficiency and media blanks

14. ORIGIN OF THE METHOD

Trichloroethylene: Method S 336, NIOSH Manual of Analytical Methods, 2nd Ed., Vol. 3 (1977)

Tetrachloroethylene: Method S 335, NIOSH Manual of Analytical Methods, 2nd Ed., Vol. 3 (1977)

Contact point: Dr P.M. Eller
Division of Physical Sciences and Engineering
National Institute for Occupational Safety and Health
4676 Columbia Parkway
Cincinnati, OH 45236, USA

METHOD 6

THE DETERMINATION OF EPICHLOROHYDRIN IN AIR

A. MacKenzie Peers

Adapted from Method 1010, NIOSH Manual of analytical Methods, 3rd Ed., Vol. 1 (1984)

1. SCOPE AND FIELD OF APPLICATION

This personal sampling pump and sorption tube method is suitable for the determination of epichlorohydrin (1-chloro-2,3-epoxypropane) over the range 2-60 mg/m³ using a 20-L air sample. In a test atmosphere containing 43 mg epichlorohydrin/m³, some breakthrough was observed (effluent concentration less than 5% of influent concentration) after sampling at 0.185 L/min for 240 min (44-L sample). The breakthrough volume is decreased by high relative humidity. The estimated limit of detection is 1 µg/sample.

2. REFERENCES

Taylor, D.G., ed. (1977) NIOSH Manual of Analytical Methods, 2nd ed., Vol. 2, U.S. Department of Health, Education and Welfare

Eller, P.M., ed. (1984) NIOSH Manual of Analytical Methods, 3rd ed., Vol. 1, U.S. Department of Health and Human Services

3. DEFINITIONS

Not applicable

4. PRINCIPLE

A known volume of air is drawn through a tube containing activated charcoal which adsorbs organic vapours. These are subsequently desorbed in carbon disulfide and determined by gas chromatography (GC), with flame-ionization detection (FID). A calibration curve is employed for quantification and a correction is applied for desorption efficiency.

5. HAZARDS

Carbon disulfide is toxic and represents an acute fire and explosion hazard. Epichlorohydrin is a strong irritant and sensitizer and may cause kidney injury. These compounds should be handled only in a ventilated hood.

6. REAGENTS[1]

Carbon disulfide	Chromatographic quality
n-Hexane	Chromatographic quality
Epichlorohydrin	Reagent grade
Nitrogen (or helium)	Purified, for GC
Hydrogen	Pre-purified, for FID
Air	Filtered, compressed, for FID
Stock standard solution	9.45 mg/mL (dissolve 80 µL of epichlorohydrin in 10 mL carbon disulfide).
Working standard solution	Prepare at least five concentrations of epichlorohydrin in carbon disulfide over the range 0.04 to 1.2 mg/mL (or range of interest). Add known amounts of stock standard solution to 10-mL volumetric flasks and dilute to the mark with carbon disulfide.
Standard spiking solution for desorption efficiency determination	Prepare an accurately known concentration (e.g., 9.45 mg/mL) of epichlorohydrin in n-hexane. Make up 10 mL in a volumetric flask.

NOTE: Epichlorohydrin is unstable and the pure compound should be stored under refrigeration. All solutions should be prepared fresh daily.

---

[1] Reference to a company and/or product is for the purpose of information and identification only and does not imply approval or recommendation of the company and/or product by the International Agency for Research on Cancer, to the exclusion of others which may also be suitable.

7. APPARATUS[1] (see INTRODUCTION, p. 173)

Air sampler tubes	Glass tubes, 7 cm x 6 mm o.d., 4 mm i.d., flame-sealed ends, containing two sections of 20/40 mesh activated (600°C) coconut-shell charcoal (front, 100 mg; back, 50 mg), separated by a 2-mm urethane foam plug. A sylilated glass-wool plug precedes the front section and a 3-mm urethane foam plug follows the back section. The pressure drop across the tube must be < 3.4 kPa at 1 L/min air flow rate.
Personal sampling pumps	0.01 to 0.2 L/min, with flexible connecting tubing.
Gas chromatograph	With FID and peak-area integrator. Column, 1.8 m x 2 mm i.d., glass, packed with 80/100 chromosorb 101.
Vials	2-mL, glass, with PTFE-lined caps
Syringes	10 µL and 100 µL
Volumetric flasks	1-mL and 10-mL
Pipette	1-mL, with bulb.

8. SAMPLING

8.1 Follow sampling instructions in INTRODUCTION, p. 173.

8.2 Sample at an accurately-known rate between 0.01 and 0.2 L/min for a total sample size of 2 to 30 L.

NOTE: 0.38 mg epichlorohydrin has been found to be stable on coconut charcoal for 6 days at ambient temperature.

---

[1] Reference to a company and/or product is for the purpose of information and identification only and does not imply approval or recommendation of the company and/or product by the International Agency for Research on Cancer, to the exclusion of others which may also be suitable.

9. PROCEDURE

9.1 Blank tests (see INTRODUCTION, p. 173)

9.1.1 Field blanks: analyse with air samples, as in 9.4-9.6.

9.1.2 Media blanks: set aside three sampler tubes from each new batch of charcoal used for sampling. Extract and analyse with air samples, as in 9.4-9.6.

9.2 Check tests

Analyse three blind spikes and three analyst spikes to check the calibration and desorption efficiency curves (see INTRODUCTION, p. 173).

9.3 Test portion

Not applicable

9.4 Sample extraction

9.4.1 Place front and back sorbent section in separate vials (discard glass-wool and foam plugs).

9.4.2 Add 1.0 mL carbon disulfide to each vial and close with crimp cap.

9.4.3 Allow to stand 30 min, with occasional agitation. Retain for GC analysis.

9.5 GC operating conditions

Column temperature	135°C
Injector temperature	175°C
Detector temperature	215°C
Carrier gas flow-rate	20 mL/min

9.6 Analyte determination

Inject 5 µL of sample extract onto GC, using solvent-flush technique (Method 1, section 12) or autosampler. Record peak area.

9.7 Calibration curve

9.7.1 Calibrate daily by injecting 5 µL aliquots of each of the working standard solutions onto the GC (analyse together with samples and blanks to minimize variations of FID response).

9.7.2 Prepare a calibration curve showing peak area *versus* concentration of analyte in mg/mL.

9.8 Desorption efficiency (DE)

9.8.1 For each batch of charcoal used for sampling, prepare three spiked tubes at each of five concentrations in the calibration range. Proceed as follows:

9.8.2 Remove and discard back sorbent section of a fresh sampler.

9.8.3 Inject a known amount of analyte onto front sorbent section with a microlitre syringe, using the standard spiking solution in *n*-hexane.

9.8.4 Cap the tube and allow to stand overnight.

9.8.5 Extract and analyse, as in 9.4-9.6.

9.8.6 Using calibration curve (9.7), calculate recovery at each spiking level and prepare a graph of DE *versus* mg analyte recovered.

10. METHOD OF CALCULATION

Using the peak area (9.6) and the calibration curve (9.7), calculate the mass concentration, ρ, of epichlorohydrin in the air sample from the equation given in Method 1, section 10.

11. REPEATABILITY AND REPRODUCIBILITY

The method S 118 (issued, 1975), of which the present is an updated version, was validated over the range 12-43 mg/m^3, using approximately 20-L sample volumes of a generated atmosphere. The coefficient of variation for the total sampling and analytical procedure was 5.7% (C.V. analytical procedure alone = 3.1%). the average DE over the cited range was 0.905.

12. NOTES ON PROCEDURE

Not applicable

13. SCHEMATIC REPRESENTATION OF PROCEDURE

Draw air sample through charcoal sampler tube
(front section, 100 mg; back section, 50 mg)

↓

Place front and back solvent sections in separate vials,
add 1.0 mL carbon disulfide and cap

↓

Stand 30 min with occasional agitation

↓

Inject 5-µL aliquot of extract onto GC

↓

Inject 5 µL aliquots of standard solutions of analytes onto GC
and prepare calibration curve

↓

Determine desorption efficiency in the calibration range,
using three spiked tubes at each of five levels.
Extract and analyse by GC, as above.

↓

Calculate mass concentration (mg/m³) of analyte in air sample,
correcting for desorption efficiency and media blanks

14. ORIGIN OF THE METHOD

Method 1010, NIOSH Manual of Analytical Methods, 3rd Ed., Vol. 1, Cincinnati, OH, National Institute for Occupational Safety and Health (1984)

Contact point: Dr P.M. Eller
Division of Physical Sciences and Engineering
National Institute for Occupational Safety and Health
4676 Columbia Parkway
Cincinnati, OH 45236, USA

METHOD 7

THE DETERMINATION OF METHYL CHLORIDE IN AIR

A. MacKenzie Peers

Adapted from Method S99, NIOSH Manual of Analytical Methods, 2nd Ed., Vol. 4 (1978)

1. SCOPE AND FIELD OF APPLICATION

This personal sampling pump and sorption tube method is suitable for the determination of methyl chloride in air in the range 0.12-1.2 g/m^3, with sample sizes of 1.5 to 0.5 L and relative humidity (RH) of ~ 80%. At a concentration of 413 mg/m^3 and RH = 82%, the breakthrough volume was found to be 3.2 L. It fell to 2.6 L at 1.4 g/m^3 and RH = 84%.

2. REFERENCES

Taylor, D.G., ed. (1978) NIOSH Manual of Analytical Methods, 2nd Ed., Vol. 4, U.S. Department of Health, Education and Welfare

Eller, P.M., ed. (1984) NIOSH Manual of Analytical Methods, 3rd Ed., Vol. 1, U.S. Department of Health and Human Services

3. DEFINITION

The breakthrough volume is reached when the amount of analyte found on the back-up charcoal tube equals 10% of that found on the front tube.

4. PRINCIPLE

A known volume of air is drawn through two tubes in series, containing activated charcoal which adsorbs organic vapours. The analyte is desorbed in methylene chloride and determined by gas chromatography (GC) with flame-ionization detection (FID). A calibration curve is employed for quantification and a correction is applied for desorption efficiency.

5. HAZARDS

Methyl chloride is toxic and flammable. Methylene chloride is moderately toxic and an eye irritant. Both compounds should be handled under a ventilated hood.

6. REAGENTS[1]

Methylene chloride	Chromatographic quality (methyl chloride-free)
Methyl chloride	95.5% (compressed gas bottle)
Nitrogen	Purified, for GC
Hydrogen	Pre-Purified, for FID
Air	Filtered, compressed, for FID
Standard solution of methyl chloride	Prepare daily at least five solutions at different concentrations over the range of interest. For each solution, add 3.0 ml of methylene chloride to a serum bottle and seal with septum and aluminum crimp cap. Add an appropriate amount of methyl chloride by bubbling it slowly into the methylene chloride, using a gas-tight syringe (see section 10.1). Shake the bottle gently after removal of the syringe needle. Record temperature and pressure. Standards should be analysed at the same time as samples, as soon as possible after preparation.

---

[1] Reference to a company and/or product is for the purpose of information and identification only and does not imply approval or recommendation of the company and/or product by the International Agency for Research on Cancer, to the exclusion of others which may also be suitable.

7. APPARATUS[1]

Air sampler tubes	The method employs two tubes in series. Both tubes are divided into two sections containing 20/40 mesh activated coconut charcoal, separated by a 2-mm plug of urethane foam. The larger (front) section of each tube is preceded by a sylilated glass-wool plug and the smaller (back) section is followed by a 3 mm plug of urethane foam. The first tube (glass, 10cm × 6 mm i.d.) contains 400 mg charcoal in the front section and 200 mg in the back section. The first tube is attached by Tygon tubing to the second (back-up) tube (glass, 7 cm × 4 mm i.d.), which contains 100 mg charcoal in the front section and 50 mg in the back. The pressure drop across the larger tube must be < 3.4 kPa at a flow rate of 0.05 L/min.
Personal sampling pump	0.02-0.1 L/min, with flexible connecting tubing. Consult INTRODUCTION, p. 173.
Gas chromatograph	With FID and peak-area integrator. Column; stainless-steel, 1.2 m × 6.4 mm o.d., packed with 80/100 mesh Chromosorb 102.
Syringes	10-μL, readable to 0.1 μL.
Gas-tight syringes	0.1, 0.25, 0.50 and 1.0 mL for preparation of standard solutions
Pipettes	3.0 mL, with bulb
Serum bottles	15-mL, glass, with 20 mm o.d. mouth
Septa	20-mm rubber septa with Teflon lining
Aluminum seals	Aluminum tear-away seals to fit serum bottles

---

[1] Reference to a company and/or product is for the purpose of information and identification only and does not imply approval or recommendation of the company and/or product by the International Agency for Research on Cancer, to the exclusion of others which may also be suitable.

Hand crimper	For sealing septa to serum bottles
Union "tee"	Glass, 8 mm o.d., with septum on one branch, for injecting methyl chloride into nitrogen stream (see 9.8).

8. SAMPLING

8.1 Immediately before sampling, break the ends of the two sampler tubes and connect the back section of the larger tube to the front section of the smaller tube, then follow the sampling procedure in the INTRODUCTION, p. 173.

8.2 For concentrations of ~ 600 mg/m³, sample at a rate of 0.1 L/min for a total volume of 0.5 L. For concentrations of ~ 200 mg/m³, sample at 0.025 L/min for a volume of 1.5 L.

8.3 Separate the tubes and seal with plastic caps. Samples are stable for ~ 7 days at ambient temperatures.

9. PROCEDURE

9.1 Blank tests (see INTRODUCTION, p. 173)

9.1.1 Field blanks: analyse with air samples, as in 9.4-9.6.

9.1.2 Media blanks: set aside three pairs of sampler tubes from each new batch of charcoal used for sampling. Extract and analyse with air samples, as in 9.4-9.6.

9.2 Check tests

Analyse three blind spikes and three analyst spikes to check the calibration and desorption efficiency curves (see INTRODUCTION, p. 173). Use only the larger of the two tubes.

9.3 Test portion

Not applicable

9.4 Sample extraction

9.4.1 Pipette 3.0 mL methylene chloride into a 15-mL serum bottle.

9.4.2 Remove plastic caps from larger tube and transfer the 200 mg portion of charcoal, then the 400 mg portion, to the serum bottle. Place septum on bottle mouth and seal with aluminum crimp cap (steps 9.4.1 and 9.4.2 should be carried out as rapidly as possible).

9.4.3 Repeat 9.4.1 and 9.4.2 with the smaller (back-up) tube.

9.4.4 Allow sealed bottles to stand 10 min, with occasional gentle agitation. Retain for GC analysis (analysis should be completed within one day after the analyte has been desorbed).

9.5 GC operating conditions

Column temperature	100°C
Injector temperature	200°C
Detector temperature	260°C
Carrier gas flow rate	40 mL/min
Injection volume	5 µL

Under the above conditions, the retention time of the analyte is ~ 3 min. The methylene chloride will be eluted after the methyl chloride.

9.6 Analyte determination

Inject 5 µL of extract (9.4.4) onto GC column, using the solvent-flush technique (see Method 1, section 12). It is not advisable to use an automatic injector for methyl chloride in methylene chloride. Record peak area.

9.7 Calibration curve

9.7.1 Inject 5 µL aliquots of each of the standard solutions onto the GC and record peak areas.

9.7.2 Prepare a curve of peak-area *versus* mg analyte/3 mL of methylene chloride solution.

9.8 Desorption efficiency (DE)

9.8.1 For each new batch of charcoal used for sampling, prepare three spiked tubes at each of five levels in the calibration range (e.g., spike with the amount of analyte contained in each standard solution). Proceed as follows:

9.8.2 Connect the union tee to a source of purified nitrogen and to the front section of a large (10 cm) sampler tube.

9.8.3 With a nitrogen flow rate of ~ 30 mL/min, inject a known amount (see 9.8.1 and 10.1) of methyl chloride into the nitrogen stream, using a gas-tight syringe. Allow nitrogen to flow for one minute after discharge of syringe through the septum, then remove syringe.

9.8.4 Seal tube with plastic (not rubber) caps and allow to stand overnight.

9.8.5 Extract and analyse, as in 9.4-9.6.

9.8.6 Prepare a curve of DE (average mass recovered/mass of spike) versus mass of spike.

10. METHOD OF CALCULATION

10.1 The mass, m, of methyl chloride injected is calculated from the volume, v, injected, using the equation

$$m = 0.81\ Pv/T \quad (mg)$$

where, P = pressure at which methyl chloride was injected (torr)
T = temperature at which methyl chloride was injected (°K)

10.2 Using the peak area (9.6) and the calibration curve (9.7.2), calculate the mass concentration, ρ, of methyl chloride in the air sample from the equation

$$\rho = 10^3(m_s - m_b)/V \quad (mg/m^3)$$

where, m_s = sum of analyte masses (corrected for DE) found in both sections of larger and smaller (back-up) tubes (mg)

m_b = average sum of analyte masses (corrected for DE) found in both sections of the pairs of media blank tubes (mg)

V = volume of air sample (L)

11. REPEATABILITY AND REPRODUCIBILITY

The coefficient of variation for the total sampling and analytical method has been found to be 5.2% for the range 122-458 mg/m³.

12. NOTES ON PROCEDURE

Not applicable

13. SCHEMATIC REPRESENTATION OF PROCEDURE

Draw air sample through two charcoal sampler tubes in series
(front tube, 600 mg; back-up tube, 150 mg)

↓

Place sorbent from front and back-up tubes in separate 15-mL bottles
containing 3 mL methylene chloride

↓

Seal with septa and aluminum caps and agitate gently
for 10 min

↓

Inject 5 μL extract onto GC and record peak area

↓

Inject 5 μL aliquots of standard solutions onto GC
and prepare calibration curve

↓

Determine desorption efficiency using 3 spiked front tubes
at each of 5 concentration levels.
Extract and analyse as above

↓

Calculate mass concentration (mg/m^3) of analyte in air sample,
correcting for desorption efficiency and media blanks

14. ORIGIN OF THE METHOD

Method S99, NIOSH Manual of Analytical Methods, 2nd Ed., Vol. 4, Cincinnati, OH, National Institute for Occupational Safety and Health (1978)

Contact point: Dr P.M. Eller
Division of Physical Sciences and Engineering
National Institute for Occupational Safety and Health
4676 Columbia Parkway
Cincinnati, OH 45236, USA

METHOD 8

THE DETERMINATION OF METHYL BROMIDE IN AIR

A. MacKenzie Peers

Adapted from Method S372, NIOSH Manual of Analytical Methods, 2nd Ed., Vol. 3 (1977)

1. SCOPE AND FIELD OF APPLICATION

This personal sampling pump and sorption tube method is suitable for the determination of methyl bromide in air over the range 20-250 mg/m^3, using an 11-L sample. The breakthrough volume depends on atmospheric composition and the capacity of the charcoal: for example, it has been found to be 18 L when sampling an atmosphere containing 161 mg analyte/m^3 at a rate of 0.19 L/min. High relative humidity decreases the trapping efficiency of the charcoal.

2. REFERENCES

Taylor, D.G., ed, (1977) NIOSH Manual of Analytical Methods, 2nd Ed., Vol. 3, U.S. Department of Health, Education and Welfare

Eller, P.M., ed. (1984) NIOSH Manual of Analytical Methods, 3rd Ed., Vol. 1, U.S. Department of Health and Human Services

3. DEFINITIONS

The breakthrough volume is reached when the concentration of analyte in the effluent reaches 5% of that in the influent.

4. PRINCIPLE

A known volume of air is drawn through two charcoal tubes in series (separated after sampling, to prevent migration to back-up tube during storage) which adsorb organic vapours. The latter are desorbed in carbon disulfide and determined by gas chromatography (GC) with flame-ionization detection (FID). Quantification is carried out using standard solutions and corrections are applied for desorption efficiency and media blanks.

5. HAZARDS

Carbon disulfide is toxic and an acute fire and explosion hazard. Methyl bromide is toxic and a strong skin irritant. Both compounds should be handled only in a ventilated hood.

6. REAGENTS[1]

Carbon disulfide	Chromatographic quality
Methyl bromide	99.5%, plus gas standards (Available from Linde (Union Carbide Co.) National specialty Gas Office) for determining concentration of stock standard solution.
_n_-Decane	Chromatographic quality, for internal standard use.
Nitrogen	Purified, for GC
Hydrogen	Pre-purified, for FID
Air	Filtered, compressed, for FID
Extraction solvent	Carbon disulfide containing sufficient internal standard (_n_-decane) to give a FID response similar to that of methyl bromide near the mean of the range of interest. The extraction procedure for 10 samples, plus blanks, desorption efficiency measurement, etc. requires ~ 150 mL of extraction solvent.

---

[1] Reference to a company and/or product is for the purpose of information and identification only and does not imply approval or recommendation of the company and/or product by the International Agency for Research on Cancer, to the exclusion of others which may also be suitable.

Stock standard solution of methyl bromide	A concentrated solution of methylbromide in carbon disulfide, which can be used to inject methyl bromide (e.g., 0.2-2.5 mg amounts) directly into sorption tubes and to prepare working standard solutions for calibration purposes. (A concentration of 20-50 g/L should be suitable.) Prepare by bubbling methylbromide gas into 50 mL carbon disulfide with a fritted glass bubbler. Determine exact concentration by comparison with gas standards. Keep refrigerated in glass container with PTFE-lined screw cap.
Working standard solutions of methyl bromide	Add known amounts of stock standard solution to the extraction solvent in a 10-mL volumetric flask and dilute to the mark. Prepare daily (shortly before use) at least five concentrations over the range of interest (e.g., 0.1-1.2 mg/mL). Keep cold in glass container with PTFE-lined screw cap.

7. APPARATUS[1] (see INTRODUCTION, p. 173)

Air sampler tubes	The method requires two tubes in series. Both are glass, 10 cm × 8 mm o.d., 6 mm i.d., flame-sealed at both ends. The front tube contains 400 mg of petroleum-based, 20/40 mesh, activated charcoal and the back-up tube 200 mg. A plug of sylilated glass wool is placed at both ends of both tubes. The pressure drop across the pair of tubes must be < 3.4 kPa at an air flowrate of 1 L/min.
Personal sampling pump	0.1 to 1 L/min, with flexible connecting tubing.
Gas chromatograph	With FID and peak-area integrator. Column: 6.1 m × 3.2 mm, stainless-steel, packed with 10% FFAP on 100/120 mesh Supelcoport.

[1] Reference to a company and/or product is for the purpose of information and identification only and does not imply approval or recommendation of the company and/or product by the International Agency for Research on Cancer, to the exclusion of others which may also be suitable.

Vials	5-mL, glass, with PTFE-lined screw caps.
Syringes	10, 100, 500 and 1000 μL
Pipettes	1-mL, graduated; 2-mL delivery, with bulbs
Volumetric flasks	10-mL
Glass containers	With PTFE-lined screw-caps, for storing standard solutions (10-mL, 50-mL) and extraction solvent.

8. SAMPLING

8.1 Immediately before sampling, break the ends of the two charcoal tubes (front and back-up) and connect them with a short piece of flexible vinyl tubing. Follow sampling instructions in INTRODUCTION, p. 173.

8.2 Sample at an accurately-known rate (e.g., 0.75 L/min) for a total sample volume of 11 L.

8.3 Separate front and back-up tubes immediately after sampling and seal with plastic (not rubber) caps.

9. PROCEDURE

9.1 Blank tests (see INTRODUCTION, p. 173)

9.1.1 Field blanks: analyse with air samples, as in 9.4-9.6.

9.1.2 Media blanks: set aside three pairs of sampler tubes from each new batch of charcoal used for sampling. Extract and analyse with air samples, as in 9.4-9.6.

9.2 Check tests

Analyse three blind spikes and three analyst spikes to check the calibration and desorption efficiency curves (see INTRODUCTION, p. 173).

9.3 Test portion

Not applicable

9.4 Sample extraction

9.4.1 Place sorbent from front and back-up tubes in separate, 5-mL vials (discard glass-wool plugs)

9.4.2 Add 2.0 mL extraction solvent (6) to each vial and seal tightly with screw cap.

9.4.3 Allow to stand 30 min, with occasional agitation. Retain for GC analysis.

9.5 <u>GC operating conditions</u>

Column temperature	65°C
Injector temperature	155°C
Detector temperature	200°C
Carrier gas flow-rate	30 mL/min
Injection volume	5 µL

9.6 <u>Analyte determination</u>

9.6.1 Inject 5 µL of sample extract (9.4.3) manually onto the GC using solvent-flush technique (see Method 1, section 12), or with auto-sampler.

9.6.2 Measure peak areas of analyte(s) and internal standard on the same chromatogram. Divide peak area of analyte by that of internal standard to obtain relative peak area.

9.7 <u>Calibration curve</u>

9.7.1 Calibrate daily by injecting 5-µL aliquots of each of the five working standard solutions of the analyte onto the GC.

9.7.2 Measure peak areas of analyte and internal standard on the same chromatogram and calculate relative peak area (9.6.2) for each concentration.

9.7.3 Prepare calibration curve showing relative peak area <u>versus</u> mg analyte/2 mL of solution.

9.8 <u>Desorption efficiency</u> (DE)

9.8.1 For each batch of charcoal used for sampling, prepare three spiked tubes at each of five levels in the calibration range (e.g., spike with the amount of analyte in 2 mL of each of the working standard solutions). Proceed as follows:

9.8.2 Inject a known amount (9.8.1) of analyte directly into the 400 mg of charcoal in a front tube, using the stock standard solution (6) and a microlitre syringe. Cap and allow to stand overnight.

9.8.3 Extract and analyse, as in 9.4-9.6.

9.8.4 Using the calibration curve, calculate the mass of analyte recovered at each level and prepare a graph of desorption efficiency *versus* mg analyte recovered.

10. METHOD OF CALCULATION

Using the relative peak area and the calibration curve, calculate the mass concentration, ρ, of methyl bromide in the air sample from the equation:

$$\rho = 10^3(m_s - m_b)/V \quad (mg/m^3)$$

where, m_s = sum of analyte masses (corrected for DE) found in front and back-up sample tubes (mg)

m_b = average sum of analyte masses (corrected for DE) found in front and back-up media blank tubes (mg)

V = volume of air sample (L)

11. REPEATABILITY AND REPRODUCIBILITY

The coefficient of variation for the total sampling and analytical method has been found to be 10.3% in the range 35-150 mg/m³.

12. NOTES ON PROCEDURE

Not applicable

13. SCHEMATIC REPRESENTATION OF PROCEDURE

Draw air sample through two charcoal tubes in series

↓

Place sorbent from front and back-up tubes in separate vials, add 2 mL carbon disulfide (+ standard) and cap

↓

Stand 30 min, with occasional agitation

↓

Inject 5 μL of extract onto GC and measure peak areas

↓

Inject 5 μL aliquots of standard solutions onto GC and prepare calibration graph

↓

Determine desorption efficiency in the calibration range using 3 spiked front tubes at each level

↓

Calculate mass concentration of analytic in air samples

14. ORIGIN OF THE METHOD

Method 1008, NIOSH Manual of Analytical Methods, 2nd Ed., Vol. 3, Cincinnati, OH, National Institute for Occupational Safety and Health (1977)

Contact point: Dr P.M. Eller
Division of Physical Sciences and Engineering
National Institute for Occupational Safety and Health
4676 Columbia Parkway
Cincinnati, OH 45236, USA

METHOD 9

MONITORING CHLOROMETHYL METHYL ETHER IN AIR

M.L. Langhorst

1. SCOPE AND FIELD OF APPLICATION

This method is applicable to the monitoring of airborne chloromethyl methyl ether (CMME) in the volume fraction range 6-900 μL/m³ (ppb, v/v) or the mass concentration range 20-3000 μg/m³ (38-3800 ng/sample). The procedure was designed for industrial hygiene personal monitoring to provide an accurate 4-hour, time-weighted average of the exposure level. Approximately 15 min are required to desorb and clean up a set of 8 sample tubes. Each chromatogram requires an additional 3 min.

2. REFERENCES

Langhorst, M.L., Melcher, R.G. & Kallos, G.J. (1981) Reactive adsorbent derivative collection and gas chromatographic determination of chloromethyl methyl ether in air. Am. Ind. Hyg. Assoc. J., 42, 47-55

Department of Labor, Occupational Safety and Health Administration (1974) Occupational Safety and Health Standards: Part III. Fed. Reg., 39, 3756-3797

3. DEFINITIONS

Yield

The yield, Y, is the amount of derivative formed relative to the amount of CMME added, and is given by,

$$Y = \frac{100\ m_2 (80.52)}{m_1\ (241.5)} \quad (\%)$$

where

m_1 = weight of CMME added or sampled (ng)
80.52 = molecular weight of CMME
241.5 = molecular weight of CMME derivative
m_2 = weight of CMME derivative formed (ng)

Desorption efficiency

Desorption efficiency is the percentage of the CMME derivative formed which is desorbed into solution.

Analytical recovery

Analytical recovery is the percentage of the desorbed CMME derivative which will be recovered through the clean-up and extraction steps for analysis by gas chromatography (GC).

Recovery independent of yield

Percent recovery independent of yield is the combined recovery from desorption efficiency and analytical recovery.

4. PRINCIPLE

A measured volume of air is drawn through a glass air-sampling tube, packed with 1.5% potassium 2,4,6-trichlorophenate on 120/140 mesh textured glass beads. The CMME is derivatized and collected on the sorbent by the reaction:

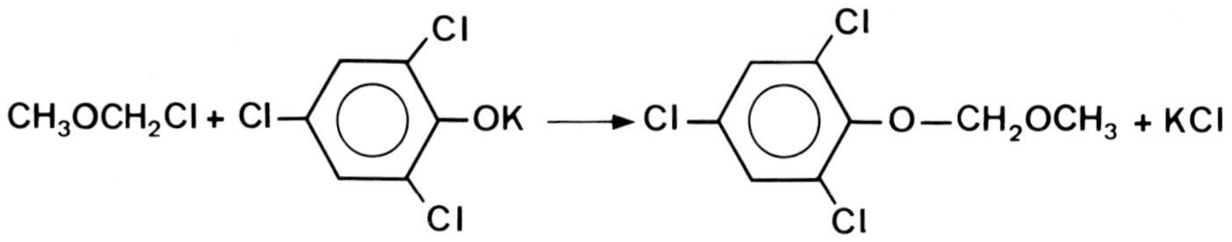

$$CH_3OCH_2Cl + Cl{-}C_6H_2Cl_2{-}OK \longrightarrow Cl{-}C_6H_2Cl_2{-}O{-}CH_2OCH_3 + KCl$$

The CMME derivative is desorbed into methanol, diluted with aqueous potassium hydroxide (KOH), and extracted with hexane. The hexane extract is analysed by GC with electron-capture detection. Quantification is based upon comparison of detector response (peak height) from sample and external standards.

5. HAZARDS

Bis(chloromethyl)ether (BCME) is known to be an impurity in CMME. Both BCME and CMME have been termed human carcinogens by the United States Department of Labor (Department of Labor, Occupational Safety and Health Administration, 1974). Extreme safety precautions should be exercised in the preparation and disposal of liquid or gas standards. The handling of CMME should be carried out in a well-ventilated hood. Always store samples in a secondary container in a cool, dry, well-ventilated area. Dispose of any gas or liquid sample by putting it into a methanol-caustic solution that is kept in the hood. A second waste bottle for disposing of gloves, contaminated paper, etc. should also be kept in the hood and sent to the burner. Use a rubber bucket and dry ice when transporting liquid samples outside the hood area.

The preparation of reagent-coated glass beads involves slurrying the beads in a solution of reagent, then evaporating the solvent under heat. Be careful to watch for splattering as the material begins to dry and cake on the bottom of the glass dish. Wear goggles and stir or swirl occasionally. Work in a hood. Avoid excessive temperatures during drying.

6. REAGENTS[1]

Hexane	Burdick and Jackson, distilled-in-glass grade
Methanol	Burdick and Jackson, distilled-in-glass grade
Water	Distilled, deionized
Potassium hydroxide pellets	Reagent grade
Chloromethyl methyl ether, 97%	Aldrich Chemical, Milwaukee, WI 53201
2,4,6-Trichlorophenol, 98%	Aldrich chemical, Milwaukee, WI 53201 Recrystallize before use (three times from hexane) to give a high-purity (>99%) white powder
GLC-100 textured glass beads	120/140 mesh, Applied Science Laboratories, Inc., State College PA 16801
Commercial silica gel air-sampling tubes	520 mg front/260 mg back section, Environmental Compliance Corp., Venetia, PA 15367

[1] Reference to a company and/or product is for the purpose of information and identification only and does not imply approval or recommendation of the company and/or product by the International Agency for Research on Cancer, to the exclusion of others which may also be suitable.

1.5% (w/w) potassium salt of 2,4,6-trichlorophenol on 120/140 mesh GLC-100	Weigh 50 g of textured glass beads into a glass evaporating dish. Add 42 mL of 2% 2,4,6-trichlorophenol solution in methanol (0.88 g in 44 mL) and 14 mL of 0.26 mol/L KOH solution in methanol (1.5 g/100 mL). Slurry glass beads in solution. Evaporate solvent slowly under an infrared heat lamp, swirling occasionally (~ 20 min). Finish drying on a hot plate or steam bath to form a free-flowing powder (~ 10 min).
Reactive sorbent air-sampling tubes	Pack 2.0 ± 0.05 of trichlorophenate-coated beads (above) into disposable glass pipettes, 7 cm × 6 mm i.d. Place a silanized glass wool plug at both ends.
Derivative standard (2,4,6-trichlorophenyl structure: Cl, Cl, Cl, $O-CH_2OCH_3$)	Synthesize from CMME and 2,4,6-trichlorophenol. Dissolve 5 g 2,4,6-trichlorophenol in 70 mL pentane. Add 2 g (2 mL) CMME. Shake to mix and let stand at room temperature for 2 h. Add 500 mL aqueous 10% sodium hydroxide solution and an additional 100 mL of pentane to extract the non-phenolic material. Separate the pentane and allow to evaporate to dryness. The product is a white crystalline material.
Standard solutions of CMME derivative for calibration	Prepare five solutions of CMME derivative in hexane at concentrations between 10 ng/mL and 150 ng/mL. These can be prepared by weighing the synthesized derivative into a vial and dissolving in hexane or by spiking CMME onto an air sampling tube for desorption (see Section 11 (b)).

7. APPARATUS[1]

Gas chromatograph	Equipped with ^{63}Ni electron-capture detector
Infrared heat lamp with variable transformer	

[1] Reference to a company and/or product is for the purpose of information and identification only and does not imply approval or recommendation of the company and/or product by the International Agency for Research on Cancer, to the exclusion of others which may also be suitable.

Portable battery-operated vacuum pump	Capable of pumping at flow rates of 5-10 mL/min. Sipin Model SP-15, Anatole J. Sipin Co., New York, NY 10016
Field sampling tube	The recommended field sampling tube, shown in Figure 1, should consist of a front tube containing 2.0 g of 1.5% potassium trichlorophenate on 120/140 mesh GLC-110 in a 7 cm x 6 mm i.d. tube. The back tube (to check for breakthrough of the derivative), should contain ~ 1 g silica gel.

8. SAMPLING

8.1 Collect samples by pulling air through the tube at 5-10 mL/min for 4 hours (1.2 to 2.4 litres). Short-term samples can be collected at 10-20 mL/min for 10-20 min. Use calibrated sampling pumps and record the time, temperature, pressure and humidity of the atmosphere sampled.

8.2 Immediately after sampling, cap the tubes. Sample tubes should be stored at room temperature (25±3°C) and have been shown to be stable for a period of at least 7 days.

FIG. 1. FIELD SAMPLING TUBE FOR CHLOROMETHYL METHYL ETHER IN AIR

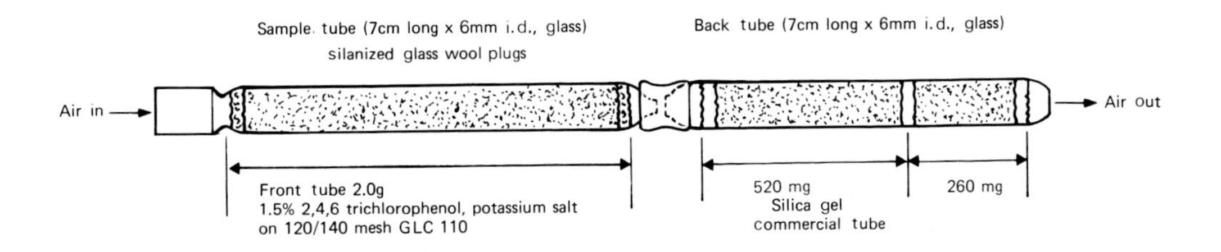

---

[1] Reference to a company and/or product is for the purpose of information and identification only and does not imply approval or recommendation of the company and/or product by the International Agency for Research on Cancer, to the exclusion of others which may also be suitable.

9. PROCEDURE

9.1 Blank test

To provide a blank, a representative sampling tube should be analysed as specified in sections 9.4–9.5 below.

9.2 Check test

Prepare a laboratory "known" by direct injection of 1-2 µL CMME solution (2.0 µg/mL) into a sorbent tube and pull air through the tube at 5-10 mL/min for 4 h.

9.3 Field confirmation test

Prepare field spikes to confirm method suitability in specific plant atmospheres. Inject a known amount of CMME standard solution into a sorbent tube and pull field air through the tube. A side-by-side field sample (not spiked) should also be collected to determine the recovery of the field spike.

9.4 Sample desorption and clean-up

9.4.1 Wash each sorbent (derivatizing sorbent and silica gel) in separate 2 mL aliquots of methanol in 3-dram vials and shake for five min. Follow steps 9.4.2 to 9.5 inclusive, for each sorbent.

9.4.2 Separate the methanol aliquot from the sorbent with a disposable pipette and transfer to a 3-dram vial.

9.4.3 Wash the sorbent with an additional 1 mL of methanol and combine the latter with the extract from 9.4.1.

9.4.4 Dilute the combined methanol extracts with 3 mL 1 mol/L aqueous KOH and extract with 2 mL hexane.

9.4.5 Transfer the hexane extract to another vial and wash the hexane extract with 2 mL aqueous 1 mol/L KOH.

9.5 Gas chromatography

Analyse the hexane layer from 9.4.5 by gas chromatography (GC) with electron-capture detection, using the following operating conditions:

Column:	2.3 m × 2 mm i.d., glass, packed with Permabond PEG-20M (T max 240°C), available from HNU Systems, Inc., Newton Highlands, MA 02161

Detector:	Electron-capture (^{63}Ni)
Carrier gas:	Nitrogen (20 mL/min)
Temperatures:	column: 140°C detector: 320°C injection port: 200°C
Injection volume:	1 µL

A typical chromatogram is shown in Figure 2.

FIG. 2. TYPICAL GAS CHROMATOGRAMS OF CMME DERIVATIVE

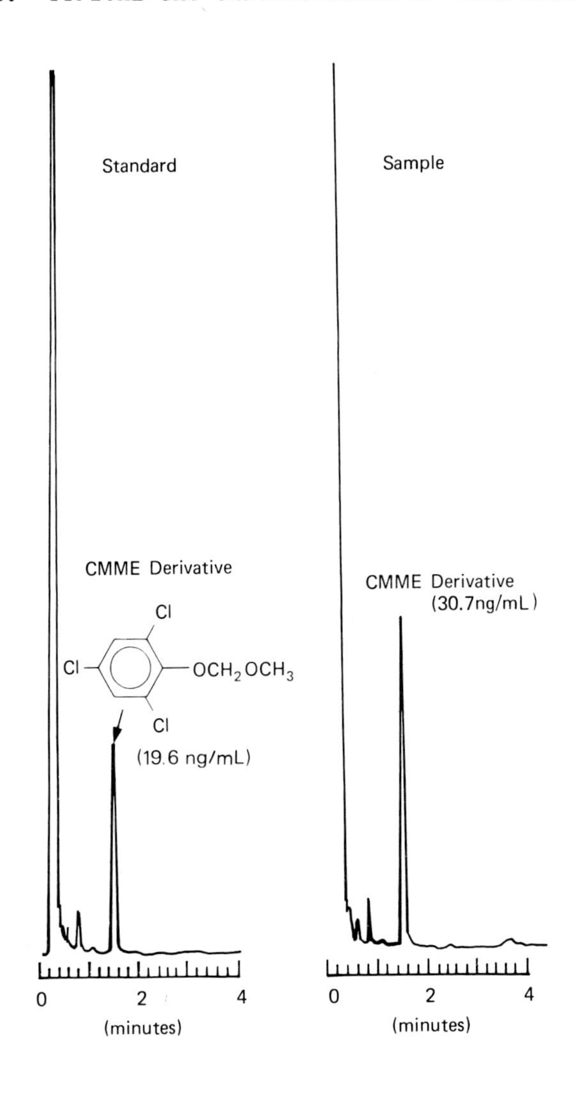

9.6 Calibration curve

Inject 1 µL aliquots of the standard solutions containing the CMME derivative (see Section 6) into the GC. Prepare a calibration curve by plotting detector response (peak height) vs concentration of derivative (ng/mL).

10. METHOD OF CALCULATION

The concentration of CMME in the original air sample can be calculated in two ways: (a) using standard solutions of the CMME derivative or (b) using standard solutions of CMME.

(a) Using a standard solution of the CMME derivative in hexane, the mass concentration, ρ_d, of the CMME derivative in the sample extract is given by

$$\rho_d = \frac{\rho_s\, h}{h_s} \text{ (ng/mL)}$$

where

ρ_s = mass concentration of CMME derivative in standard solution (ng/mL)
h = peak height of derivative in sample chromatogram (mm)
h_s = peak height of derivative in standard chromatogram (mm)

The corresponding mass concentration, ρ, of CMME in the extract is then given by

$$\rho = \frac{100\rho_d}{Y}\left[\frac{80.52}{241.5}\right] \text{ (ng/mL)}$$

where,

80.52 = molecular weight of CMME
241.4 = molecular weight of the CMME derivative
Y = yield of derivative from CMME (%) (from validation data, Y= 56%)

The mass concentration, ρ_{air}, of CMME in the air sample is given by

$$\rho_{air} = \frac{\rho V_1}{V_2} \text{ (µg/m}^3\text{)}$$

where,

V_1 = volume of hexane extraction 9.4.5 (mL)
V_2 = volume of air sampled in 8.2 (L)

The volume fraction, ϕ, of CMME in the air sample is given by

$$\phi = \left[\frac{24.45}{80.52}\right]\rho_{air} \quad (\mu L/m^3)$$

(b) The alternative method of calculation, using standard solutions of CMME, requires spiking a known weight of CMME onto an air sampling tube, then pulling a small volume of air through the tube. The derivative peak height is then determined by GC, as described in sections 9.4-9.5

The weight of CMME collected in field samples can then be determined by direct comparison of the peak height from the field sample chromatogram with the peak height from the spiked tube chromatogram.

11. REPEATABILITY AND REPRODUCIBILITY

11.1 Derivative yields for CMME with volume fractions between 6 and 900 $\mu L/m^3$ are 56 ± 10% (2σ). Percent recoveries independent of yield were 96 ± 6% (2σ). These results were obtained from direct injection simulated air samples (20 determinations).

11.2 Percent yield for samples prepared by pulling CMME in from SARAN[1] bags was 54 ± 14% for volume fractions from 3 to 700 $\mu L/m^3$ in air collected at 5-10 mL/min for 4-8 h.

11.3 The analytical limit of detection is ~ 0.5 $\mu L/m^3$ for a 1.2 L air sample, but additional validation data would be recommended before reporting volume fractions below 6 $\mu L/m^3$.

11.4 For 4 h collection at 5-10 mL/min, the volume fraction limit for the CMME derivative appears to be greater than 900 mm^3/m^3. Under those conditions, no significant breakthrough of CMME was detected on the back tube (silica gel).

11.5 The recovery of the CMME derivative from silica gel back tubes was 96-100% (desorption efficiency).

12. NOTES ON PROCEDURE

12.1 It is recommended that procedure (a), using CMME derivative standards, be employed, because it is very difficult to obtain and store a good analytical standard of CMME (> 90%). The compound tends to decompose at room temperature to HCl, methylal and methanol, and more dilute standard stolutions cannot be trusted several days after preparation. In addition, the use of derivative standards eliminates the need for handling CMME standards, which are toxic.

12.2 The alternative procedure, (b), has the advantage of automatically determining derivative yield and would correct for any variations in adsorbent batches.

12.3 Effect of humidity, sample storage duration and temperature and potential interferences.

12.3.1 Eleven samples collected at 80% relative humidity (21°C) showed 56 ± 12% yield and 96 ± 4% recovery for volume fractions between 9 and 200 $\mu L/m^3$ CMME in air.

12.3.2 Eight samples stored 7 days without refrigeration showed 54 ± 10% yield and 95 ± 6% recovery.

12.3.3 Eight excursion samples collected for 10-20 min at 10-20 mL/min showed 57 ± 14% yield and 97 ± 4% recoveries for CMME volume fractions from 25-300 $\mu L/m^3$.

12.4 The percent yield obtained is dependent upon flow rate. Maximum yield is obtained between 5 and 20 mL/min. Higher flow rates significantly decrease yield.

12.5 Based on one experiment at 0°C (1.2 L), the sorbent tube method appears to be not significantly affected by changes in temperature.

12.6 This method was not affected by the presence of humidity, HCl vapours, and trimethylamine vapours.

---

[1] Trademark of The Dow Chemical company

13. SCHEMATIC REPRESENTATION OF PROCEDURE

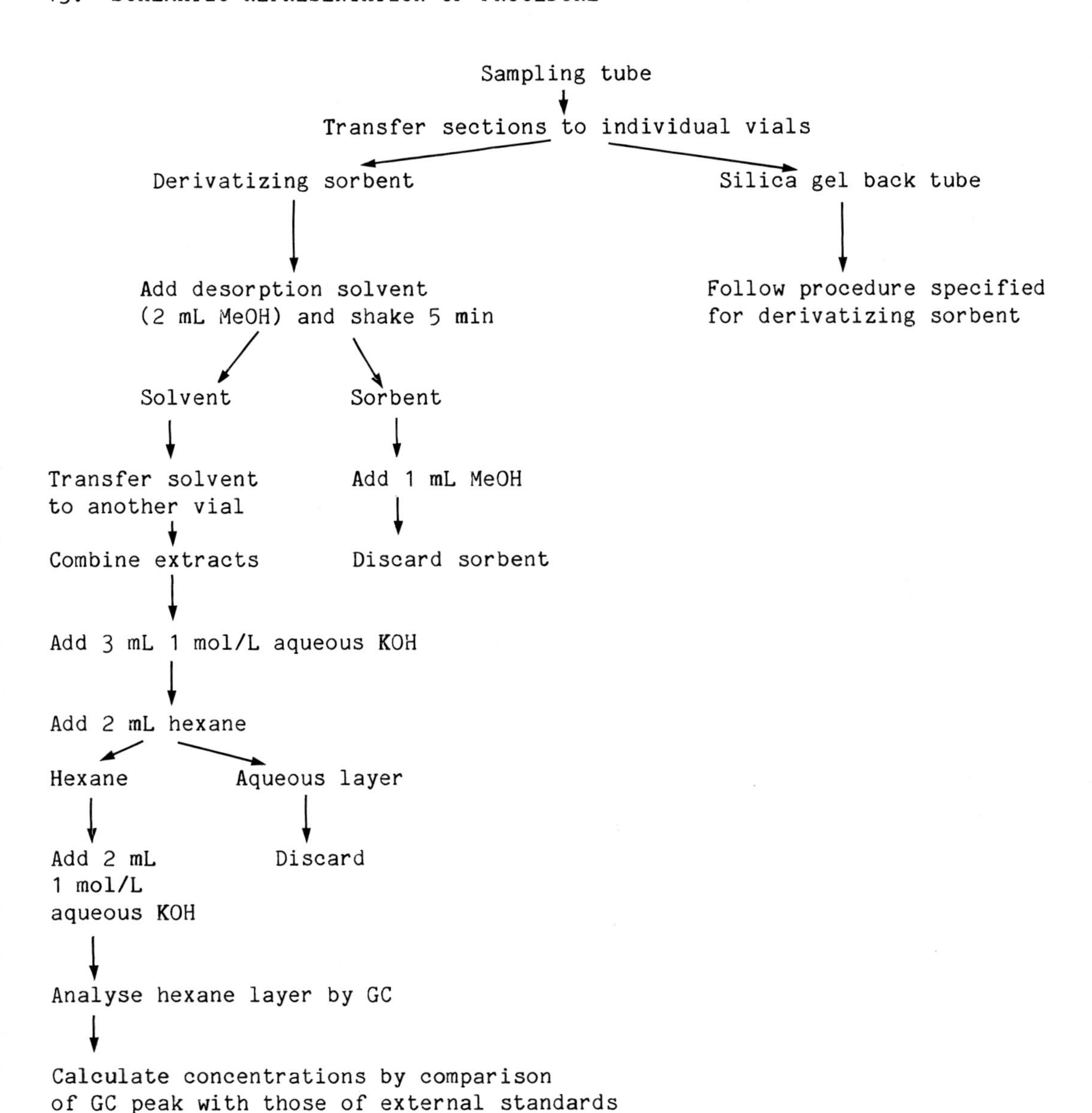

14. ORIGIN OF THE METHOD

The Dow Chemical Company
Analytical Laboratory, 574 Building
Midland, MI 48640

Contact point: M.L. Langhorst
The Dow Chemical Company
Analytical Laboratory
574 Building
Midland, MI 48640, USA

METHOD 10

GC-EC DETERMINATION OF BIS(CHLOROMETHYL)ETHER IN AIR

J.W. Russell

1. SCOPE AND FIELD OF APPLICATION

This method is applicable to the determination of bis(chloromethyl)ether in air in the weight range of 5-2400 nanograms, equivalent to a volume fraction of 0.1-50 µL/m³ in a 10-liter sample. No common airborne interferences are known. The time required for a single sample determination (exclusive of sampling or calibration) is about one hour.

NOTE: µL/m³ = ppb (v/v).

2. REFERENCES

Department of Labor, Occupational, Safety and Health Administration (1974) Occupational Safety and Health Standards: Part III. Fed. Regist., 39, 3756-3797

Lovelock, J.E. (1973) United States Patent 3,725,009, Detection of Trace Gases Utilizing an Electron Capture Detector.

Solomon, R.A. & Kallos, G.J. (1975) Determination of chloromethyl methyl ether and bis-chloromethyl ether in air at the part per billion level by gas chromatography. Anal. Chem., 47, 955-957

Tou, J.C. & Kallos, G.J. (1974) Kinetic study of the stabilities of chloromethyl methyl ether and bis(chloromethyl)ether in humid air. Anal. Chem., 46, 1866-1869

Yao, C.C., Zollinger, H., Kallos, G.J., Solomon, R.A. & Tou, J.C. (1979) Exchange of comments: Analytical methods of bis(chloromethyl)ether in air. Anal. Chem., 51, 299-302

3. DEFINITIONS

Not applicable

4. PRINCIPLE

A measured volume of air is drawn through an air sampling tube packed with Chromosorb 101 porous polymer gas-chromatographic stationary phase. The bis(chloromethyl)ether concentrated in the porous polymer is thermally desorbed with a stream of nitrogen into a derivatizing solution. (see Notes on Procedure, 13.1). Quantification is based on comparison of detector responses (peak heights) obtained with sample and with an external standard.

5. HAZARDS

Bis(chloromethyl)ether has been classified as a human carcinogen by the United States Department of Labor. Extreme safety precautions must be exercised in the preparation, handling, storage and disposal of liquid or gas standards. All handling of bis(chloromethyl)ether should be performed in a verified high-efficiency fume hood. Store solutions in secondary containers in the hood and dispose of them by addition to methanolic sodium hydroxide, followed by incineration.

Rubber gloves and goggles must be worn. Dispose of used gloves and contaminated paper, etc., in a waste bottle stored in the hood. Dispose of the waste bottle by incineration.

Handle reagents as recommended by the supplier. Be particularly aware of the corrosiveness and flammability of sodium methoxide.

6. REAGENTS[1]

Hexane	Distilled-in-glass
Methanol	Distilled-in-glass
Water	Distilled, deionized
Sodium methoxide	Reagent. Keep tightly sealed to minimize contact with humid air. Replace with fresh reagent after 1 year.
2,4,6-Trichlorophenol	Reagent. Vacuum distill and recrystallize three times from hexane. Resulting product must be free of impurities that would interfere in the gas chromatographic separation.

---

[1] Reference to a company and/or product is for the purpose of information and identification only and does not imply approval or recommendation of the company and/or product by the International Agency for Research on Cancer, to the exclusion of others which may also be suitable.

Bis(chloromethyl)ether	
Nitrogen	Prepurified grade
Helium	High-purity grade
Chromosorb 101 porous polymer	80/100 mesh
Sodium hydroxide, 1 mol/L aqueous concentrate	"Dilut-it" brand from J.T. Baker
0.5 mol/L sodium hydroxide	Dilute the 1 mol/L nominal sodium hydroxide concentrate to 2 liters with deionized water.
Derivatizing reagent, sodium trichlorophenate in methanol	Dissolve 12.0 g of purified 2,4,6-trichlorophenol and 3.2 g of sodium methoxide in 750 mL of methanol. Minimize exposure of methoxide powder and solution to air to minimize hydrolysis. Discard reagent after 4 months.
Silanized glass wool	
Bis(chloromethyl)ether standard solution	Using a 10-µL syringe, accurately add 2.0 µL of bis(chloromethyl)ether to 10.0 mL of hexane previously pipetted into a 7-dram vial. Cap and mix. This solution contains 266 ng/µL based upon a specific gravity of 1.33 for bis(chloromethyl)ether. Dilute to 26.6 ng/µL by accurately pipetting 1 mL into a 10-mL volumetric flask and diluting to volume with hexane. Transfer to a 3-dram vial.

7. APPARATUS[1]

Air sampler tubes	Stainless steel tubes (14 cm length x 6.4 mm o.d.) are rinsed in hexane followed by methanol, dried and gravity packed, using gentle tapping, with Chromosorb 101 porous polymer. The packing is held in place with 10 mm long plugs of silanized glass wool at each end. See Figure 2.

[1] Reference to a company and/or product is for the purpose of information and identification only and does not imply approval or recommendation of the company and/or product by the International Agency for Research on Cancer, to the exclusion of others which may also be suitable.

Gas chromatograph	Equipped with a ^{63}Ni electron-capture detector.
Recorder	1 mV, 25 cm (full-scale)
Portable battery operated vacuum pump	Capable of pumping air at a constant rate between 10 and 100 mL/min. Sipin Model SP-1, Anatole J. Sipin Co., New York, NY 10016
Heating tape	1.2 m × 1.3 cm, in glass
Temperature controller with thermocouple	Capable of controlling at 110°C
Variable transformer	0-130 VAC
Sparger	Fabricated from glass as shown in Figure 1. Substitute designs may affect recovery of bis(chloromethyl)ether. Frit is Ace "coarse" porosity, Ace Glass Inc., 1430 Northwest Blvd, Vineland, NJ 08360
Tubing	Stainless steel, 6.4 mm o.d., 5.3 mm i.d.
Tube fittings	6.4 mm stainless steel nuts, ferrules, caps, tees, unions; 6.4 mm ferrules of Teflon polymer for sparger
10-Microliter syringes	

8. SAMPLING

8.1 Condition packed air-sampling tubes overnight prior to each use by heating at 190°C with 30 mL/min helium flow. After cooling to ambient temperature, remove from helium flow and immediately cap ends.

8.2 When sampling air, use a pump which has been calibrated while attached to a sampling tube.

8.3 Draw air through a sampling tube at a constant rate (between 10 and 100 mL/min) up to a total volume of 10 liters. Record the time, temperature, pressure and humidity of the atmosphere sampled. Cap tubes immediately after sampling.

8.4 If desired, prepare a field spike by injecting a known amount of bis-(chloromethyl)ether standard solution into the front of a tube prior to sampling. An unspiked field sample should also be collected side-by-side with the spiked sample to allow determination of recovery of the latter.

FIG. 1. SPARGER DESIGN

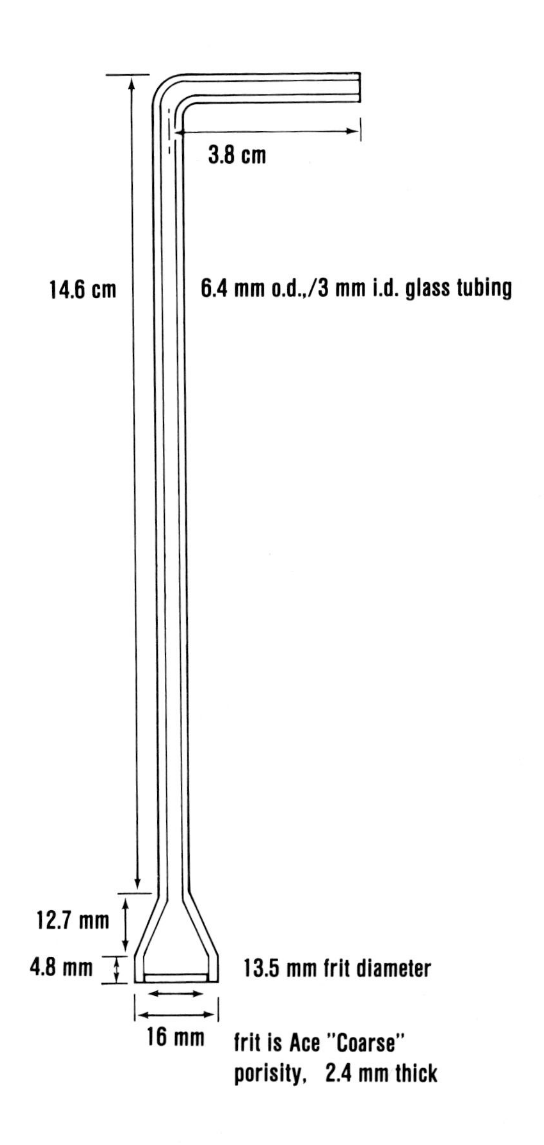

8.5 Store tubes at ambient temperature. Sampling tubes should be analysed within 1 week.

9. PROCEDURE

9.1 Blank test

Assemble the apparatus as shown in Figure 2 and analyse a conditioned sampling tube as described in sections 9.4.2 to 9.6.3, inclusive.

9.2 Check test

Conduct a test of bis(chloromethyl)ether desorption and derivatization efficiency by accurately injecting 5.0 µL of the 26.6 ng/µL bis(chloromethyl)ether in hexane solution into the injection T (Fig. 2) with two conditioned sampling tubes and the sparger in place. Procede as described in sections 9.4.2 to 9.6.4, inclusive. Compare response obtained to that of the calibration standard. (Expected recovery of the spike is 85-95%.)

9.3 Test portion

Not applicable.

9.4 Sample desorption, derivatization and clean-up

9.4.1 Disconnect the sampling tube from the pump and connect the end through which the air entered to a fresh, conditioned (clean-up) tube which is connected to the sparger (Fig. 2).

9.4.2 Attach the temperature controller thermocouple to the clean-up tube using glass adhesive tape. Tightly wrap tubes and unions with heating tape, avoiding gaps.

9.4.3 Pipette 10.0 mL of derivatizing reagent into a 7-dram vial and place under the sparger. The sparging surface should be 3 mm from the bottom of the vial and at a slight angle (5° from horizontal) to prevent pooling of nitrogen.

9.4.4 Start prepurified nitrogen flow through the tubes at 100 mL/min, heating the tubes to 110°C for 15 min (see Notes on Procedure, 12.2).

9.4.5 Separate vial from sparger. Cap loosely and place on a steam bath for 5 min. Vial contents should warm sufficiently to cause reflux of methanol, but without ebullition (Notes on Procedure, 12.3).

FIG. 2.

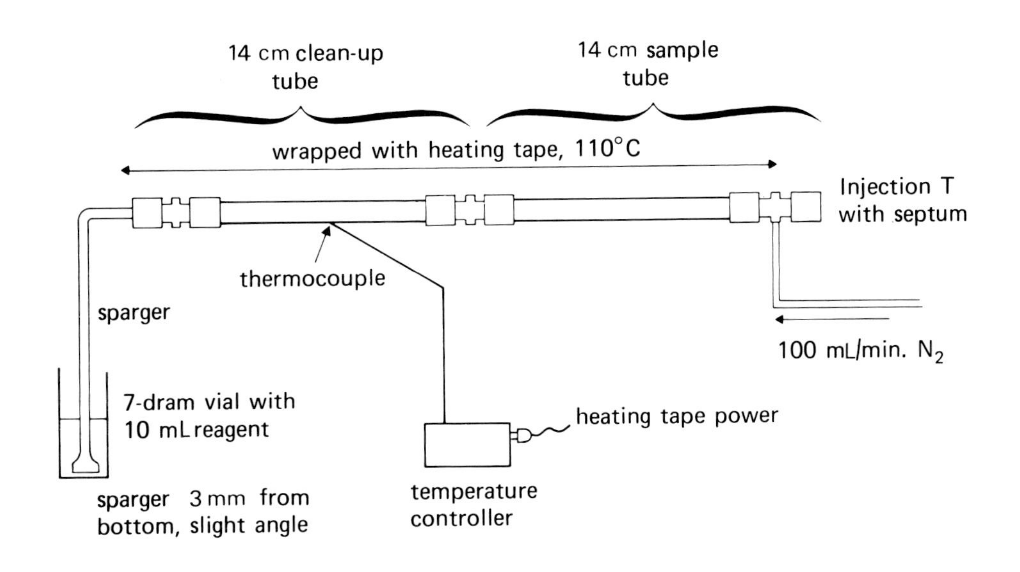

9.4.6 Cool to ambient temperature and add by pipette 10.0 mL of 0.5 mol/L sodium hydroxide and 2.0 mL hexane. Cap and shake vigorously for 5 min to extract.

9.4.7 After phase separation, transfer the upper (hexane) layer to a 2-dram vial. Discard the aqueous layer.

9.4.8 Add 1.0 mL of 0.5 mol/L sodium hydroxide to the hexane, cap and shake vigorously for 30 s.

9.5 GC operating conditions

Column	1.2 m × 6.4 mm o.d., 2 mm i.d. glass, packed with 0.5% OV-17 phenylmethyl silicone on Chromosorb W HP 80/100 mesh. The first 6 cm of the column should be free of packing and glass wool. (Notes on Procedure, 12.4) Condition overnight at 220°C with 35 mL/min prepurified nitrogen flow.
Carrier gas	Nitrogen at 35 mL/min
Oven temperature	220°C

Detector temperature 300°C

Injection port On-column design, 220°C

9.6 Analyte determination

9.6.1 Immediately after calibration injections (9.7.6), inject a 2 µL aliquot of the washed hexane extract (upper layer) from 9.4.8 onto the GC (Notes on Procedure, 12.5).

9.6.2 Measure bis(chloromethyl)ether derivative peak height to the nearest mm. (Typical chromatograms are shown in Figure 3.)

9.6.3 Repeat injection until peak heights agree to within 2%.

9.6.4 Reinject standard (9.7.4) to check calibration.

9.7 Calibration

9.7.1 Pipette 10.0 mL of derivatizing reagent into a 7-dram vial.

9.7.2 Using a 10-µL syringe, accurately add 5.0 µL of 26.6 ng/µL bis(chloromethyl)ether standard. Cap and mix. Final standard contains 133 ng of bis(chloromethyl)ether. (Notes on Procedure, 12.6).

9.7.3 Loosen cap and place on steam bath for 5 min. Cool. Proceed as in 9.4.6 to 9.4.8, inclusive.

9.7.4 Inject a 2-µL aliquot of the hexane (upper) layer into the GC.

9.7.5 Measure the bis(chloromethyl)ether derivative peak height to the nearest mm.

9.7.6 Repeat 9.7.4 and 9.7.5, until peak heights agree to within 2%.

10. METHOD OF CALCULATION

10.1 The mass, m_e, of bis(chloromethyl)ether found in the sample hexane extract is given by

$$m_e = \frac{h_e m_s}{h_s} \quad \text{(ng)}$$

where

h_e = sample peak height obtained in step 9.6.3 (mm)
h_s = standard peak height obtained in step 9.7.6 (mm)
m_s = mass of bis(chloromethyl)ether in final standard (9.7.2) (ng)

10.2 The volume fraction, ϕ, of bis(chloromethyl)ether in the air sample (at 25°C and 1 atmosphere pressure) is given by

$$\phi = \frac{24.45\ m_e}{115\ V} \quad (\mu L/m^3)$$

where

V = volume of air sampled, corrected to 25°C and 1 atmosphere (L)
and m_e (ng) is defined in 10.1.

10.3 Values of ϕ should be expressed to two significant figures when $\geq$ 1.0 $\mu L/m^3$ and one significant figure when $<$ 1.0 $\mu L/m^3$.

11. REPEATABILITY AND RECOVERY

11.1 Repeatability for determination of 6 $\mu L/m^3$ bis(chloromethyl)ether in 5 liters of air, prepared in a gas bag, was ± 6% at the 95% confidence level.

11.2 Recovery for 0.1 to 50 $\mu L/m^3$ bis(chloromethyl)ether in 10 liters of air prepared in a gas bag was 90 ± 4% at the 95% confidence level.

11.3 With an air flow rate of 100 mL/min and 20% relative humidity at 21°C, breakthrough of bis(chloromethyl)ether from a 14 cm sampling tube occurred at 40-43 liters. Higher humidity may, and higher temperature will, decrease this volume. The effect of relatively low temperatures (below 5°C) should be investigated if such sampling conditions are necessary.

FIG. 3. CHROMATOGRAMS OF BIS(CHLOROMETHYL)ETHER DERIVATIVE

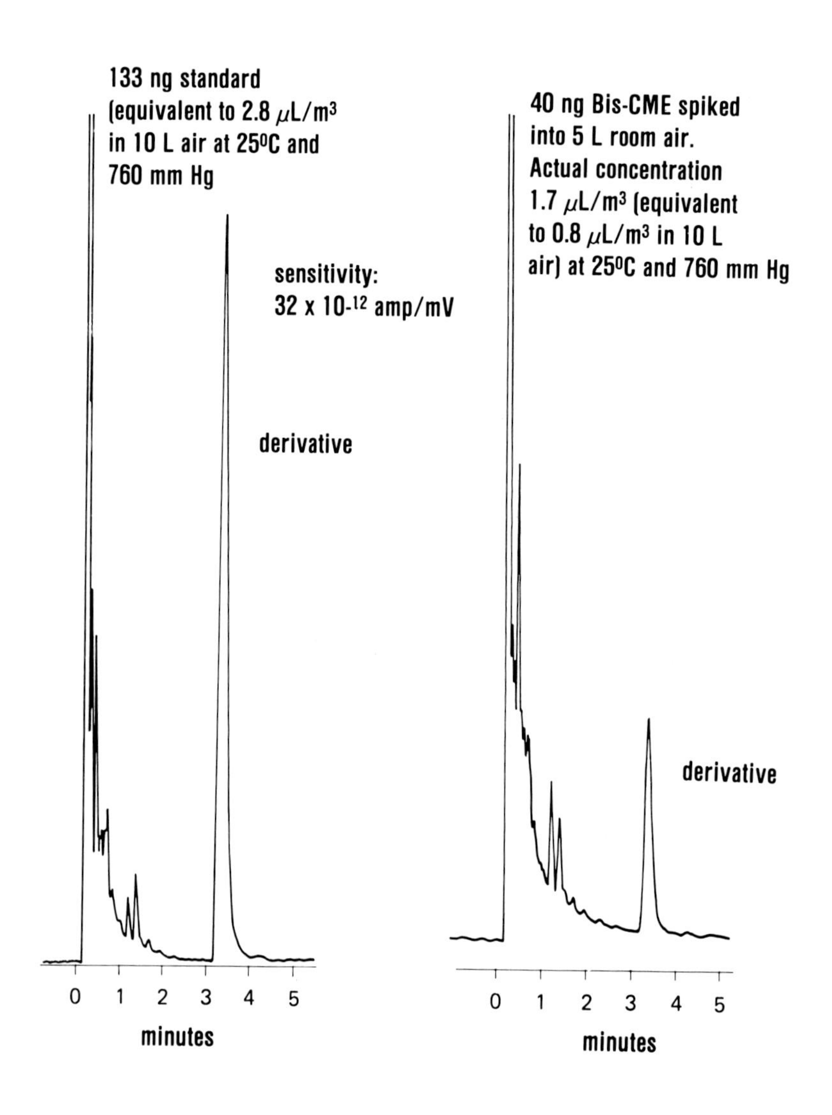

12. NOTES ON PROCEDURE

12.1 The chemistry of the derivatization is as follows:

$ClCH_2OCH_2Cl$ + Na–(2,4,6-trichlorophenyl) $\xrightarrow{CH_3OH}$

1) $CH_3OCH_2OCH_2OCH_3$

2) $CH_3OCH_2OCH_2$ O–(2,4,6-trichlorophenyl)

3) Cl–(2,4,6-trichlorophenyl)–OCH_2OCH_2O–(2,4,6-trichlorophenyl)–Cl

The third derivative is determined in this method because the greatest specificity is assumed. The ratio of the quantity of this derivative formed to the quantity of bis(chloromethyl)ether added is constant for constant conditions (in particular, the same batch of derivatizing reagent).

12.2 The optimum transformer output voltage is determined by trial to obtain fast heating, without excessive temperature overshoot.

12.3 Insufficient heating may result in incomplete reaction.

12.4 Column lifetime may be increased by replacing the first 2.5 cm of column packing with 3% OV-17, instead of 0.5%, and by cooling the column when not in use.

12.5 Avoid injecting any of the aqueous layer, which would damage the column.

12.6 The standard concentration is equivalent to 2.8 $\mu L/m^3$ bis(chloromethyl)-ether in 10 liters of air. For sampling higher concentrations, a standard of proportionately higher concentration should be prepared or smaller air volumes should be collected.

12.7 Clean-up tubes will require reconditioning at intervals governed by the contaminants in sampled air. Reconditioning after 3 analyses may suffice when relatively uncontaminated air is sampled.

13. SCHEMATIC REPRESENTATION OF PROCEDURE

Sampling tube

Thermal desorption at 110°C with 100 mL/min nitrogen flow into derivatizing solution

↓ → Recondition sampling tube (Sections 8.1 and 12.7)

Heat solution 5 min on steam bath and cool

↓

Extract with 10 mL 0.5 mol/L NaOH and 2 mL hexane

↓ → Discard aqueous layer

Transfer hexane (upper layer) to another vial

↓

Extract with 1 mL 0.5 mol/L NaOH

↓

Inject 2 μL of hexane layer into GC

↓

Measure peak height

↓

Calculate concentration using external standard calibration

14. ORIGIN OF THE METHOD

The Dow Chemical Company
Analytical Laboratories
574 Building
Midland, MI, 48667, USA

Contact point: J.W. Russell
The Dow Chemical Company
Analytical Laboratories
574 Building
Midland, MI, 48667, USA

METHOD 11

DETERMINATION OF DIBROMOCHLOROPROPANE AND ETHYLENE DIBROMIDE IN AIR

S. Fredrickson, T. Jackson, C.L. Liao & K.T. Maddy

1. SCOPE AND FIELD OF APPLICATION

This method is suitable for the analysis of 1,2-dibromo-3-chloropropane (DBCP) and 1,2-dibromoethane (ethylene dibromide, EDB) in air.

The range of the method is from 1.0 to 130 $\mu L^3/m^3$, with a precision of ± 10% (based on a 60 L air sample). Higher concentrations may be measured by taking smaller samples. The analysis is simple and rapid.

2. REFERENCES

NIOSH (1977) NIOSH Manual of Analytical Methods. Vol. 1, Organic Solvents in Air, P & CAM 127

NIOSH (1978) NIOSH Manual of Analytical Methods. U.S. Department of Health, Education and Welfare, publication No. 78-175

NIOSH (1977) NIOSH Manual of Sampling Data Sheets. U.S. Department of Health, Education and Welfare publication No. 77-159

U.S. Department of health and Human Services, Food and Drug Administration (1983) Pesticide Analytical Manual, Vol. 1, Methods which detect multiple residues

EPA (1979) Manual for Analytical Quality Control for Pesticides and Related Compounds in Human and Environmental Samples. U.S. Environmental Protection Agency Publication, EPA-600, 1-79-008

3. DEFINITIONS

Not applicable

4. PRINCIPLE

DBCP or EDB is adsorbed from the air onto activated charcoal-filled tubes. The analytes are desorbed from the charcoal with toluene and ethyl acetate, respectively, and the solutions are analysed using gas chromatography with electron-capture detection.

5. HAZARDS

Both DBCP and EDB are potent animal carcinogens and cause adverse effects on the reproductive system in certain test animals. Exposure to DBCP has been shown to cause lowered sperm counts in man. All operations involving handling of DBCP or EDB should be carried out in a fume hood and disposable surgical gloves should be worn. Safe and legal disposal of all solutions is mandatory.

6. REAGENTS[1]

Toluene	Pesticide grade (must be tested for interfering GC peaks)
Ethyl acetate	Pesticide grade (must be tested for interfering GC peaks)
EDB	Analytical grade standard (> 95% purity)
DBCP	Analytical grade standard (> 95% purity)
Activated charcoal tube	SKC 226-09 (11 cm × 8 mm o.d.) SKC Inc., R.D.1 No. 395 Valley View Road, Eightyfour, PA 15330, USA
Stock standard solution of EDB	1 g/L in ethyl acetate. Store in freezer
Stock standard solution of DBCP	1 g/L in toluene. Store in freezer

---

[1] Reference to a company and/or product is for the purpose of information and identification only and does not imply approval or recommendation of the company and/or product by the International Agency for Research on Cancer, to the exclusion of others which may also be suitable.

Working standard solution	Dilute aliquots of the stock standard solution with the appropriate solvent. Prepare concentration range from 1 μg/L to highest concentration on linear range of chromatographic system. Determine the linear range analytically by injecting identical volumes of the working standards. Determine the required working standard concentrations for routine use (from air sample volume, anticipated concentrations of compounds of interest and linear range of GC).

7. APPARATUS[1]

Usual chromatography-laboratory syringes and glassware, plus the following items:

Personal sampling pump	MSA Monitair Sampler, model S or TD, with adjustable flow rate, properly calibrated (see INTRODUCTION, p. 173)
Developing (desorption) vials	5-mL glass, screw-top, with foil-lined lids
Small triangular file	To score charcoal tubes
Rotator or agitator	
Gas chromatograph	Hewlett-Packard, or equivalent, equipped with ^{63}Ni electron-capture

8. SAMPLING

Techniques for correct air sampling can be found in the NIOSH Manual of Sampling Data Sheets (Section 2). Note that high humidity may affect trapping efficiency.

8.1 To monitor worker exposure by inhalation, connect a Tygon intake hose to a charcoal tube properly attached to the collar of the worker. Follow the sampling procedure in INTRODUCTION, p. 173.

---

[1] Reference to a company and/or product is for the purpose of information and identification only and does not imply approval or recommendation of the company and/or product by the International Agency for Research on Cancer, to the exclusion of others which may also be suitable.

8.2 Begin sampling at the desired flow rate (usually 1 L/min). Note atmospheric pressure and temperature.

NOTE: Samples should be kept cool during transport to the laboratory. If using wet ice, seal in a plastic or glass jar to prevent water contamination during transport. Samples should not be stored in an automobile trunk and should be kept away from gasoline and automobile exhaust to avoid contamination by EDB. Blank tubes should be opened, immediately capped, and transported in the same manner as the samples.

8.3 Store samples at -5°C, or colder, if they are not to be extracted on the same day. Samples should be extracted within a week.

9. PROCEDURE

9.1 <u>Blank tests</u>

See NOTE above, and 9.5.2

9.2 <u>Check test</u>

Not applicable

9.3 <u>Test portion</u>

Not applicable

9.4 <u>Sample desorption</u>

9.4.1 With a sharp file, score each charcoal tube in front of the first section of charcoal.

9.4.2 Break open the tube. Remove and discard the glass wool.

9.4.3 Transfer the first section of charcoal to a labelled desorption vial containing 4.0 mL pesticide-grade toluene (for DBCP) or ethyl acetate (for EDB).

9.4.4 Remove and discard the separating section of foam from the tube.

9.4.5 Transfer the second section of charcoal to a second labelled desorption vial containing 4.0 mL of the appropriate solvent.

9.4.6 Place the desorption vials on the agitating apparatus for one hour. Retain for GC analysis.

9.4.7 If the sample extract is not to be analysed on the day of preparation, transfer an aliquot to another developing vial, label, and freeze (-5°C or colder) until needed. The chromatography should be completed within a week, unless samples are frozen in sealed ampules.

9.5 Determination of desorption efficiency

9.5.1 Using the standard solutions (section 6) and a syringe, inject known amounts of DBCP or EDB (preferably covering the range of expected results) into at least 5 freshly-opened charcoal tubes and cap the tubes. (These tubes must have the same lot number as those used for the sampling.)

9.5.2 Allow the 5 tubes, plus another tube[1] for a blank determination, to stand overnight (or longer) to assure complete adsorption. Retain for analysis (see Section 9.8).

9.6 Gas chromatography (GC) determination of DBCP

9.6.1 GC conditions:

Column:	Nickel or S.S. tubing, 6.1 m x 2.0 mm i.d., packed with 10% SP-2100 on 100/120 mesh Chromosorb W-HP
Carrier gas:	Ar:5% methane; flow-rate, 30 mL/min
Column temperature:	90°C
Injector temperature:	200°C
Detector temperature:	350°C

9.6.2 GC procedure

9.6.2.1 Flush a 10 µL syringe several times with the sample extract (9.4.6), then inject 5 µL (or other appropriate volume[2]) of the sample extract into the GC[3].

---

[1] Treated in precisely the same way as the other 5 tubes, except that no DBCP (or EDB) is injected.

[2] The minimum detectable amount on the GC should be 0.01 ng.

[3] The "solvent flush technique" may also be employed and is probably somewhat more accurate in experienced hands. See Method 1, section 12, for details.

9.6.2.2 Measure the peak area (DBCP elutes in 7 min under the above conditions).

9.6.3 Calibration curve

Inject known amounts of DBCP standard solutions (at least three) to give peak areas above, below and close to that of the sample.

A calibration curve may be constructed by plotting peak area against the amount of standard injected (check linearity).

NOTE: Duplicate injections of each sample and standard should be made to check repeatability. Samples and standards should be analysed on the same day, within as short a time-span as possible.

9.7 Gas chromatography determination of EDB

9.7.1 GC conditions:

Column:	Nickel or S.S. tubing, 6.1 m x 2.0 mm i.d., packed with 10% FFAP on 100/120 mesh Supelcoport.
Carrier gas:	Ar:5% methane flowrate, 30 mL/min
Column temperature:	130°C
Injector temperature:	200°C
Detector temperature:	350°C

9.7.2 GC procedure
Proceed as in 9.6.2. EDB elutes in 5 min under the above conditions.

9.8 Extraction (desorption) efficiency

9.8.1 Extract the 5 spiked tubes and the blank tube from 9.5.2 as described in steps 9.4.1 to 9.4.6, inclusive.

9.8.2 Carry out GC determination as in 9.6 (or 9.7), omitting step 9.6.3.

9.8.3 Prepare 5 standard solutions by injecting into labelled desorption vials, the same amounts of DBCP (or EDB) standard solutions that were injected into the tubes in 9.5.1. Bring the volume in each vial up to 40 mL with the appropriate solvent. (Alternatively, begin by adding 4.0 mL of solvent to each vial, then remove a volume equal to the volume of standard solution to be added. Use the same syringe for the two operations.)

9.8.4 Inject standard solutions 9.8.3 into the GC, using the same injection volume as in 9.8.2.

For each tube, the extraction efficiency, E, (≤ 1) is calculated from the peak areas as follows,

$$E = \frac{\text{area spiked tube - area blank}}{\text{area standard}}$$

where the appropriate spiked tube and standard are those containing the same amount of DBCP or EDB.

10. METHOD OF CALCULATION

10.1 Determine the amount of analyte injected into the GC, by comparison of the peak area of the sample extract with those of standard solutions (calibration curve, 9.6.3). Sum the values found for the two sections of the tube and subtract the blank volume.

10.2 The mass, m_1, of analyte extracted from the sample tube is given by,

$$m_1 = \frac{wV}{v} \ (\mu g)$$

and the mass collected by the sample tube is

$$m_2 = m_1/\langle E\rangle \ (\mu g)$$

where

w = amount of analyte injected into GC (ng)
v = volume of analyte injected into GC (µL)
V = volume of extraction solvent in 9.4.3 (mL)
<E>= average extraction efficiency from 9.8.

10.3 The volume of air sampled, V_1, is converted to standard (industrial hygiene) conditions of 25°C and 760 mm Hg by

$$V_2 = \frac{PV_1}{760}\left(\frac{298}{T}\right)$$

where

V_2 = volume of air at STP (L)
V_1 = volume of air measured under sampling conditions (L)
P = barometric pressure measured under sampling conditions (mm Hg)
T = temperature of sampled air (°K)

10.4 The mass concentration, ρ, of analyte in the sampled air is given by,

$$\rho = \frac{m_2}{V_2} \quad (mg/m^3)$$

where

m_2 (μg) and V_2 (L) are defined above.

10.5 The volume fraction, ϕ, of analyte in the sampled air is given by,

$$\phi = \frac{24.45\ \rho}{M} \quad (mL/m^3)$$

where M = molecular weight of analyte and ρ is defined in 10.4.
(M of DBCP = 236, M of EDB = 188)

11. REPEATABILITY AND REPRODUCIBILITY

Results for 10 samples ranging from 0.05 to 100 micrograms/sample indicated 90% ± 5% recovery.

12. NOTES ON PROCEDURE

Not applicable

13. SCHEMATIC REPRESENTATION OF PROCEDURE

Pump air sample (1 L/min) through charcoal tube

↓

Transfer charcoal from first section of tube into 5.0 mL vial containing 4.0 mL toluene (for DBCP) or 4.0 mL ethyl acetate (for EDB)

↓

Repeat preceeding instruction with charcoal from second section

↓

Agitate vials for 1 hour (store aliquot of extract at -5°C)

> Prepare 5 spiked tubes covering expected range of analyte concentration, plus 1 blank tube, and allow to stand overnight
>
> Extract analyte from spiked tubes + blank as described above for sample tubes
>
> Prepare 5 standard solutions containing same amounts of analytes as the 5 spiked tubes
>
> Inject equal amounts of spiked-tube extracts and standard solutions onto appropriate GC column, calculate extraction efficiencies from peak areas

↓

Inject extract aliquot onto appropriate GC column and determine peak areas for both tube sections

↓

Repeat preceeding instruction with standard solutions to obtain calibration curve

↓

Calculate analyte concentration in total sample extract and in air sample, correcting for blank and extraction efficiency

14. ORIGIN OF METHOD

California Department of Food and Agriculture[1]
Worker Health and Safety Unit, and
Chemistry Laboratory Services
3292 Meadowview Road
Sacramento, California 95832, USA

Contact point: G. Tichelaar
Chief of Chemistry Laboratory Services

---

[1] Reference to a company or product is for the purpose of information and identification only and does not imply approval or recommendation of the company or product by the California Department of Food and Agriculture, to the exclusion of others which may also be suitable.

METHOD 12

DETERMINATION OF VOLATILE ORGANIC COMPOUNDS IN AMBIENT AIR USING TENAX ADSORPTION AND GAS CHROMATOGRAPHY/MASS SPECTROMETRY

R.M. Riggin

1. SCOPE AND FIELD OF APPLICATION

This method is suitable for the collection and determination of volatile organic compounds which can be captured on Tenax GC (poly(2,6-diphenylphenylene oxide)) and determined by gas chromatography/mass spectrometry (GC/MS), following thermal desorption. Organic compounds which are non-polar and have boiling points in the range of approximately 80°C-200°C can be determined by this method. However, not all compounds falling into this category can be determined. Table 1 gives a list of compounds for which the method has been used. The limits of detection are generally of the order of 0.1-1.0 ng/L. approximately one hour is required for sample analysis, once the instrument is calibrated.

2. REFERENCES

Annual Book of ASTM Standards, Part 11.03, Atmospheric Analysis, American Society for Testing and Material, Philadelphia, Pennsylvania, 1984

Grob, K., Jr, Grob, G. & Grob, K. (1978) Comprehensive standardized quality test for glass capillary columns. J. Chromatogr., 156, 1-20

Kebbekus, B.B. & Bozzelli, J.W. (1982) Collection and Analysis of Selected Volatile Organic Compounds in Ambient Air. In: Proceedings Air Pollut. Control Assoc., Paper No. 82-65.2. Air Pollut. Control Assoc., Pittsburgh, Pennsylvania

Krost, K.J., Pellizzari, E.D., Walburn, S.G. & Hubbard, S.A. (1982) Collection and analysis of hazardous organic emissions. Anal. Chem., 54, 810-817

Pellizzari, E.O. & Bunch, J.E. (1979) Ambient Air Carcinogenic Vapors-Improved Sampling and Analytical Techniques and Field Studies, EPA-600/2-79-081, U.S. Environmental Protection Agency, Research Triangle Park, North Carolina

Riggin, R.M. (1983) Technical Assistance Document for sampling and Analysis of Toxic Organic Compounds in Ambient Air, EPA-600/4-83-027, U.S. Environmental Protection Agency, Research Triangle Park, North Carolina

Walling, J.F., Berkley, R.E., Swanson, D.H. & Toth, F.J. (1982) Sampling Air for Gaseous Organic Chemical-Applications to Tenax, EPA-600/7-54-82-059, U.S. Environmental Protection Agency, Research Triangle Park, North Carolina

Table 1. Breakthrough volume estimates for compounds on Tenax

Compound	Estimated breakthrough volume at 38°C (L/g)
Benzene	19
Toluene	97
Ethyl benzene	200
Xylene(s)	~200
Cumene	440
n-Heptane	20
1-Heptene	40
Chloroform	8
Carbon tetrachloride	8
1,2-Dichloroethane	10
1,1,1-Trichloroethane	6
Tetrachloroethylene	80
Trichloroethylene	20
1,2-Dichloropropane	30
1,3-Dichloropropane	90
Chlorobenzene	150
Bromoform	100
Ethylene dibromide	60
Bromobenzene	300

3. DEFINITIONS

Breakthrough volume = the sample volume at which the amount of analyte of interest found on a back-up cartridge reaches 20% of the amount found on the front cartridge (see 9.2).

4. PRINCIPLE

Ambient air is drawn through a cartridge containing 1-2 grams of Tenax to trap volatile organic compounds. The cartridge is placed in a heated chamber and purged with an inert gas which transfers the volatile compounds from the cartridge onto a cold trap and subsequently onto a high-resolution (capillary) GC column, which is held at low temperature (e.g., -70°C). The column temperature is then increased (temperature programmed) and the components eluting from the column are identified and quantified by mass spectrometry. Component identification is normally accomplished by a library search routine, using GC retention times and mass spectral characteristics.

5. HAZARDS

Many halogenated organic compounds are proven or suspect carcinogens. Extreme safety precautions must be exercized in the preparation, handling, storage and disposal of liquid or gaseous standards. Such chemicals should be handled in a verified, high-efficiency fume hood. If the user elects to prepare compressed gas cylinder standards in the laboratory, the filling operation must be conducted by trained personnel according to strict safety standards.

6. REAGENTS[1]

Acetone	Pesticide quality or equivalent
Methanol	Pesticide quality, or equivalent
Pentane	pesticide quality or equivalent
Helium	Ultra pure, compressed gas (99.9999%)
Nitrogen	Ultra pure, compressed gas (99.9999%)
Liquid nitrogen	
Perfluorotributylamine (FC-43)	
Analytes of interest	Highest purity available, for use as standards (see 9.5.2 and Table 1)

[1] Reference to a company and/or product is for the purpose of information and identification only and does not imply approval or recommendation of the company and/or product by the International Agency for Research on Cancer, to the exclusion of others which may also be suitable.

Granular activated charcoal	For preventing contamination of Tenax cartridges during storage
Tenax GC	60/80 mesh (2,6-diphenylphenylene oxide polymer). Purify before use as described in section 12.2.
Glass wool	Silanized

7. APPARATUS[1]

Gas chromatograph/mass spectrometer	System should be capable of subambient temperature programming. Unit mass resolution or better up to 800 amu. Capable of scanning 30-440 amu region every 0.5-1 second. Equipped with data system for instrument control as well as data acquisition, processing and storage (see 12.3).
Thermal desorption unit	Designed to accommodate Tenax cartridges in use. See Figures 1a or b
Sampling system	Capable of accurately drawing an air flow of 10-500 mL/min through the Tenax cartridge (see Figures 2a and b).
Vacuum oven	Connected to water aspirator vacuum supply
Stopwatch	
Pyrex disks	For drying Tenax
Glass jar	Capped with Teflon-lined screw cap. for storage of purified Tenax
Powder funnel	For delivery of Tenax into cartridges
Culture tubes	To hold individual glass Tenax cartridges

[1] Reference to a company and/or product is for the purpose of information and identification only and does not imply approval or recommendation of the company and/or product by the International Agency for Research on Cancer, to the exclusion of others which may also be suitable.

FIG. 1. TENAX CARTRIDGE DESORPTION MODULES

(a) Glass cartridges (compression fit)

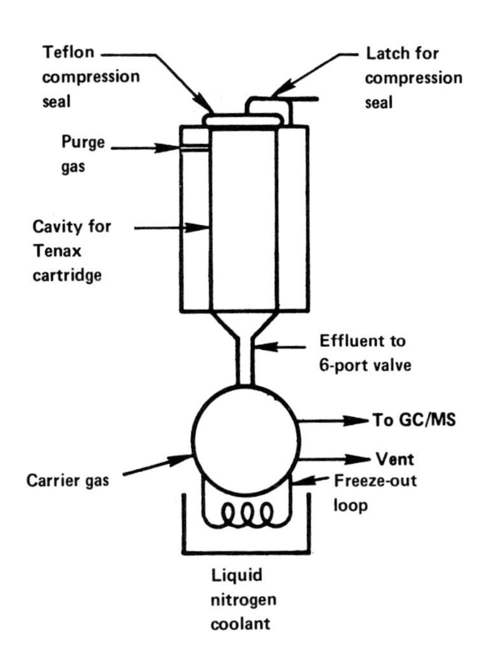

(b) Metal cartridges (Swagelok fittings)

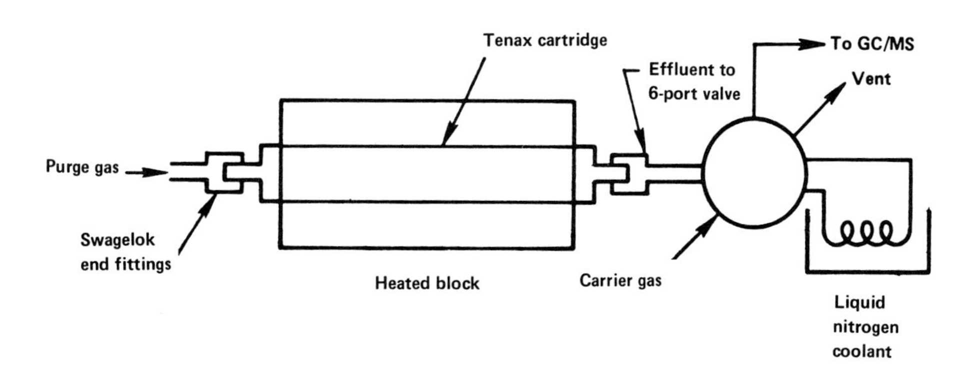

FIG. 2. TYPICAL SAMPLING SYSTEM CONFIGURATIONS

(a) Mass flow control

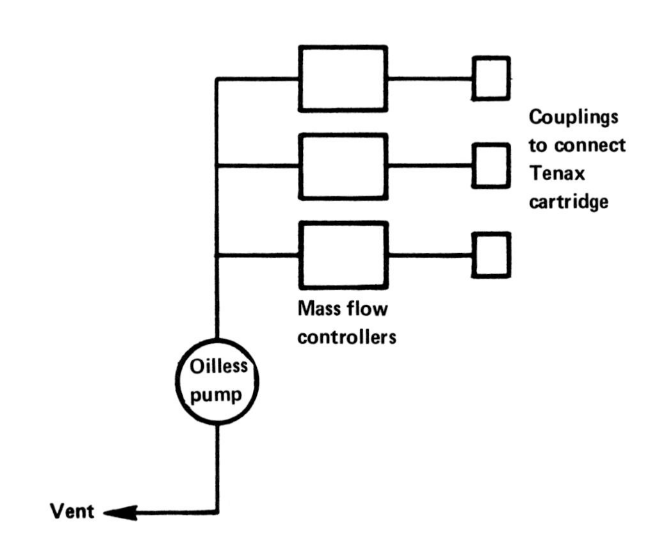

(b) Needle-valve control

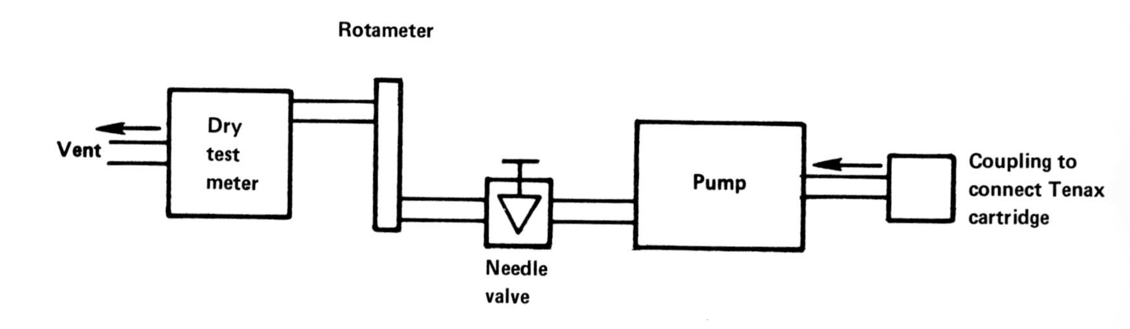

Friction-top can (paint can)	To hold clean Tenax cartridges
Filter holder	Stainless steel or aluminum (to accommodate 1 inch diameter filter). Other sizes may be used if desired. (Optional)

Dilution bottle	2-L, with septum cap for preparation of standards
Gas-tight glass syringes	10-500 μL, with stainless steel needles
Microliter syringes	5-50 μL for injecting liquid standards into dilution bottle
Oven (60 ± 5°C)	For equilibrating dilution flasks
Magnetic stirrer	With 1-inch, Teflon-covered stirring bar
Heating mantle	
Variable-voltage transformer (Variac)	
Soxhlet extraction apparatus	With glass thimbles for purifying Tenax
Infrared lamp	For drying Tenax
GC column	SE-30 or alternative coating, glass capillary or fused silica.
Empty Tenax cartridges	Glass or stainless steel (see Fig. 3a and b and Notes on Procedure, 12.1).

FIG. 3. TENAX CARTRIDGE DESIGNS

(a) Glass cartridge

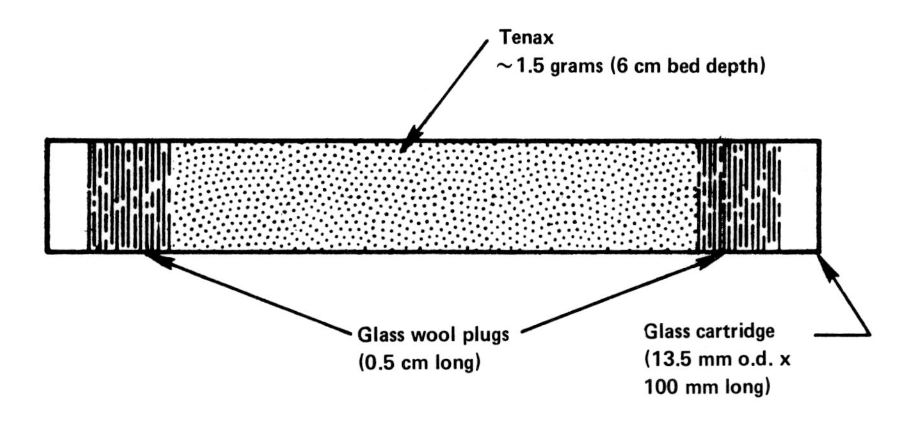

FIG. 3. TENAX CARTRIDGE DESIGNS

(b) Metal cartridge

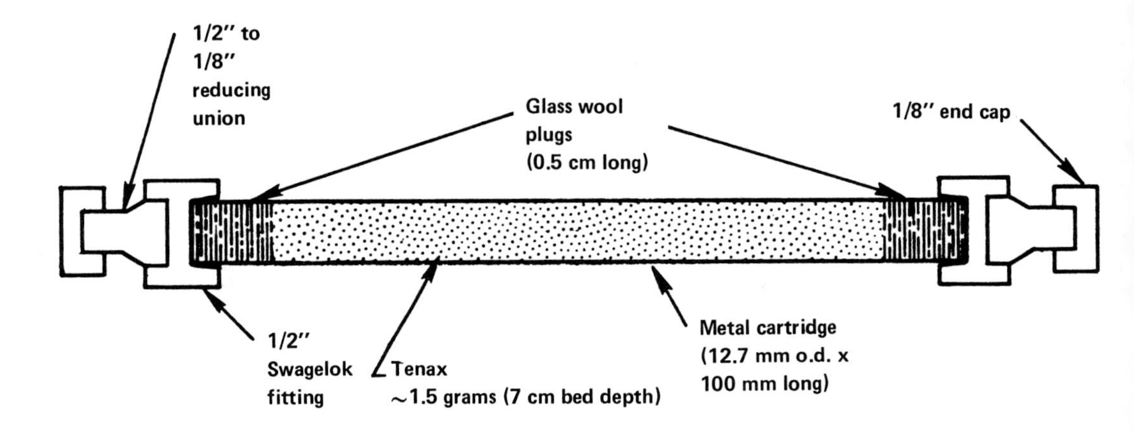

Polyester gloves	For handling glass Tenax cartridges
Glass fiber filter	One inch diameter, to fit in filter holder. (Optional)

8. SAMPLING

8.1 Cartridge preparation and pretreatment

NOTE: If the glass cartridge is employed, it should be handled with polyester gloves.

8.1.1 Place a 0.5-1 cm glass-wool plug in the base of the cartridge, then fill with Tenax (see 12.2) to within approximately 1 cm of the top. Place 0.5-1 cm glass-wool plug on top of the Tenax.

8.1.2 Thermally condition the cartridges by heating for four hours at 270°C under an inert gas (helium) purge (100-200 mL/min).

8.1.3 Immediately place glass cartridges (without cooling) in clean culture tubes with Teflon-lined screw caps and a glass-wool cushion at the top and the bottom. Shake each tube to ensure that the cartridge is held firmly in place. (Metal cartridges are allowed to cool to room temperature under inert gas purge and are then closed with stainless-steel plugs.)

8.1.4 Label the cartridges and place in a tightly-sealed metal can (e.g., paint can or similar friction-top container). Cartridges should be used for sampling within 2 weeks after preparation and analysed within two weeks after sampling. If possible, the cartridges should be stored at -20°C in a clean freezer (i.e., no solvent extracts or other sources of volatile organic compounds).

8.2 Flow rate and total volume selection

8.2.1 Each compound has a characteristic breakthrough volume which must not be exceeded (see Walling *et al.*, 1982). Approximate breakthrough volumes at 38°C are provided in Table 1. These volumes are given only as rough guidance and are subject to considerable variability, depending on cartridge design as well as sampling parameters and atmospheric conditions.

8.2.2 To calculate the maximum total volume of air which can be sampled, use the equation,

$$V_{max} = V_b m_T / 1.5$$

where,

V_{max} = maximum total volume of air sample (L)

V_b = breakthrough volume (Table 1) for the least-retained compound of interest (L/g Tenax)

m_T = mass of Tenax in the cartridge (g)

1.5 = safety factor appropriate for temperatures in the range 25-30°C. (If higher temperatures are encountered, the factor should be increased.)

8.2.3 To calculate maximum flow rate use the equation,

$$F_{max} = V_{max} / t$$

where

F_{max} = maximum volumetric flow rate (L/min)

t = desired sampling time (min)

8.2.4 The maximum flow rate should yield a linear flow velocity of 50-500 cm/min. Calculate the linear velocity corresponding to the maximum flow rate using the equation,

$$v = 10^3 F_{max} / r^2$$

where

v = linear flow velocity (cm/min)

r = internal radius of the cartridge (cm)

If v is greater than 500 cm/min, either V_{max} or F_{max} should be reduced. If v is less than 50 cm/min, F_{max} should be increased.

8.2.5 The flow rate calculated as described above is the maximum flow rate allowed. In general, one should collect additional samples in parallel for the same time period, but at lower flow rates. this practice yields a measure of quality control (see Walling et al., 1982). In general, flow rates 2- to 4-fold lower than the maximum flow rate should be employed for the parallel samples. In all cases, a constant flow rate should be achieved for each cartridge, since accurate integration of the analyte concentration requires that the flow be constant over the sampling period.

8.3 Sample collection

NOTE: For the collection of an accurately known volume of air, the use of mass flow controllers, rather than conventional needle valves or orifices, is highly recommended, especially at low flow velocities (e.g., < 0.1 L/min). Figure 2a shows a sampling system using mass flow controllers, which readily allows for collection of parallel samples. Figure 2b shows a commercially-available system using needle-valves.

8.3.1 Prior to sample collection, insure that the sampling system has been calibrated over a range including the flow rate to be used for sampling, with a representative Tenax cartridge in place. Generally calibration is accomplished using a soap-bubble flow meter or calibrated wet test meter. The flow calibration device is connected to the flow exit, assuming the entire flow system is sealed. ASTM Method D3686 describes an appropriate calibration scheme, not requiring a sealed flow system downstream of the pump.

8.3.2 To collect an air sample, remove the cartridges from the sealed container just prior to initiation of the collection process. If glass cartridges are employed, they must be handled only with polyester gloves and should not contact any other surfaces.

8.3.3 Place a particulate filter and holder on the inlet to the cartridges and connect the exit end to the sampling apparatus. (The use of a filter is not necessary if only the total concentration of a component is desired.) Glass cartridges are connected using Teflon ferrules and Swagelok (stainless steel or Teflon) fittings.

8.3.4 Start the pump and record the following parameters: date, sampling location, time, ambient temperature, barometric pressure, relative humidity, dry gas meter reading (if applicable) flow rate, rotameter reading (if applicable), cartridge number and dry gas meter serial number.

8.3.5 Check the flow rate before and after each sample collection. If the sampling interval exceeds four hours, check the flow rate at an intermediate point during sampling as well.

8.3.6 At the end of the sampling period, record the parameters listed in 8.3.4. If the flow-rates at the beginning and end of the sampling period differ by more than 10%, the cartridge should be marked as of doubtful value.

8.3.7 Remove the cartridges (one at a time) and place in the original container. Seal the cartridges or culture tubes in the friction-top can (containing a layer of charcoal) and ship immediately to the laboratory for analysis. Store cartridges at reduced temperature (e.g., -20°C) before analysis.

8.3.8 Calculate and record the average sampling rate for each cartridge.

8.3.9 Calculate and record the total volume sampled for each cartridge at the average temperature and pressure recorded.

Calculate the total volume at standard conditions, V_s, (25°C and 760 mm Hg) from the equation,

$$V_s = \frac{298\ V_m P}{760T}$$

where,

V_m = total volume sampled at average pressure and temperature (L)

P = average barometric pressure during sampling (mm Hg)

T = average ambient temperature during sampling (°K)

9. PROCEDURE

9.1 Blank tests

9.1.1 Check each batch of Tenax cartridges (prepared as described in 8.1) for contamination by analysing one cartridge immediately after preparation. The blank cartridge should be demonstrated to contain less than one fourth of the minimum level of interest for each analyte. For most compounds, the blank level should be less than 10 ng per cartridge. If a cartridge does not meet this criterium the entire lot should be rejected.

9.1.2 For each group of samples to be taken at a given site, send at least one clean cartridge to the field and back to the laboratory (without being used) to serve as a field blank. If the blank level is greater than 25% of the sample level, data for the analyte concerned must be considered to be of doubtful value.

9.2 Check test

In addition to the field blank(s), attach a second cartridge in series with one or more of the sampling cartridges. These back-up cartridges should trap less than 20% of the amount of analytes of interest found in the front cartridges, or the same amounts as the field blank, whichever is the greater. The frequency of use of back-up cartridges should be increased if increased flow rate is shown to yield reduced component levels for parallel sampling (8.3). This practice will help to detect the occurrence of breakthrough of the compound of interest during sampling.

9.3 Test portion

Not applicable

9.4 GC/MS operating conditions

9.4.1 Prior to instrument calibration or sample analysis, assemble the GC/MS system as shown in Figure 4. Set helium purge flows (through the cartridge) and carrier flow at approximately 10 mL/min and 1-2 mL/min, respectively. If applicable, set the injector sweep flow at 2-4 mL/min.

9.4.2 Once the column and other system components are assembled and the various flows established, increase the column temperature to 250°C for approximately four hours (or overnight if desired) to condition the column.

9.4.3 Set MS and data system according to the manufacturer's instructions. Use electron-impact ionization (70 eV) and an electron multiplier gain of approximately 5×10^4. Once the entire GC/MS system has been set up, calibrate as described in 9.5. Prepare a detailed standard operating procedure (SOP) describing this process for the particular instrument being used.

9.5 Instrument calibration

9.5.1 Carry out tuning and mass standardization of the MS system according to manufacturer's instructions and relevant information from the user-prepared SOP. Employ perfluorotributylamine for this purpose and introduce it directly into the ion source through a molecular leak. Adjust the instrumental parameters to give the relative ion abundances shown in Table 2, as well as acceptable resolution and peak shape. In the event that the instrument employed cannot achieve these relative ion abundances, but is otherwise operating properly, adopt another set of relative abundances as performance criteria. However, these alternative values must be repeatable on a day-to-day basis.

FIG. 4. BLOCK DIAGRAM OF ANALYTICAL SYSTEM

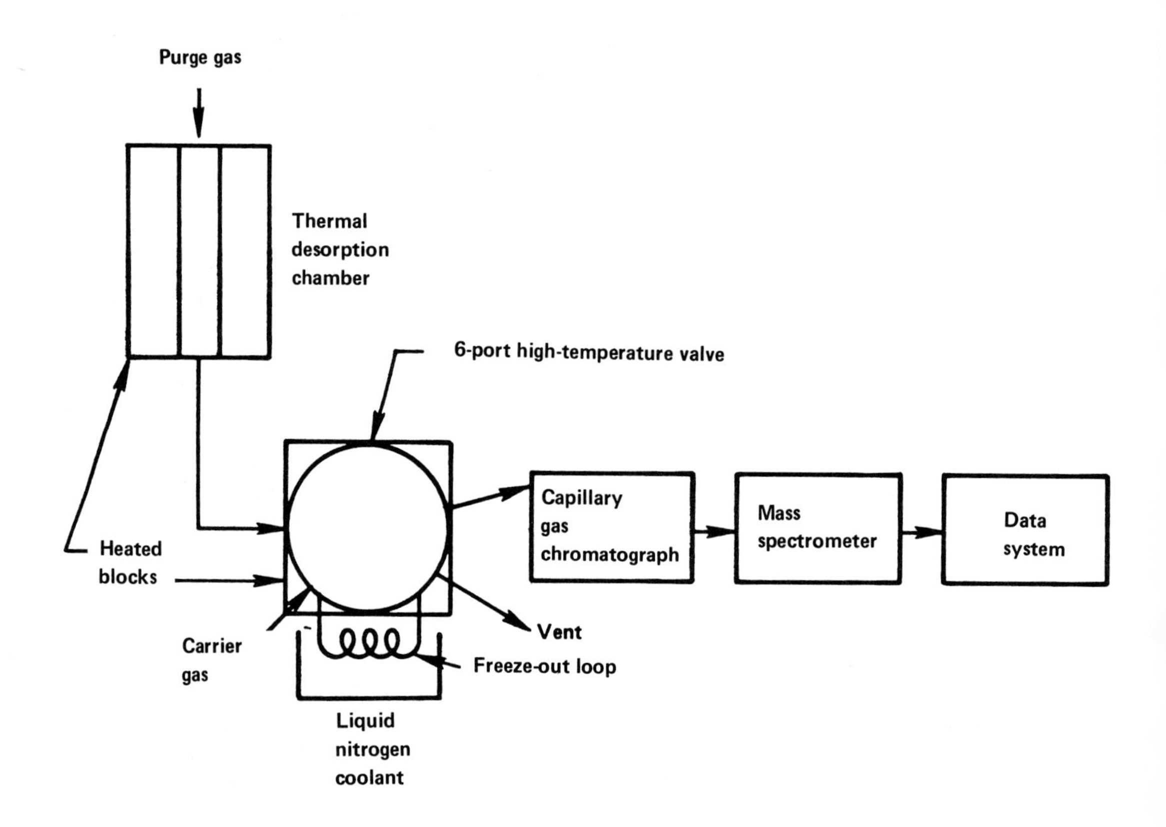

Table 2. Suggested performance criteria for relative ion abundances from FC-43 mass calibration

Ion (m/z)	Relative abundance
51	1.8 ± 0.5
69	100
100	12.0 ± 1.5
119	12.0 ± 1.5
131	35.0 ± 3.5
169	3.0 ± 0.4
219	24.0 ± 2.5
264	3.7 ± 0.4
314	0.25 ± 0.1

9.5.2 After the mass standardization and tuning process has been completed and the appropriate values entered into the data system, calibrate the entire system by introducing known quantities of the analytes of interest into the system. One quantity near the mid-point of the range of interest should be chosen for injection every day, to determine response reproducibilty. One of three procedures may be employed for the calibration process: (1) direct syringe injection of dilute vapour-phase standards, prepared in a dilution bottle, onto the GC column, (2) injection of dilute vapour-phase standards into a carrier gas stream directed through the Tenax cartridge, and (3) introduction of permeation or diffusion tube standards onto a Tenax cartridge. These procedures are described in sections 12.4-12.6. The following paragraphs describe the instrument calibration process for each procedure.

9.5.3 Calibration by direct injection of gaseous standards.

Prepare a standard in a dilution bottle as described in 12.4. Cool the GC column to -70°C (or manually cool a portion of the column inlet with liquid nitrogen). The electron-emission filament should be cold during the initial 2-3 min of the run, while oxygen and other highly-volatile components elute. Inject an appropriate volume (< 1 mL) of the gaseous standard onto the GC using an accurately-calibrated gas-tight syringe. Maintain the column at -70°C for 2 min, then rapidly increase the column temperature to the desired initial temperature (e.g., 30°C). Start the temperature program at a fixed time (e.g., four min) after injection, switch on electron-emission filament and initiate data acquisition. Repeat 9.5.3 for each analyte of interest, using different injection volumes, as desired. (Ideally, each analyte of interest should be determined at three or more calibration levels daily.)

9.5.4 Calibration by analysis of spiked Tenax cartridges

Prepare a set of cartridges as described in 12.5 or 12.6. Prior to analysis, store the cartridges as described in section 8.1. Cool the GC column to -70°C, immerse the collection loop in liquid nitrogen and maintain the desorption module at 250°C. With the inlet valve in the desorb mode, place the standard cartridge in the desorption module, making certain that no leakage of purge gas occurs. Purge the cartridge for 10 min, place the inlet valve in the inject mode and remove the liquid nitrogen from the collection trap. Maintain the column at -70°C for 2 min, then continue as described in 9.5.3. After the process is complete, remove the cartridge from the desorption module and store for subsequent use as described in 8.1.

NOTE: Data processing for instrument calibration involves determining retention times and integrated characteristic-ion intensities for each of the compounds of interest. In addition, for at least one chromatographic run, the individual mass spectra should be inspected and compared to reference spectra to ensure proper instrumental performance.

9.6. Sample analysis

Analyse samples and blanks by following the procedure described in 9.5.4 for the analysis of spiked Tenax cartridges.

NOTE: Sample data processing generally involves (1) qualitatively determining the presence or absence of each compound of interest (presence or absence of characteristic ion at correct retention time) using a reverse-search software routine, (2) quantifying each compound by integrating the intensity of a characteristic ion and comparing the peak area to that of the calibration standard, and (3) tentative identification of other compounds observed using a forward (library) search software routine.

10. METHOD OF CALCULATION

10.1 Calibration response factors

10.1.1 Use data from calibration standards to calculate a response factor for each compound of interest. Determine each response factor (area/ng injected) from the linear least-squares fit of a plot of nanograms injected versus area for the characteristic ion. If data for at least three calibration levels is not available for a given day, pool calibration data from consecutive days to yield a response factor, provided that analysis of replicate standards of the same concentration are shown to agree within 20% on the consecutive days.

10.1.2 If substantial non-linearity is present in the calibration curve, fit the data to the equation,

$$A = \alpha + \beta m + \gamma m^2$$

where

A = GC peak area for characteristic ion

m = mass of compound (ng)

α, β and γ are numerical coefficients

10.2 Analyte concentrations

10.2.1 Calculate analyte quantities on a sample cartridge from the equation,

$$A_a = \alpha + \beta m_a + \gamma m_a^2$$

where

A_a = GC peak area of the analyte characteristic-ion for the sample cartridge

m_a = mass of analyte on the sample cartridge (ng)

α, β and γ = constants calculated from the calibration curve (10.1.2)

NOTE: If instrument response is essentially linear over the concentration range of interest, a linear equation ($\gamma = 0$) can be employed.

10.2.2 Calculate the mass concentration of analyte, ρ_a, in the original air sample from the equation,

$$\rho_a = m_a/V_s \text{ (ng/L)}$$

where V_s and m_a have been defined in 8.3.9 and 10.2.1, respectively.

NOTE: If the response factor (10.1) is obtained using spiked-cartridge calibration standards (9.5.4), ρ_a includes an automatic correction for recovery of analyte from the cartridge.

11. REPEATABILITY AND REPRODUCIBILITY

The performance of this method has not been rigorously tested. However, repeatability is estimated to be ± 25% (relative standard deviation). Recovery is a function of the component retention (breakthrough) volume, sampling volume, ambient condition, and related factors. Under ideal conditions, overall recoveries (trapping + desorption) of > 75% can be achieved for many compounds of interest.

12. NOTES ON PROCEDURE

12.1 The glass cartridge minimizes contact of the sample with metal surfaces, which can lead to decomposition in certain cases. However, a disadvantage is the need to avoid contamination of the outside portion of the cartridge, since it is exposed to the gas stream during the desorption process. The metal cartridge has the advantage that only the interior is purged.

12.2 Prior to use, subject the Tenax resin to the following treatment (all glassware used in Tenax purification and all cartridge materials should be thoroughly cleaned by water rinsing, followed by an acetone rinse, and dried in an oven at 250°C). Place bulk Tenax in a glass extraction thimble and hold in place with a plug of clean glass wool. Place the thimble in the soxhlet apparatus and extract for 16-24 h with methanol, then for 16-24 h with pentane, at approximately 6 cycles/h. (Glass wool for cartridge preparation should be cleaned in the same manner.) Immediately place the extracted Tenax in an open glass dish and heat under an infrared lamp for two hours in a fume hood (do not over-heat), then place in a vacuum oven (evacuated using a water aspirator) without heating for one hour. (Use an inert gas purge of 2-3 mL/min for the removal of solvent vapours.) Then increase the oven temperature to 110°C, maintaining inert gas flow, and hold for one hour. Allow the oven to cool to room temperature and, before opening, pressurize slightly with nitrogen to prevent contamination with ambient air. Sieve the Tenax through a 40/60 mesh sieve (acetone-rinsed and oven-dried) into a clean glass vessel. If the Tenax is not to be used immediately for cartridge preparation it should be stored in a clean glass jar with a Teflon-lined screw cap and placed in a desiccator.

12.3 A block diagram of the typical GC/MS system required for analysis of Tenax cartridges is depicted in Figure 4. Exposure of the sample to metal surfaces should be minimized and only stainless steel or nickel should be employed. The volume of tubing and fittings leading from the cartridge to the GC column must be minimized and all areas must be well-swept by helium carrier gas. The GC column inlet should be capable of being cooled to -70°C and subsequently heated rapidly to approximately 30°C. The GC column and temperature program employed will depend on the compounds of interest. Appropriate conditions are described in the literature (see Krost et al., 1982; Pellizzari & Bunch, 1979; Kebbekus & Bozzelli, 1982). In general, a non-polar stationary phase (e.g., SE-30, OV-1) programmed from 30°C to 200°C at 8°/min will be suitable. Fused-silica bonded-phase capillary columns are preferable to glass columns, since they can be inserted directly into the MS ion source. Column dimensions of 50 m × 0.3 mm i.d. are generally appropriate.

12.4 Direct injection of gaseous calibration standards

12.4.1 Place fifteen 3-mm diameter glass beads and a one-inch Teflon stirbar in a clean 2-L glass, septum-capped bottle and determine the exact volume by weighing the bottle before and after filling with deionized water. Rinse with acetone and dry at 200°C.

12.4.2 Calculate the mass of each standard to be injected into the bottle using the equation,

$$m_b = m_i V_b / V_i$$

where

m_b = mass of analyte to be injected into the bottle (mg)

m_i = desired mass of analyte to be injected onto GC or spiked cartridge (ng)

V_i = GC or cartridge injection volume (μL) (should not exceed 500 μL)

V_b = volume of dilution bottle determined in 12.4.1 (L)

12.4.3 Determine the volume of the liquid standard to be injected into the dilution bottle using the equation,

$$V_l = m_b / d$$

where

V_l = volume of liquid standard to be injected into bottle (μL)

d = mass density of the liquid standard (g/mL) and

m_b is defined in 12.4.2

12.4.4 Deliver V_l μL of the liquid standard through the septum using a microliter syringe and place the bottle in a 60°C oven for at least 30 min prior to removal of a vapour-phase standard.

12.4.5 To withdraw a standard for GC injection, remove the bottle from the oven and stir for 10-15 s. Insert a gas-tight syringe (warmed to 60°C) through the septum cap and pump three times slowly. Draw the appropriate volume of sample (approximately 25% larger than the desired injection volume) into the syringe and adjust the volume to the exact value desired, then immediately inject over a 5-10 s period onto the GC.

12.5 Preparation of spiked cartridges by vapour-phase injection

12.5.1 Prepare a dilution bottle containing the desired concentrations of the compound(s) of interest as described in 12.4.

12.5.2 Assemble a helium purge system wherein the helium flow (20-30 mL/min) is passed through a stainless-steel "tee" fitted with a septum injector. Connect the clean Tenax cartridge downstream of the tee, using appropriate Swagelok fittings, and inject the desired volume of vapour standard (taken from the dilution bottle as described in 12.4.5). Flush the syringe several times by alternately filling with carrier gas and displacing the contents into the gas stream, without removing the syringe from the septum. Maintain gas flow through the cartridge for approximately 5 min after injection.

12.6 Preparation of spiked cartridges using permeation or diffusion tubes

12.6.1 Generate a flowing stream of inert gas containing known amounts of each compound of interest according to ASTM Method D3609(6) (see Annual Book of ASTM Standards, 1984). Note that a method which maintains the temperature within ± 0.1°C is required and the system usually must be equilibrated for at least 48 h before use.

12.6.2 Draw an accurately known volume of the standard gas stream (usually 0.1-1.0 L) through a clean Tenax cartridge, using the sampling system shown in Figure 2. If mass-flow controllers are employed they must be calibrated for the carrier gas used in 12.6.1 (usually nitrogen).

13. SCHEMATIC REPRESENTATION OF PROCEDURE

Purify Tenax GC and prepare sampling cartridges

↓

Sample known volume of air with Tenax cartridge

↓

Calibrate GC/MS system and obtain response factors (calibration curves) for analytes of interest, using direct injection of standards or spiked cartridges

↓

Analyse sample cartridges and blanks

↓

Identify and quantify analytes, using retention times, GC peak areas (characteristic ion) and response factors (calibration curves)

14. ORIGIN OF THE METHOD

A number of research organizations have employed variations of the method described herein. This presentation was prepared by:

Dr Ralph M. Riggin
Battelle Columbus Laboratories
505 King Avenue
Columbus, Ohio 43201-2693
USA

Contact point: Dr Ralph M. Riggin

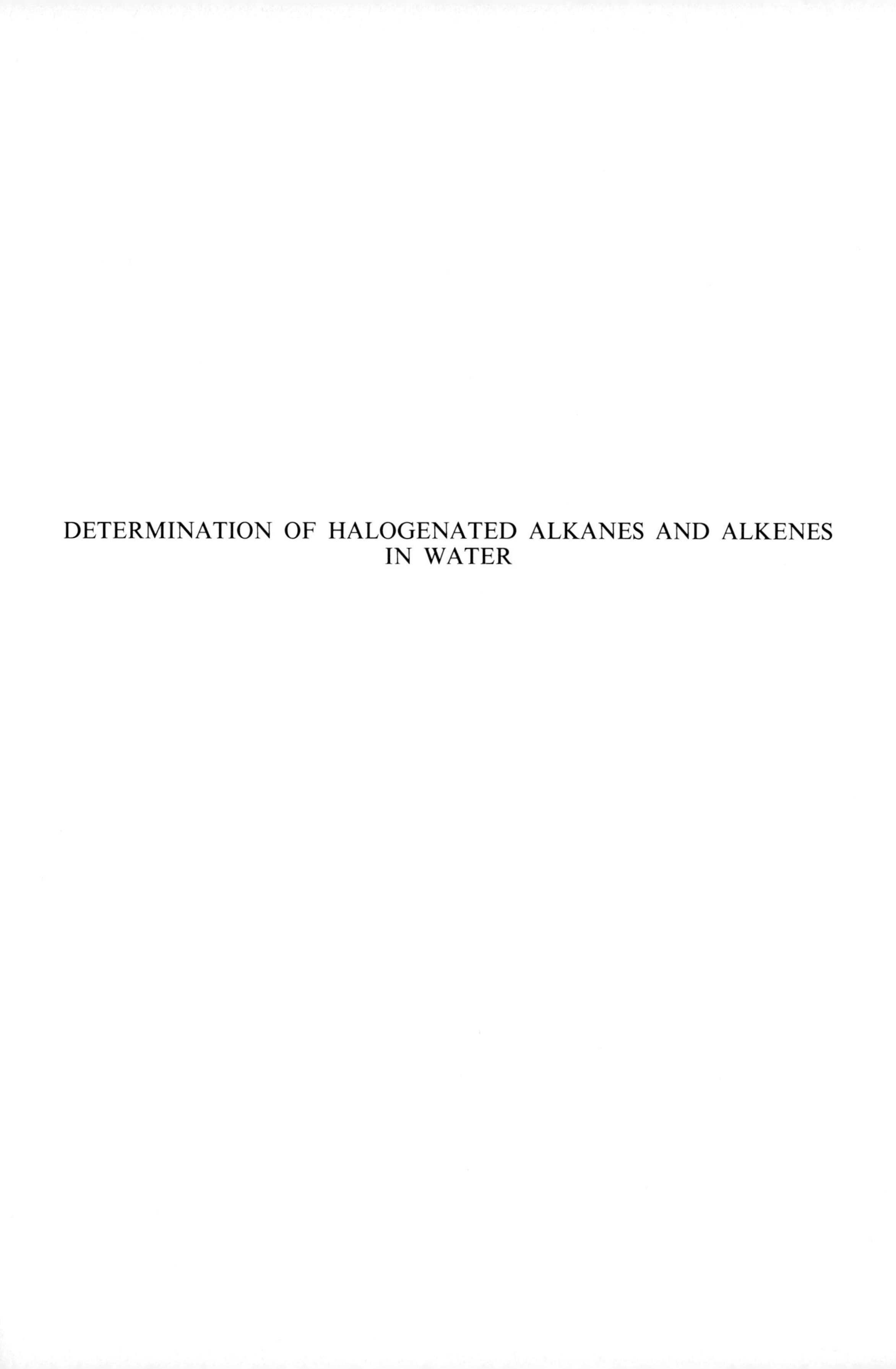

DETERMINATION OF HALOGENATED ALKANES AND ALKENES IN WATER

WATER SAMPLING

A. MacKenzie Peers

INTRODUCTION

Experience indicates that the ultimate purpose of any analytical programme is best served when the analyst is involved as far as possible in all aspects of the sampling process, and particularly in those of collection and preservation prior to analysis. It is axiomatic that the finest analysis is of little use if the sample is not representative and it should therefore be emphasized that sample collection and preservation are critical steps in the overall analytical process.

Water supplies are analysed for a wide variety of reasons, and these largely determine the choice of site, sampling programme and type of sample required. Detailed statistical and other considerations pertaining to the establishment of adequate sites and sampling programmes, however, are beyond the scope of this chapter, which will deal in a general way with the more salient aspects of sample collection and preservation. The variability of water quality is such that detailed procedures in any given case must be left to the judgement of the analyst. The following exposition has been condensed from the references cited at the end of this Introduction.

Sample collection

Sampling requirements may differ considerably between rivers, lakes and underground waters. Variability is generally greatest in rivers and the determinand range will be greater, the closer its source to the site of collection. An 'integrated' sample (see below) is preferable in the case of rivers or streams. If only a single 'grab' sample can be collected, it is best taken in the middle of the stream, at mid-depth.

Lakes and reservoirs exhibit seasonal thermal stratification and are also subject to variations from causes such as rainfall and wind. All these factors, and the purpose of the investigation, must be taken into account when choosing sampling locations, frequencies and types.

When sampling water from wells, the water should be pumped sufficiently to ensure that the sample represents the groundwater that feeds the well.

Distribution systems should likewise be adequately flushed to ensure that the sample is representative of the supply.

The detailed selection of sampling points is critical in the case of rivers and lakes, as determinand concentrations may vary considerably with numerous factors, such as distance from shore, type and rate of flow, depth below surface, temperature, turbidity, etc. The concentrations of volatile determinands (e.g., certain halocarbons) are particularly sensitive to temperature and distance from source. Particulate matter can adsorb some halocarbons and other determinands. If necessary, it may be removed by filtration or centrifugation and analysed separately, as recommended by Wegman and Oake (1983).

Blank samples should be prepared (using distilled, 'organic-free' water) at the time the water samples are collected, and should be submitted to the entire procedure (collection, preservation, transport, storage and analysis) to which the water samples are subjected.

Frequency and time of sampling

To characterize a water source, it is often necessary to determine mean, maximum and minimum values over a period of time. The closeness of the measured values to the true values will depend on the variability of the determinand and on the number of samples taken. The variations with time may be random (storms, industrial discharges, accidents, etc.) or cyclic, the latter with periods ranging from a few hours to a few months (seasonal) or years, so that careful consideration must be given to time and frequency of sampling. It may be noted that the reliability of a mean value is not proportional to the number of samples taken, but to the square root of that number.

Type of sample

(a) A 'spot', 'grab' or 'catch' sample can only represent the water composition at the time and place at which it is taken. Single grab samples are therefore useful only when the determinand is known to vary little with time and/or location.

(b) A 'composite' sample consists of a mixture of a series of grab samples taken at different times at a given point. If the grab samples are of equal volume, the composite is "time-based" (or -weighted). If the volumes are proportional to the flow rate at the time of sampling, the composite is termed "flow-based". Such samples are used to give average values for the time interval employed. Composite samples cannot be used if the determinand is subject to significant, unavoidable changes on storage.

(c) An 'integrated' sample consists of a mixture of grab samples taken at the same time at a number of different locations. If it is desired, for example, to determine the average composition (or total loading) of a stream, grab samples should be taken at different points on a cross-section of the stream. If the sample volumes are equal, each sample must be analysed separately and the average determinand concentration calculated only after each result has been multiplied by the volume flow rate at the point of sampling. Multiple analyses may be avoided, however, if the volume of each grab sample is proportional to the flow-rate at the sampling point.

Containers

Sample bottles are normally of glass or plastic and must be capable of being tightly sealed. In the case of organic determinands, glass bottles are to be preferred, being less likely to contaminate the sample or adsorb the determinand.

For the determination of halocarbons, it is recommended that the bottles be rinsed with acetone and heated two or three hours at 150°C before use. The bottle should be completely filled, leaving only a minimal air-space when capped.

A sample volume of two litres is sufficient for most analyses.

Preservation

Once a sample has been taken, changes in composition will begin. These can, at best, only be retarded by conservation techniques. The most appropriate technique in a given case will depend on the nature of the determinand, as well as on the overall composition of the sample. Microbiological activity, for example, may reduce residual chlorine to chloride. Such activity can be greatly retarded by keeping the sample in the dark and at low temperature (4°C), which will also reduce the rates of other physical and chemical reactions. The addition of chemical preservatives is generally less desirable than storage at low temperatures in the absence of light, since they render the sample unfit for a number of determinations. When determining halocarbons, however, free chlorine should be removed by the addition of a reducing agent, such as sodium sulfite or ascorbic acid, since chlorine can react with organic materials to form halogenated compounds. It can also oxidize bromide and iodide ions to species capable of participating in reactions resulting in the formation of halomethanes (Wegman & Oake, 1983).

Whatever method of preservation is employed, the time interval between sample collection and analysis should be kept as short as possible.

REFERENCES

American Public Health Association, American Water Works Association, Water Pollution Control Federation (1976) Franson, M.A., ed., Standard Methods for the Examination of Wastewater, 14th Ed., Washington, American Public Health Association, pp. 38-45

UNEP/WHO/UNESCO/WMO (1978) GEMS: Global Environmental Monitoring System: GEMS/Water Operational Guide, Sampling procedures, Geneva, World Health Organization, Chap. 2

Wegman, R.C.C. & Oake, R.J. (1983) Guidelines for the determination of extractable organic halogen in waters (EOH). In: Commission of the European Communities, Directorate General for Science, Research & Development, Concerted Action "Analysis of Organic Micropollutants in Water (COST 64B-bis), doc. OMP/37/83, Part I, pp. 22-26

DETERMINATION OF ORGANIC-BOUND HALOGEN IN WATER SAMPLES

DETERMINATION OF ORGANIC-BOUND HALOGEN IN WATER SAMPLES

INTRODUCTION

P.A. Greve & R.C.C. Wegman

Organohalogen compounds are produced in large quantities in many countries and are used for a wide variety of purposes. It is not surprising, therefore, that many organohalogen compounds have been found in the environment and that the possible pollution of ground-, drinking- and surface-waters by these compounds is a matter of concern.

Specific methods which detect individual organohalogen compounds are available (see this volume, Methods 15 and 16), but the wide variety of organohalogen compounds which can occur in principle makes a complete coverage by compound-specific methods virtually impossible. It is therefore desirable to have general methods which determine, by means of a "sum-parameter", the group of organohalogen compounds as a whole. Comparison of the sum-parameter with the sum of identified organohalogen compounds will indicate whether or not unidentified compounds are also present. The sum-parameter can also be used to detect substantial deviations from "normal" values.

The analytical background of the determination of organic halogen has been discussed by Wegman (1981). From this discussion, it can be concluded that one parameter covering all possible organohalogen compounds in a water cannot be given; instead, at least three parameters must be defined, accounting for:

(i) the more-volatile, sparingly water-soluble fraction;
(ii) the less-volatile, sparingly water-soluble fraction;
(iii) the water-soluble (polar) fraction.

For present purposes, only fractions (i) and (ii) are of interest.

Fraction (i) mainly consists of compounds with a Henry coefficient of more than 0.05 at 25°C; e.g., dichloromethane, chloroform, 1,2-dichloroethane, etc. It is currently designated as the fraction containing "Volatile Organic-bound Halogen" (VOH). A purge-and-trap technique is used as a concentration step prior to the determination of VOH, hence it is sometimes also called "Purgeable Organic-bound Halogen" (POH, or, using the symbol **X** for halogen, POX).

Fraction (ii) consists of the other sparingly water-soluble organohalogen compounds (in practice, compounds with a solubility in water of more than 1% at room temperature contribute little to this parameter). It is determined in a petroleum ether (or hexane, or pentane) extract of the sample and is therefore currently designated as "Extractable Organic-bound Halogen" (EOH or EOX). More polar solvents have been used for the extraction, but, mainly because of their questionable purity (presence of organohalogen compounds), none of them has found wide application (Oake, 1983).

Both EOH and VOH are determined by microcoulometry. This method is sensitive and relatively cheap, but does not differentiate between the four halogens. From data supplied by Wegman & Greve (1977) it can be calculated that the average relative response for the four halogens I/Br/Cl/F = 0.4/0.7/1.0/0.0. In practice, both VOH and EOH are expressed as chlorine. Recoveries and occurrence of a great number of organohalogen compounds are given by Oake (1983).

Analytical procedures for the determination of EOH and VOH are given in Methods 13 and 14, which follow.

REFERENCES

Oake, R.J. (1983) Methods for determination of extractable organo-halide. In: Commission of the European Communities, Directorate-General for Science, Research and Development, Concerted Action "Analysis of Organic Micropollutants in Water" (COST 64B-bis), doc. OMP/37/83, Part II, pp. 38-60

Wegman, R.C.C. (1981) Determination of organic halogens; a critical review of sum parameters. In: Bjørseth, A. & Angeletti, G., eds, Proceedings Second European Symposium on the Analysis of Micropollutants in Water, Killarney, 17-19 November 1981; D. Reidel Publishing Comp., pp. 249-263

Wegman, R.C.C. & Greve, P.A. (1977) The micro-coulometric determination of extractable organic halogen in surface water; application to surface waters of the Netherlands. Sci. Total Environ., 7, 235-245

METHOD 13

THE DETERMINATION OF EXTRACTABLE ORGANIC-BOUND HALOGEN (EOH) IN WATER

P.A. Greve & R.C.C. Wegman

1. SCOPE AND FIELD OF APPLICATION

The method is suitable for the determination of Extractable Organic-bound Halogen in water samples. The detection limit is approximately 0.2 µg chlorine/L and the time required for analysis is about 90 min.

2. REFERENCES

Wegman, R.C.C. & Greve, P.A. (1977) The micro-coulometric determination of extractable organic halogen in surface water; application to surface waters of the Netherlands. Sci. Total Environ., 7, 235-245

Wegman, R.C.C. & Oake, R.J. (1983) Guidelines for the determination of extractable organic halogen in waters (EOH). In: Commission of the European Communities, Directorate-General for Science, Research and Development, Concerted Action "Analysis of Organic Micropollutants in Water" (COST 64B-bis), doc. OMP/37/83, Part I, pp. 22-26

3. DEFINITION

Extractable Organic-bound Halogen (EOH) is defined as the halogen, measured by microcoulometry and expressed as chlorine, formed by total combustion of a petroleum ether extract of a water sample. It is of importance to stress that the parameter is defined by the analytical procedure, so that the method described must be strictly followed.

4. PRINCIPLE

The water sample is extracted twice with petroleum ether. The extract is dried with sodium sulfate and concentrated to a small volume. *n*-Hexadecane is added as a "holder" and the solution is concentrated further at room temperature. An aliquot of the concentrated extract is injected into a combustion furnace. The hydrohalogens formed are passed into a cell and determined by microcoulometry.

5. HAZARDS

As the microcoulometric determination of halogen ions is carried out in concentrated acetic acid solution, the titration cell should be placed in a separate, well-ventilated housing.

The procedure involves the use of oxygen at high temperatures and the usual precautions for working under these conditions should be taken.

Aldrin is highly toxic by ingestion and inhalation and is absorbed by the skin.

6. REAGENTS[1]

Petroleum ether, b.r. 40-60°C	Freshly-distilled. Check for the absence of halogen-containing impurities
Sodium sulfate	Heat at 500°C for 3 hours
n-Hexadecane	Fluka, art.nr. 52210, or equivalent. Check for the absence of halogen-containing impurities
Glacial acetic acid, A.R.	AnalaR, art.nr. 10001, or equivalent
Sodium perchlorate monohydrate	Baker Analysed Reagent, art.nr. 2815, or equivalent
Aldrin	
Oxygen	
Argon	
Helium	ultra-pure
Concentrated sulfuric acid	AnalaR, art.nr. 10276, or equivalent
Generator electrolyte	Dissolve 10 g sodium perchlorate monohydrate in approx. 200 mL aqua bidest, add 780 g glacial acetic acid and make up to 1 000 mL with aqua bidest.

[1] Reference to a company and/or product is for the purpose of information and identification only and does not imply approval or recommendation of the company and/or product by the International Agency for Research on Cancer, to the exclusion of others which may also be suitable.

Aldrin standard solution	Prepare 50 to 100 mL at a concentration of 45-55 µg/L in petroleum ether: n-hexadecane (9:1, v/v). Store in refrigerator and replace after one year.
Sodium chloride standard solution	Approximately same concentration (in water) as Aldrin standard solution.

7. APPARATUS[1]

Filtration system	Millipore (Waters Ass. Milford, Mass., USA)
Kuderna-Danish evaporator	
Centrifuge	For 1 L sample, 800-900 x g
Combustion furnace, equipped with automatic injection device and coupled to a microcoulometer with integrator, <u>via</u> a trap containing concentrated sulfuric acid	The complete system, based on the principles described by Wegman and Greve (1977), is produced by the "Keuringsinstituut voor Waterleidingartikelen" (KIWA), Churchilllaan 273, 2280 AB Rijswijk, Netherlands, and by Euroglas B.V., Buitenwatersloot 341, 2614 GS Delft, Netherlands. A flow diagram of the system is given in Figure 1.
Glass bottles	1 L, with screw cap. (Blood bottles with Teflon-lined rubber septa are also suitable.)

8. SAMPLING

8.1 Prior to sampling, rinse the bottles with acetone and heat them for 2-3 hours at 100°C.

8.2 Drinking water samples may contain free chlorine, which has to be removed; to this end, add approximately 0.1 g of ascorbic acid to each water sample as it is collected.

---

[1] Reference to a company and/or product is for the purpose of information and identification only and does not imply approval or recommendation of the company and/or product by the International Agency for Research on Cancer, to the exclusion of others which may also be suitable.

FIG. 1. FLOW DIAGRAM FOR THE DETERMINATION OF EOH

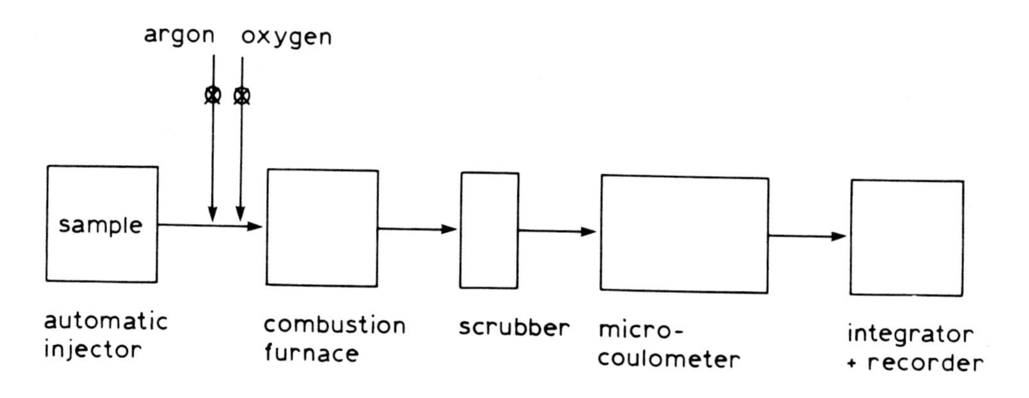

8.3 Prepare a blank sample (see 9.1) and treat in all respects in the same way as the water sample. Fill the bottles so as to leave only a minimal air-space.

8.4 Store samples in the dark at about 4°C, in a contaminant-free area.

Analysis should be carried out as soon as possible after sampling, since the maximum time of safe storage (without decomposition or artefact formation) depends on the nature of the water to be analysed.

9. TEST PROCEDURES

9.1 Blank test

Blank tests must be carried out in order to ensure that the reagents used do not contain halogens in amounts which would interfere with the determination. Blank samples can be prepared from distilled water by passing it through a Millipore system, followed by purging with ultra-pure helium for 48 h at 95°C.

Care must be taken that the air of the laboratory, which can also contain halogenated compounds, does not give rise to erratic results.

9.2 Check test

The performance of the combustion and titration system must frequently be checked by determining the yield obtained with standard substances. Aldrin is often used for this purpose; the yield for this compound should be at least 90%. As for the unknown extracts, the solution must be prepared in petroleum ether:n-hexadecane (9:1, v/v). Typically, 100 µL of solution containing 4.66 ng aldrin is injected into the combustion furnace.

The performance of the titration cell must be checked by the direct addition of standard solutions of sodium chloride.

The generator anode should be regularly re-plated with silver, as described in the manual provided with the apparatus.

The sulfuric acid in the scrubber must be replaced regularly.

Small particles (e.g., of partly burned materials) in the combustion furnace can substantially decrease the yield of the combustion. The tubes must therefore be kept scrupulously clean.

9.3 Test portion

100 mL, representative water sample

9.4 Determination of EOH

9.4.1 Assemble the combustion/titration system as described in the manual. The following settings are advised:

argon flow:	80 mL/min
oxygen flow:	250 mL/min
temperatures:	injection port, 350°C
	scrubber inlet, 70°C
	furnace, 850°C
bias voltage:	-310 mV (using the Ingold Ag/AgCl electrode, type 373 M5 NS, as the reference electrode)
gain:	4 mA/V
injection speed:	0.2 mL/sec

9.4.2 Extract 1000 mL water with 2 × 100 mL petroleum ether (See Sections 12.1 and 12.2). Shake for at least 5 min at each extraction.

9.4.3 Combine the organic phases and dry with sodium sulfate.

9.4.4 Concentrate the dried solution in a Kuderna-Danish evaporator to approximately 5 mL.

9.4.5 Add 0.5 mL n-hexadecane and concentrate further to exactly 1 mL by means of a gentle stream of nitrogen.

9.4.6 Inject 100 μL into the furnace, using the automatic injection device.

9.4.7 Record the number of millicoulombs (mC) used for the titration (given by the integrator).

10. METHOD OF CALCULATION

The mass concentration, ρ, of EOH is given by

$$\rho = \frac{0.368\ QV_e}{V_i V_s}\ (\mu g/L)$$

where

0.368 = conversion factor derived from Faraday's law (μg/mC)
Q = quantity of electrical charge used for titration (mC)
V_s = volume of water sample analysed (L)
V_i = volume injected (L)
V_e = final volume of extract in 9.4.6 (L)

11. RECOVERY, REPEATABILITY AND REPRODUCIBILITY

The performance of the system is best checked by determining the recovery for aldrin as described in section 9.2. Typically, a recovery of 92% (standard deviation, 2%) is found for repeated determinations on one day. The coefficient of variation over a two-year period in the authors' laboratory was 4%.

12. NOTES ON PROCEDURE

12.1 If the sample contains appreciable amounts of particulate matter, the following procedure is recommended before the extraction, step (9.4.2).

12.1.1 Centrifuge water sample at 800-900 × g for 10 min.

12.1.2 Agitate (e.g., in an ultrasonic bath) 1 g particulate matter with 5 mL acetone.

12.1.3 Centrifuge, decant the supernatant and repeat the extraction with another 5 mL portion of acetone.

12.1.4 Combine the acetone extracts and add 100 mL water. Extract with 2 × 100 mL petroleum ether and shake for at least 5 min at each extraction. Proceed as described in 9.4.3 to 9.4.7 inclusive.

12.2 It can be of interest to differentiate between basic, neutral and acid organohalogen compounds in the water sample under investigation. To this end, extraction (9.4.2) is first carried out at pH 4 after acidifying with 4 mol/L hydrochloric acid and the aqueous layer is subsequently re-extracted at pH 7 and 10 (after addition of 4 mol/L sodium hydroxide solution). Each organic extract is analysed separately.

13. SCHEMATIC REPRESENTATION OF THE METHOD

Separate, if necessary, particulate matter from the water phase by centrifugation

Particulates
↓
Extract 1 g with 2 × 5 mL acetone
↓
Combine the organic phases, add 100 mL water, extract with 2 × 100 mL petroleum ether, combine the organic phases and dry with sodium sulfate
↓
Concentrate each extract to a few milliliters in a Kuderna-Danish evaporator, add 0.5 mL n-hexadecane, concentrate further to exactly 1 mL with a gentle stream of nitrogen at room temperature
↓
Inject 100 µL of each extract in the combustion furnace and titrate. Record the responses of the integrator
↓
Calculate the EOH-concentrations in the samples using the equation of section 10.

Supernatant
↓
Extract 1000 mL with 2 × 100 mL petroleum ether
↓
Combine the organic phases and dry with sodium sulfate

14. ORIGIN OF THE METHOD

Laboratory for Organic Chemistry
National Institute of Public Health and Environmental Hygiene
P.O. Box 1
3720 BA Bilthoven, Netherlands

Contact points: P.A. Greve
R.C.C. Wegman

METHOD 14

THE DETERMINATION OF VOLATILE ORGANIC-BOUND HALOGEN (VOH) IN WATER

P.A. Greve & R.C.C. Wegman

1. SCOPE AND FIELD OF APPLICATION

The method is suitable for the determination of Volatile Organic-bound Halogen (VOH) in water samples. The detection limit is approximately 0.05 µg chlorine/L and the time required for analysis is about 1 hour.

2. REFERENCES

Wegman, R.C.C. & Hofstee, A.W.M. (1979) The microcoulometric determination of volatile organic halogen in water samples. In: Commission of the European Communities, Directorate-General for Science, Research and Development, Concerted Action: "Analysis of Organic Micropollutants in Water" (COST 64B bis), doc. OMP/29/82, Vol. 2, pp. 245-253

Piet, G.J. & Wegman, R.C.C. (1983) The determination of Purgeable Organic Halogen (POH). In: Commission of the European Communities, Directorate-General for Science, Research and Development, Concerted Action: "Analysis of Organic Micropollutants in Water" (COST 64B bis), doc. OMP/37/83, part I., pp. 9-21

3. DEFINITION

Volatile Organic-bound Halogen (VOH) is defined as the halogen, measured by microcoulometry and expressed as chlorine, formed by total combustion of a concentrate prepared by passing an inert gas through the water sample and adsorbing the volatile compounds on a Tenax column. It is of importance to stress that the parameter is defined by the analytical procedure, which must therefore be strictly followed.

4. PRINCIPLE

The volatile halogen-containing compounds are stripped from the water sample by passing a stream of helium through the sample and the vapours are adsorbed on Tenax. The Tenax is subsequently heated and the vapours are led into a combustion furnace. The hydrohalogens formed are passed into a cell and halogen is determined by microcoulometry.

5. HAZARDS

As the microcoulometric determination of halogen ions is carried out in concentrated acetic acid solution, the titration cell should be placed in a separate, well-ventilated housing.

The procedure involves the use of oxygen at high temperatures and the usual precautions for working under these conditions should be taken.

6. REAGENTS[1]

Adsorbent, Tenax-GC, 60-80 mesh.	Chrompack, art.nr. 245, or equivalent. Condition new batches of Tenax by heating *in situ* (Fig. 1) to 200-210°C for 1 h under a stream of helium.
Sodium perchlorate monohydrate	Baker Analysed Reagent, art.nr. 2815, or equivalent
Helium	
Oxygen	
Dichloromethane	
Concentrated sulfuric acid	AnalaR, art.nr. 10276, or equivalent
Generator electrolyte	Dissolve 10 g sodium perchlorate monohydrate in approx. 200 mL aqua bidest, add 780 g glacial acetic acid and make up to 1 000 mL with aqua bidest.
Dichloromethane standard solution	4.5-5.5 mg/L in methanol or ethanol
Sodium chloride standard solution	Approximately same concentration (in water) as dichloromethane standard solution

---

[1] Reference to a company and/or product is for the purpose of information and identification only and does not imply approval or recommendation of the company and/or product by the International Agency for Research on Cancer, to the exclusion of others which may also be suitable.

7. APPARATUS[1]

A flow diagram of the VOH-determination is shown in Figure 1 and consists of the following items:

Liquid sample concentrator with thermostated waterbath	Tekmar, LSC-1
Tenax column (300 mg)	
Microcoulometer with combustion furnace and gas valves	Dohrmann, DE-20
Integrator	Becker 763
Recorder	Varian A-25
Glass bottles	1-L, with screw cap. (Blood bottles with Teflon-lined rubber septa are also suitable.)

FIG. 1 FLOW DIAGRAM FOR THE DETERMINATION OF VOH

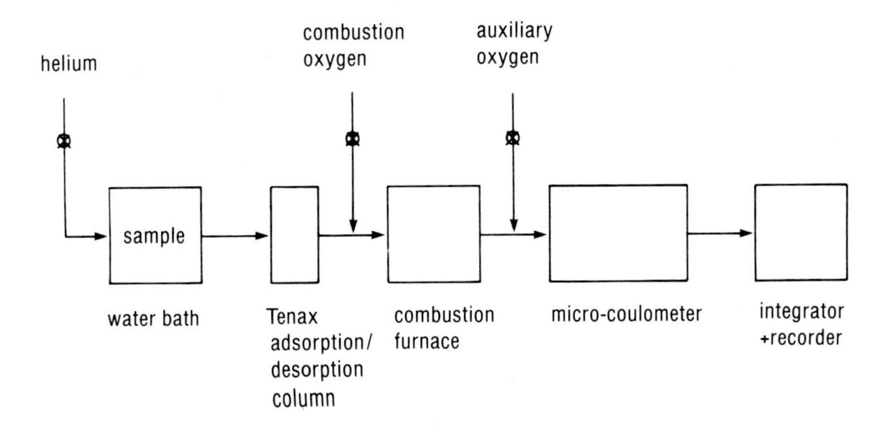

8. SAMPLING

8.1 Prior to sampling, rinse the bottles with acetone and heat them for 2-3 hours at 150°C.

8.2 Drinking water samples may contain free chlorine which has to be removed; to this end, add approximately 0.1 g of ascorbic acid to each water sample as it is collected.

8.3 Fill the bottles with sampled water so that only a small air space remains when they are capped.

8.4 At the same time, prepare a blank sample using distilled water passed through a Millipore system and purged with ultra-pure helium for 48 h at 95°C. Treat the blank in all respects in the same way as the water sample.

8.5 Transport the bottles upside-down and store in the dark at about 4°C, in a contaminant-free area. Minimise agitation. Analysis should be carried out as soon as possible after sampling, since the maximum time of safe storage (without decomposition or artefact formation) depends on the nature of the water to be analysed.

9. TEST PROCEDURES

9.1 Blank test

9.1.1 Analyse the blank (see 8.4) in the same way as the water sample.

9.1.2 The adsorption/desorption system (Fig. 1) must be checked for the absence of halogen in amounts which would interfere with the determination (Piet & Wegman, 1983). Careful conditioning of new batches of adsorbent and thorough desorption after the analysis of each sample are especially important.

---

[1] Reference to a company and/or product is for the purpose of information and identification only and does not imply approval or recommendation of the company and/or product by the International Agency for Research on Cancer, to the exclusion of others which may also be suitable.

9.2 Check test

The performance of the system must frequently be checked by determining the recovery obtained with standard substances. Dichloromethane is often used for this purpose; the recovery for this compound should be at least 80%. The standard solution is prepared in methanol or ethanol. Typically, 5 µL of a standard solution of dichloromethane in ethanol (4.5-5.5 mg/L) is added to 50 mL water, which is subsequently analysed as described in section 9.4.

The performance of the titration cell must be checked by the direct addition of standard solutions of sodium chloride.

Small particles (e.g., of partly burned materials) in the combustion furnace can substantially decrease the yield of the combustion. The tubes must therefore be kept scrupulously clean.

9.3 Test portion

50 mL of representative water sample.

9.4 Determination of VOH

9.4.1 Assemble the adsorption/desorption system as shown in Figure 1.

amount of adsorbent: 300 mg

stripping/carrier gas: helium, 40 mL/min

The following settings are advised for the combustion/titration system:

combustion gas: oxygen, 200 mL/min

auxiliary gas: oxygen, 50 mL/min

furnace temperature: 850°C

bias potential: 262 mV

range: 500 Ohm

integrator settings: sensitivity: 60 mV sec
output: 1 mV

9.4.2 Add 50 mL water sample to the 60 mL stripping vessel (see Section 12.1) in a thermostated water bath at 60°C.

9.4.3 Set the system to the adsorption mode and pass helium through the sample for 20 min.

9.4.4 Heat the Tenax tube quickly to 190°C (system in desorption mode).

9.4.5 Observe the integrator signal and continue heating the adsorbent until the pen is back at baseline.

9.4.6 Stop heating and cool the tube by switching on the fan.

9.4.7 When the temperature of the tube is below 100°C, reset the system for adsorption and repeat the adsorption/desorption procedure as described above (9.4.3 to 9.4.5, inclusive. See section 12.2).

9.4.8 Read the amount of millicoulombs used during the two titration procedures (given by the integrator).

10. METHOD OF CALCULATION

The concentration, ρ, of VOH is given by

$$\rho = \frac{0.368\ Q}{V} \quad (\mu g/L)$$

where

0.368 = conversion factor derived from Faraday's law (µg/mC)

Q = total quantity of electrical charge used during the two desorption steps (mC)

V = volume of water sample taken in step 9.4.2 (L)

11. RECOVERY, REPRODUCIBILITY AND REPEATABILITY

The performance of the system is best checked by determining the recovery for dichloromethane, as described in section 9.2. Typically, a recovery of 86% (standard deviation 3.5%) is found for repeated determinations on one day. The coefficient of variation over a two-year period in the authors' laboratory was 8%.

12. NOTES ON PROCEDURE

12.1 If a high VOH content is expected, the sample size can be reduced to 5 mL, in which case the 6 mL stripping vessel must be used. The detection limit is thereby increased to 0.5 µg/L.

12.2 The repetition of the adsorption/desorption step is necessary in order to achieve satisfactory yields for the less-volatile compounds, such as 1,2-dichloroethane, trichloroethylene and dibromochloromethane (Wegman & Hofstee, 1979). If the first purging procedure is prolonged for more than 20 min, losses will occur for volatile compounds such as dichloromethane.

13. SCHEMATIC REPRESENTATION OF THE METHOD

Strip the water sample (5 or 50 mL) at 60°C with helium for 20 min and collect the vapours on a Tenax-GC column at room temperature

↓

Desorb the vapours on the Tenax column at 190°C, using helium as a carrier gas, and pass them into a combustion furnace coupled to a microcoulometer

↓

When the integrator signal is back to base-line, cool the Tenax column to room temperature and repeat the adsorption/desorption procedure

↓

Add the integrator response obtained for the two adsorption/desorption procedures and calculate the VOH-concentration in the water sample using the equation of section 10.

14. ORIGIN OF THE METHOD

Laboratory for Organic Chemistry
National Institute of Public Health and Environmental Hygiene
P.O. Box 1
3720 BA Bilthoven, Netherlands

Contact points: P.A. Greve
R.C.C. Wegman

DETERMINATION OF VOLATILE ORGANIC HALOGEN COMPOUNDS IN WATER SAMPLES BY HEAD-SPACE GAS CHROMATOGRAPHY

DETERMINATION OF VOLATILE ORGANIC HALOGEN COMPOUNDS IN WATER SAMPLES BY HEAD-SPACE GAS CHROMATOGRAPHY

INTRODUCTION

G.J. Piet, W.C.M.M. Luijten & P.C.M. van Noort

Volatile organic halogen compounds are defined in this chapter as halogen-containing compounds with three or less carbon atoms. Some of these compounds are used in industrial processes, some are produced on chlorination of water (leading to haloform formation) and others are metabolites with halogen atoms.

Several very volatile halogen compounds are toxic, carcinogenic, mutagenic or teratogenic and have been selected for further study by the Commission of the European Communities (CEC) and the Environmental Protection Agency (USA) as priority pollutants. Compounds such as 1,2-dichloroethane (widely used for the production of vinyl chloride monomer), 1,1-dichloroethene, tetrachloroethane and trichloromethane are known to be harmful to human health, and widely used solvents, such as trichloroethene and tetrachloroethene, are suspected to be so as well. These compounds may be persistent in the aquatic environment and, in industrialized areas, they endanger the subsoil aquifer and can pass into the drinking-water supply.

It is obvious that a standardized analytical method is needed for environmental control of volatile organic halogen compounds. This method must be suitable for a broad spectrum of compounds, even at the sub-μg/L level.

Depending on the sensitivity requirements, the size of the sample and the number of samples to be analysed, one of two procedures can be employed; these are referred to as "static" and "dynamic" head-space procedures, and are used with high-resolution gas chromatography and electron-capture, flame-ionization, or mass-spectrometric detection.

The "dynamic" head-space technique uses a purge-and-trap method, which gives rise to an enrichment factor.

When gas chromatography with electron-capture detection is used, the "static" head-space procedure is suitable for analysis at the μg/L level. Compounds with three or more halogen atoms can be determined at sub-μg/L concentrations. The head-space methods have been applied to the compounds listed below. The "static" procedure is not suitable for the compounds preceded by an asterisk, nor for the higher brominated methanes because of the relatively high detection limits.

Compound	CAS No.	Compound	CAS No.
Bromochloromethane	75-27-4	*(E)-1,2-Dichloroethene	156-60-5
*Bromoethane	74-83-9	*Dichloromethane	75-09-2
*Chloromethane	74-87-3	Hexachloroethane	67-72-1
*3-Chloropropene	107-05-01	Tetrachloroethane	127-18-4
Dibromochloromethane	124-48-1	Tetrachloromethane	50-23-5
1,2-Dibromo-3-chloro-propane	96-12-8	1,1,1-Trichloroethane	71-55-6
		Trichloroethene	79-01-6
*1,2-Dibromoethane	106-93-4	Trichlorofluoromethane	75-69-4
Dichlorodifluoromethane	75-71-8	Trichloromethane	67-66-3
*1,2-Dichloroethane	187-06-02	1,1,1-Trifluoro-2-bromo-2-chloroethane	151-67-7
*1,1-Dichloroethene	75-35-4		
*(Z)-1,2-Dichloroethene	156-69-2		

Other purgeable organic halogenated compounds (Henry coefficient > 0.05 at 25°C can also be determined with head-space methods.

The "static" and "dynamic" head-space techniques are described in methods 15 and 16, below.

The analysis of compounds subject to decomposition in water, such as bischloromethyl ether, chloromethylmethyl ether and α-epichlorohydrin, is not dealt with.

REFERENCES

International Agency for Research on Cancer (1979) IARC Monographs on the Evaluation of the Carcinogenic Risks of Chemicals to Humans, Vol. 20, Some Halogenated Hydrocarbons, Lyon, International Agency for Research on Cancer

Middleditch, B.S., Missler, S.R. & Hines, H.B. (1981) Mass Spectrometry of Priority Pollutants, New York and London, Plenum Press

Nunez, A.J., Gonzales, L.F. & Janak, J. (1984) Pre-concentration of headspace volatiles for trace organic analysis by gas chromatography. J. Chromatogr., 300, 127-162

METHOD 15

"STATIC" HEAD-SPACE DETERMINATION OF VOLATILE ORGANIC HALOGEN COMPOUNDS IN WATER

G.J. Piet, W.C.M.M. Luijten & P.C.M. van Noort

1. SCOPE AND FIELD OF APPLICATION

The automated and non-automated static head-space method is suitable for the determination of purgeable (volatile, apolar) halogen compounds (see Introduction). The electron-capture detection limit depends on the number of halogen atoms in the compound. The method can be applied to the analysis of water at the µg/L level. The analytical procedure takes 30 min or less. The method can be used as a screening technique prior to the use of the purge-and-trap procedure (Method 16).

2. REFERENCES

Murray, D.A.J. (1980) Analysis of headspace gases for parts per billion concentration of volatile organic contaminants in water samples by gas chromatography. Environ. Sci. Res., 80, 207-216

Piet, G.J., Slingerland, P., de Grunt, F.E., van de Heuvel, M.P.M. & Zoeteman, B.C.J. (1978) Determination of very volatile halogenated organic compounds in water by means of direct headspace analysis. Anal. Lett., A11, 5, 437-448

Umbreit, G.R. & Grob, R.L. (1980) Experimental application of gaschromatographic headspace analyses to priority pollutants. J. Environ. Sci. Health, Part A, 15, 429-466

U.S. Environmental Protection Agency (1979) Handbook for analytical quality control in water and wastewater laboratories, EPA 600/4-79-019, Cincinnati OH, EMSL, Office of Research and Development

3. DEFINITIONS

Not applicable

4. PRINCIPLE

A standard 9-mL vessel, which can be closed with a PTFE-lined septum cap, is filled with 5 mL of the sample at 4°C. The vial is immediately capped and equilibrated in a water bath for two hours (30°C) under exclusion of direct sunlight. An aliquot of the gas phase is injected onto a thick-film, wide-bore gas chromatographic (GC) capillary column, connected to a tandem electron-capture/flame-ionization detector system, or only to an electron-capture detector. Standard aqueous solutions are employed for quantification purposes (see also 12.1). Compound identification is based on accurate GC retention data for calibration substances.

5. HAZARDS

Suitable precautions must be taken when handling dangerous calibration compounds (see EPA Handbook, section 2). When hydrogen is used as the GC carrier gas, a flow controller must be installed or a GC protection unit must be used to avoid explosions.

6. REAGENTS[1]

Gases for GC	Hydrogen, nitrogen, air
Methanol	Merck. A.R.
Calibration compounds (see Section 11)	
gases	Merck, laboratory gases quality
liquids	Merck, A.R.
Stock standard solutions of gaseous compounds	Weigh the 25-mL glass device (Fig. 1), add 25 mL methanol and weigh again. Cool device to 0°C and connect to gas cylinder containing the calibration compound. Dissolve about 25 mg of compound, then weigh the closed vessel. Store at -20°C and do not keep longer than two weeks.

[1] Reference to a company and/or product is for the purpose of information and identification only and does not imply approval or recommendation of the company and/or product by the International Agency for Research on Cancer, to the exclusion of others which may also be suitable.

Stock standard solutions of liquid compounds	Weigh a 100-mL bottle with screw-cap, add ~ 99 mL methanol and weigh again. Using a syringe, add 100 µL of the analyte of interest at 0°C. Weigh again to nearest 0.1 mg. Store at -20°C and keep not longer than one month. (In case of frequent use, store solution in 5 bottles of 20 mL each.) If the mass density of the compound at 0°C is not known, determine it with a pycnometer.
Contaminant-free water	Purge tap-water at 90°C for 4 h, using GC-grade nitrogen and the purging system described in section 7. Check before use by GC analysis, using the standard procedure for water samples.

7. APPARATUS[1]

Gas chromatograph (GC)	High-resolution, equipped with a 50 m x 0.35 mm i.d. thick-film OV-225 glass capillary column, connected to a tandem electron-capture/flame-ionization detector (ECD/FID) or to an ECD. Head-space samples can be injected manually with a gas-tight syringe or, preferably, by means of an auto-sampler (Carlo Erba Fractovap 2900 Series, with sampler module 250, or equivalent). With automated instruments, overnight analyses are facilitated by using data systems such as the Apple II, equipped with a printer/plotter. Precision is improved by using systems with syringes which puncture the vial system while a solenoid valve is opened to equalize the head space pressure and the pressure at the head of the column.
Glass bottles	1 000 mL, with silicon septum and screw-cap.
Bottles	20, 100 and 1000 mL, with PTFE-lined screw-cap

[1] Reference to a company and/or product is for the purpose of information and identification only and does not imply approval or recommendation of the company and/or product by the International Agency for Research on Cancer, to the exclusion of others which may also be suitable.

25-mL glass device	For preparing methanol stock solutions of gaseous compounds (see Fig. 1).

FIG. 1. GLASS DEVICE FOR PREPARATION OF METHANOL STOCK SOLUTIONS OF GASEOUS COMPOUNDS

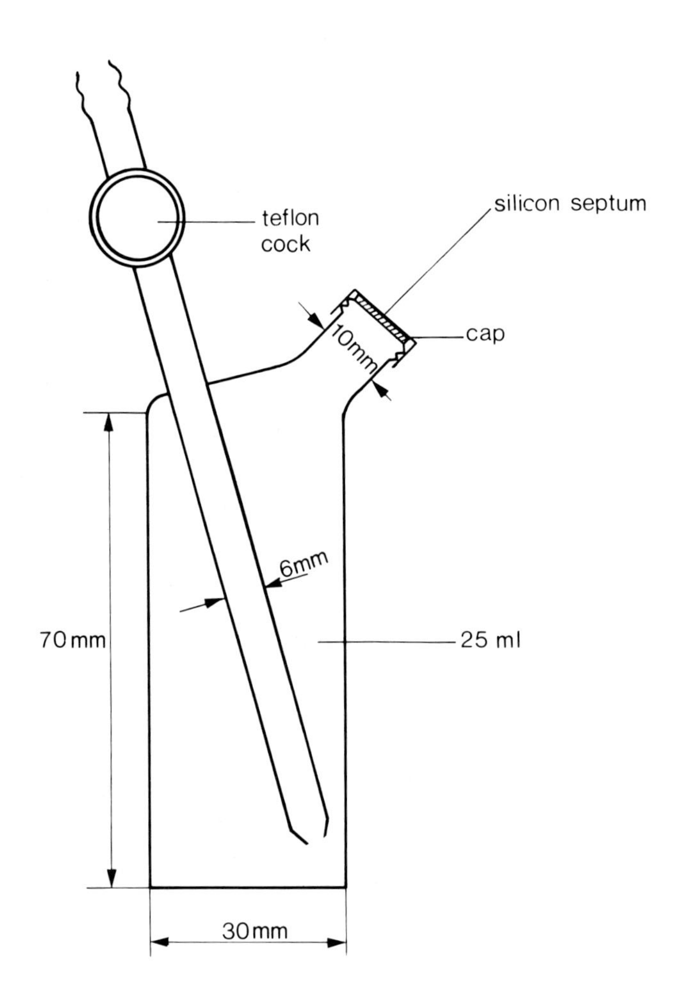

Head-space vials	9-mL, crimp-top, with PTFE-lined septum and aluminum cap (Chrompack)

Syringes	10, 50, 100 mL
Gas-tight syringes	50, 100, 500, 1 000 µL
Water/air bath (30.0°C)	
Freezer (-20°C)	
Analytical balance	100 g, 0.1 mg accuracy
Purging system	For preparing contaminant-free water; consists of a 1-3 L glass vessel that can be heated to 90°C, connected to a condensor. The water is purged with purified nitrogen (40 mL/min) through a glass frit.

8. SAMPLING

8.1 Sampling can be carried out in the 9-mL vials when appropriate filling in the field can take place. Otherwise, use one-litre glass bottles (with gas-tight PTFE-lined screwcap), pre-heated to 150°C. Fill using a glass device to introduce the sample at the bottom of the flask. Close the bottle after flushing with three times its volume, leaving about 0.5 mL head-space.

8.2 When free chlorine is present, add about 100 mg ascorbic acid to the sample.

8.3 With each series of 3-10 samples (depending on analyte concentration variability), prepare a blank by pouring contaminant-free water (section 6) into a sampling bottle (8.1) under conditions of sample collection.

8.4 With each series of 5 samples, prepare a calibration sample as follows: weigh a 1-L glass bottle with screw-cap, add 1 000 mL of contaminant-free water, cap and weigh again. Inject 10-100 µL stock standard solution of each analyte of interest through septum and mix thoroughly (the concentrations of the standard compounds should be comparable with those of the analytes in the water sample). For alternative quantification procedure, see 12.1.

NOTE: The blank, the internal standard sample and the water samples must all be subjected to the same procedures of transport, storage and analysis.

8.5 Transport samples upside-down in the dark, store at 4°C in the dark and analyse within 48 h (not later than 72 h).

9. PROCEDURE

9.1 Blank test

9.1.1 See 8.3 above. Proceed as in 9.4.1 to 9.4.4

9.1.2 Test the laboratory air regularly by adding 5.0 mL of contaminant-free water to 9-mL vials in the laboratory atmosphere and analysing under standard conditions (9.4.1-9.4.4).

9.2 Check tests

9.2.1 Check instrument performance every day by analysing calibration samples (8.4). The system response should be linear.

9.2.2 Check the analytical system for memory effects by analysing contaminant-free water after every ten samples (8.3). When the concentration level between successive analyses varies by more than an order of magnitude, such a blank must be analysed after three samples.

9.2.3 No condensation of water is allowed in the syringe. The injected gas volume and the back-pressure in the GC injection port must be constant. The performance of the syringe and possible septum leakage must be checked every day at least, particularly when autosamplers are used.

9.3 Test portion

Five mL of water sample at 4°C in 9-mL vial.

9.4 Analyte determination

9.4.1 Add 5.0 mL sample to a clean 9-mL vial at 4°C with a clean syringe, cool all devices to 4°C before use). Close the vial immediately.

9.4.2 Place the vial in a bath at 30.0°C and let it equilibrate for 2 h, away from direct sunlight.

9.4.3 Set GC conditions as follows:

injection port temperature:	180°C
detector temperature	220°C
column temperature programme	initial temperature 35-60°C (depending on the analytes), increase at 4 to 8°C/min to final temperature of 120-160°C (depending on the analytes)

9.4.4 Inject 0.05-1.0 mL head-space vapour and start the GC programme, after selection of detector attenuation factors.

9.4.5 After every 5 samples, analyse a calibration sample (8.4), as in 9.4.1 to 9.4.4, inclusive.

9.4.6 Analyse a blank sample (8.3) after 3 to 10 samples (more frequently when the analyte concentration varies appreciably).

10. METHOD OF CALCULATION

10.1 The mass concentration, ρ_a (µg/L), of an analyte in a water sample is given by,

$$\rho_a = \rho_c \left(\frac{R_a - R_b}{R_c - R_b} \right)$$

where

R_a = detector response for the analyte in water sample (9.4.4)

R_c = detector response for the analyte in the calibration sample (9.4.5)

R_b = detector response for blank sample (9.4.6)

ρ_c = concentration of analyte in calibration sample, 8.4 (µg/L)

10.2 The above value of ρ_c is given by,

$$\rho_c = V_s \rho_s / V_w$$

where

V_s = volume of stock standard solution of analyte used in 8.4 (µL)

ρ_s = concentration of the analyte in the stock standard solution (g/L)

V_w = volume of water (determined by weight) in calibration sample, 8.4 (L)

10.3 The mass concentration, ρ_s, of the calibration compound in a stock standard solution is given by,

$$\rho_s = V_c \, d_c / 10^3 V_m \qquad \text{for liquid compounds}$$

and W_c / V_m for gaseous compounds,

where

V_c = volume of added calibration compound (section 6) at 0°C (μL)

d_c = mass density of calibration compound at 0°C (g/L)

V_m = volume of the methanol stock standard solution (mL)

W_c = weight of added calibration gas (mg).

11. REPEATABILITY AND REPRODUCIBILITY

Under good laboratory conditions (EPA Handbook, section 2), the following results have been obtained:

Compound class	Concentration range (μg/L)	Analytical error (95% confidence level)	Detection limit (μg/L)
A	0.1 - 1	> 10%	0.1 - 0.5
	1 - 10	< 10%	
	10 - 100	< 5%	
B	0.1 - 1	not applicable	1 - 5
	1 - 10	< 100%	
	10 - 100	< 10%	
C	0.1 - 1	not applicable	3 - 15
	1 - 10	< 100%	
	10 - 100	< 20%	

A: dichlorodifluoromethane, tetrachloroethane, tetrachloromethane, 1,1,1-trichloroethane, trichloroethene, trichlorofluoromethane, trichloromethane

B: dibromodichloromethane, bromomethane, dibromochloromethane, 1,2-dibromo-3-chloropropane, 1,2-dibromoethane, hexachloroethane, 1,1,1-trifluoro-2-bromo-2-chloroethane

C: chloromethane, 3-chloropropene, 1,2-dichloroethane, 1,1-dichloroethene, 1,2-dichloroethene (Z/E), dichloromethane

12. NOTES ON PROCEDURE

12.1 Calibration samples (8.4) are employed for the check test (9.2.1) and for quantification. Alternatively, the method of standard addition may be employed for quantification. In that case, proceed as in 8.4, except that the standard solution of the analyte(s) of interest is added to a water sample (not to contaminant-free water). The amount of analyte added should be 2-4 times the amount originally present in the sample. The detector response of this "spiked" sample is compared with that of an identical water sample containing no added standard and the analyte concentration in the latter is readily calculated.

12.2 In most cases the water matrix does not substantially affect the analytical results. The laboratory air in many laboratories, however, is not clean enough to analyse water at the low- and sub-μg/L level.

12.3 The precision of the method depends on the analyte, the number of blanks and the number of calibration samples analysed for a given series of water samples.

12.4 Internal standards must have an analyte concentration similar to that of the samples and must be analysed under identical conditions. The injected vapour phase volume must be standardized for a sample series and no calculations should be made using different amounts of injected vapour phase.

13. SCHEMATIC REPRESENTATION OF PROCEDURE

Collect water samples and prepare blank sample
and calibration sample
Transport and store under identical conditions

↓

Check laboratory air for purity
Check analytical system for memory effects
Check instrument performance by analysing
calibration sample (as below)

↓

Add 5.0 mL sample to 9-mL vial
Cap and equilibrate for 2 h at 30°C

↓

Inject 0.05-1.0 mL head-space vapour onto GC
and record detector responses

↓

Repeat preceeding two steps with blanks
and with calibration samples

↓

Calculate analyte concentration in water sample
by comparison of detector responses in water and
in calibration samples for same analyte
(corrected for blank response)

14. ORIGIN OF THE METHOD

National Institute of Public Health and Environmental Hygiene
P.O. Box 1
3720 BA Bilthoven, The Netherlands

Contact points: G.J. Piet
W.C.M.M. Luijten

METHOD 16

"DYNAMIC" HEAD-SPACE DETERMINATION OF VOLATILE ORGANIC HALOGEN COMPOUNDS IN WATER

G.J. Piet, W.C.M.M. Luijten & P.C.M. van Noort

1. SCOPE AND FIELD OF APPLICATION

This method is suitable for the determination of purgeable organic compounds. The detection limit for organic halogen compounds in water depends on the nature of the analytes and can be decreased substantially by optimizing analytical conditions. All purgeable organic halogen compounds can be measured at the sub-μg/L level (see Section 11). The method can be applied to all water samples and can be semi-automated if the purging step is off-line.

2. REFERENCES

Dreisch, F.A. & Munson, T.D. (1983) Purge and trap analysis using fused silica capillary column, GC/MS. J. Chromatogr. Sci., 21, 111-118

Nunez, A.J., Gonzales, L.F. & Janak, J. (1984) Pre-concentration of headspace volatiles for trace organic analysis by gaschromatography. J. Chromatogr., 300, 127-162

Trussel, A.R., Moncur, J.G., Lieu, F.Y., Clark, R.R. & Leong, L.Y.C. (1983) New developments in dynamic headspace analysis of halogenated organics. Water Chlorination: Environ. Impact Health Eff., Vol. 5, No. 1, Ann Arbor Sci. Publ., pp. 583-592

3. DEFINITIONS

Not applicable

4. PRINCIPLE

Volatile compounds are removed from water by a gas-stripping process at 80°C. The volatiles are trapped on a Tenax-GC cartridge, after passing through a condenser at 10°C. If analytes with a wide volatility range (including methyl chloride) have to be determined, Porapak N or Ambersorb is used.

The analytes are thermally desorbed from the Tenax cartridge and the compounds are trapped in a 'mini-trap' containing 2-5 mg Tenax-GC at -30°C. The trap is flash-heated to inject the compounds onto a high-resolution gas chromatographic (GC) column. A tandem electron-capture/flame-ionization detector (ECD/FID), or an ECD alone, may be employed, but mass-spectrometric (MS) detection is preferable. A screening technique using "static" headspace analysis is used to determine whether dilution of samples prior to purge-and-trap is necessary to avoid overloading of the cartridge.

5. HAZARDS

Suitable precautions must be taken when handling dangerous calibration compounds. When hydrogen is used as a GC carrier gas, a flow controller must be installed or a GC protection unit must be used to avoid explosions.

6. REAGENTS[1]

Gases for GC	Hydrogen, nitrogen, air
Methanol	Merck, A.R.
Perdeuterated toluene	for MS internal standard (see section 11)
Calibration compounds	
gases	Merck, laboratory gases quality
liquids	Merck, A.R.
Tenax-GC, 80 Mesh	Chrompack, Middelburg, Netherlands
Porapak N or Ambersorb	
Stock standard solutions of gaseous componds	Weigh the 25-mL glass device (Fig. 1, Method 15), add 25 mL methanol and weigh again. Cool device to 0°C and connect to gas cylinder containing the calibration compound. Dissolve about 25 mg of compound, then weigh the closed vessel. Store at -20°C and do not keep longer than two weeks.

---

[1] Reference to a company and/or product is for the purpose of information and identification only and does not imply approval or recommendation of the company and/or product by the International Agency for Research on Cancer, to the exclusion of others which may also be suitable.

Stock standard solutions of liquid compounds	Weigh a 100-mL bottle with screw-cap, add ~ 99 mL methanol and weigh again. Using a syringe, add 100 µL of the analyte of interest at 0°C. Weigh again to nearest 0.1 mg. Store at -20°C and keep not longer than one month. (In case of frequent use, store solution in 5 bottles of 20 mL each.) If the mass density of the compound at 0°C is not known, determine it with a pycnometer.
Contaminant-free water	Purge tap-water at 90°C for 4 h, using GC-grade nitrogen and the purging system described in section 7. Check before use by GC analysis, using the standard procedure for water samples.

7. APPARATUS[1]

Gas chromatograph	High-resolution, equipped with an OV-225, 50 m × 0.22 mm i.d., capillary column, a detector system (ECD/FID, ECD or, preferably, MS) and an injection system able to flash-heat Tenax cartridges (an automatic system for cartridges, such as the Perkin-Elmer ATD-50 System, is preferred for standard conditions). A data system such as the Apple II, with a printer/plotter, facilitates overnight analyses. In case of GC/MS analysis, the Perkin-Elmer ATD-50 is connected to the Finnigan 1020 OWA automated GC/MS system. MS systems must be used for very complex environmental samples.
Off-line purge-and-trap unit	see Figure 1.

---

[1] Reference to a company and/or product is for the purpose of information and identification only and does not imply approval or recommendation of the company and/or product by the International Agency for Research on Cancer, to the exclusion of others which may also be suitable.

FIG. 1. OFF-LINE PURGE-AND-TRAP UNIT

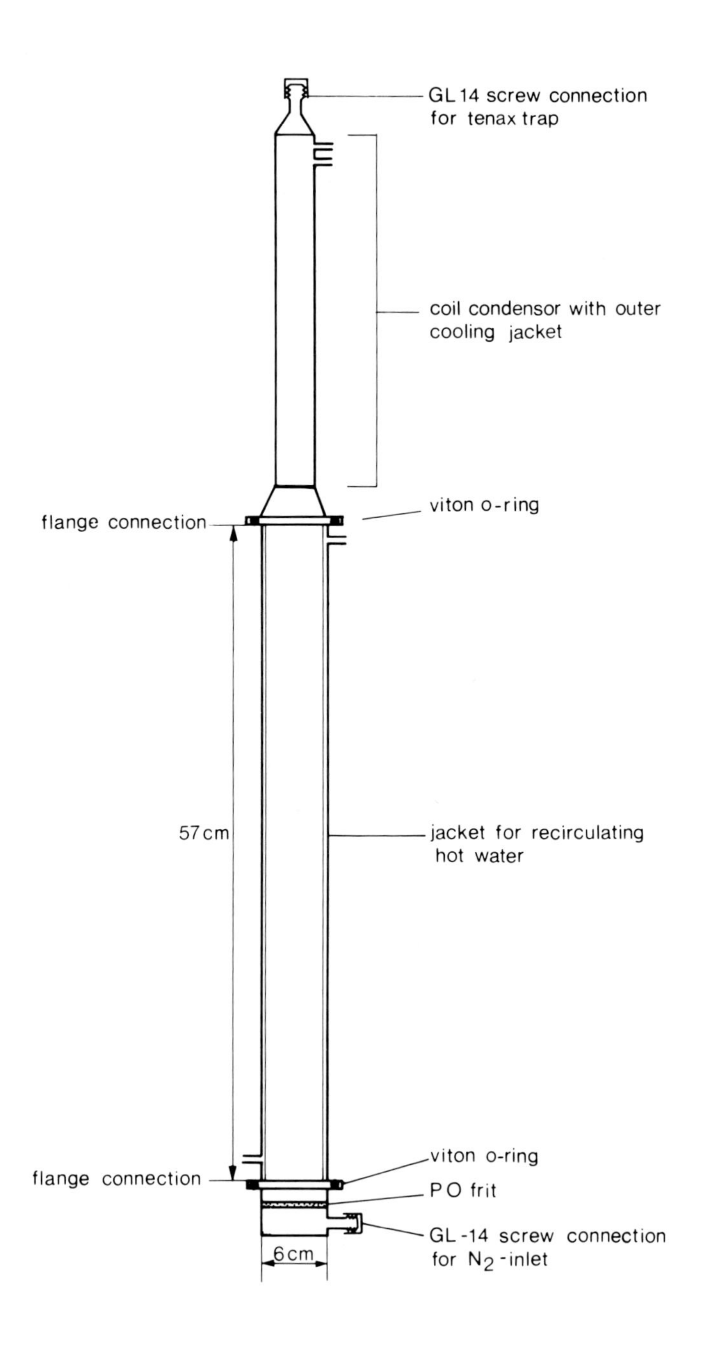

Tenax-GC sorbent cartridges	60 mm x 5 mm i.d., (see Fig. 2) containing 200 mg Tenax. If fresh Tenax is used, the loaded cartridge must be heated to 250°C for 24 h under a helium flow of 20 mL/min (see Notes on Procedure, 12.5). Other Tenax cartridges may be used if recoveries and confidence limits are known.
Heating system for purge unit	Haake N_3 thermostat, 3 150 VA, for 5 purge units

FIG. 2. A TENAX-GC SORBENT CARTRIDGE

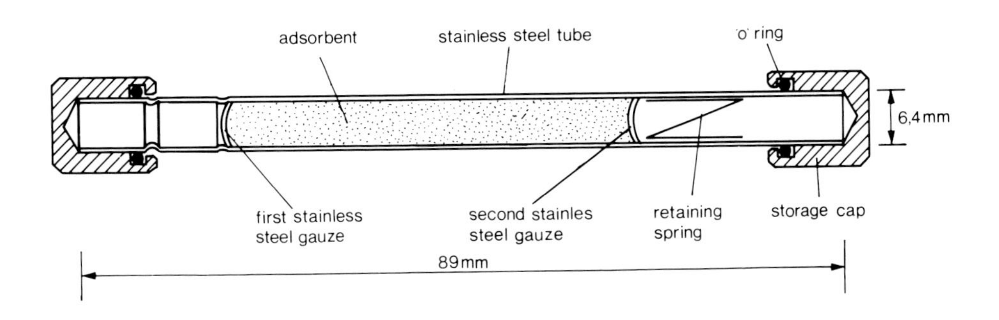

Glass bottles	1 000-mL, with silicon septum and screw-cap
Bottles	20, 100 and 1 000-mL, with PTFE-lined screw-cap
25-mL glass device	For preparing stock solution of gaseous compounds (see Fig. 1, Method 15)
Head-space vials	9-mL
Syringes	10, 50 and 100-mL
Gas-tight syringes	50, 100, 500 and 1 000-μL
Water/air bath (30.0°C)	
Freezer (-20°C)	
Analytical balance	100 g, 0.1 mg accuracy

Purging system	For preparing contaminant-free water; consists of a 1-3 L glass vessel that can be heated to 90°C, connected to a condenser. The water is purged with purified nitrogen (40 mL/min) through a glass frit

8. SAMPLING

8.1 Use one-litre pre-heated (150°C) glass bottles (with gas-tight PTFE-lined screw-cap) and fill using a glass device to introduce the sample at the bottom of the flask. Close the bottle after flushing with three times its volume, leaving about 0.5 mL head-space (if MS detection is used, add a known amount of internal standard, such as perdeuterated toluene).

8.2 When free chlorine is present, add about 100 mg ascorbic acid to the sample.

8.3 With each series of 3-10 samples (depending on analyte concentration variability), prepare a blank by pouring contaminant-free water (section 6) into a sampling bottle (8.1) under conditions of sample collection.

8.4 With each series of 5 samples, prepare a calibration sample as follows: weigh a 1-L glass bottle with screw-cap, add ~1000 mL of contaminant-free water, cap and weigh again. Inject 10-100 µL stock standard solution of each analyte of interest through septum and mix thoroughly (the concentrations of the standard compounds should be comparable with those of the analytes in the water sample). Add a known amount of perdeuterated toluene to calibration sample if MS is used. The alternative "standard addition" procedure (Method 15, section 12.1) can also be employed here.

8.5 Transport samples upside-down in the dark, store at 4°C in the dark and analyse within 18 h (not later than 72 h).

NOTE: The blank, the spiked sample and the water samples must all be subjected to the same procedures of transport, storage and analysis.

9. PROCEDURE

9.1 Blank tests

9.1.1 See 8.3 above. Proceed as in 9.4.1 to 9.4.7.

9.1.2 Test the laboratory air regularly by adding 5.0 mL of contaminant-free water to 9-mL vials in the laboratory atmosphere and analysing under standard conditions (9.4.1-9.4.7).

9.1.3 Before each purge-and-trap procedure, check the Tenax cartridges by flash-heating to 250°C in the injection system desorption unit, then flash-heating the -30°C mini-trap. Use GC conditions specified in 9.4.6.

9.1.4 Prepare a blank cartridge by desorbing a used cartridge under the desorption conditions obtaining during analysis (9.4.5); hold at 250°C for 5 min.

9.2 Check tests

9.2.1 Check instrument performance every day by analysing calibration samples (8.4) with analyte concentrations similar to those of the samples. The system response should be linear.

9.2.2 Check the analytical system for memory effects by analysing contaminant-free water after every ten samples, using a blank (clean) Tenax cartridge.

9.3 Test portion

One litre of water sample at 4°C.

9.4 Analyte determination

9.4.1 Disconnect the condenser and fill the clean purging vessel with 1 L of sample at 4°C, with a minimum of exposure to the laboratory air (see Notes on Procedure, 12.7).

9.4.2 Install the condenser with two clean Tenax-GC cartridges on top of it and start the purge flow (40 mL nitrogen/min) with the condenser at 10°C (see Notes on Procedure, 12.4).

9.4.3 Heat the sample to 80°C and continue purging for a set time (20-30 min, depending on the analytes).

9.4.4 Stop heating and purging, cap the cartridges on both sides with a minimum exposure to laboratory air. If cartridges are to be stored before analysis, it is necessary to store under helium atmosphere at 4°C, and to store one cleaned (flash-heated) cartridge under the same conditions to check storage effects.

9.4.5 Insert the uncapped cartridge in the injection system desorption unit (with a minimum exposure to laboratory air). Desorb by flash-heating to 250°C and trap the analytes at -30°C. Maintain a minimum carrier flow of 10 mL/min through the cartridge.

9.4.6 Set the GC conditions as follows:

detector temperature	220°C
column temperature programme	initial temperature (hold 5 min) varies from 5°C (vinylchloride) to 25°C (depending on analytes); raise at 5°C/min to final temperature of 220°C

9.4.7 Inject sample onto GC by flash-heating the -30°C mini-trap to 250°C

9.4.8 After every 10 samples, analyse a blank cartridge (9.1.4), as in 9.4.5 to 9.4.7, inclusive

9.4.9 Analyse blank (8.3) and calibration sample (8.4), as in 9.4.1 to 9.4.7, inclusive.

10. METHOD OF CALCULATION

10.1 The mass concentration, ρ_a (μg/L), of an analyte in a water sample is given by,

$$\rho_a = \rho_c \left(\frac{R_a - R_b}{R_c - R_b} \right)$$

where

R_a = detector response for the analyte in water sample (9.4.7)

R_c = detector response for the analyte in the calibration sample (9.4.9)

R_b = detector response for blank sample (9.4.8)

ρ_c = concentration of analyte in calibration sample, 8.4 (μg/L)

10.2 The above value of ρ_c is given by,

$$\rho_c = V_s \rho_s / V_w$$

where

V_s = volume of stock standard solution of analyte used in 8.4 (μL)

ρ_s = concentration of the analyte in the stock standard solution (g/L)

V_w = volume of water (determined by weight) in calibration sample 8.4 (L)

10.3 The mass concentration, ρ_s, of the calibration compound in a stock standard solution is given by,

$\rho_s = V_c d_c / 10^3 V_m$ for liquid compounds

and W_c / V_m for gaseous compounds,

where

V_c = volume of added calibration compound (section 6) at 0°C (μL)

d_c = mass density of calibration compound at 0°C (g/L)

V_m = volume of the methanol stock standard solution (mL)

W_c = weight of added calibration gas (mg).

11. REPEATABILITY AND REPRODUCIBILITY

No data are available for repeatability and reproducibility. Under optimum conditions the following results have been obtained for recoveries and detection limits:

Compound	% Recovery (in tap-water)	Detection limit (µg/L)		
		Tap-water		Surface-water
		ECD	MS	MS
Bromodichloromethane	> 90	0.05	< 0.05	0.1
Bromomethane	> 90	0.05	< 0.05	0.1
Chloromethane	> 95	0.05	0.01	0.05
3-Chloropropene	> 95	0.1	0.01	0.05
Dibromochloromethane	> 90	0.05	< 0.05	0.1
1,2-dibromo-3-chloropropane	> 90	0.05	< 0.05	0.1
1,2-dibromoethane	> 90	0.05	< 0.05	0.1
Dichlorodifluoromethane	> 95	0.01	0.01	0.05
1,2-Dichloroethane	> 90	0.1	< 0.05	0.1
(Z)-1,2-Dichloroethane	> 95	0.1	< 0.05	0.1
(E)-1,2-Dichloroethene	> 95	0.1	< 0.05	0.1
Dichloromethane	> 95	0.1	< 0.05	0.1
Hexachloroethane	> 95	0.01	0.01	0.05
Tetrachloroethene	> 95	0.01	0.01	0.05
Tetrachloromethane	> 95	0.01	0.01	0.05
1,1,1-Trichloroethane	> 95	0.01	0.01	0.05
Trichloroethene	> 95	0.01	0.01	0.05
Trichlorofluoromethane	> 95	0.01	0.01	0.05
Trichloromethane	> 95	0.01	0.01	0.05
1,1,1-Trifluoro-2-bromo-2-chloroethane	> 90	0.05	< 0.05	0.1

12. NOTES ON PROCEDURE

12.1 In most cases, the precision of the method is different for tap-water, surface-water and effluents, due to the complexity of the sample and changing concentration levels.

12.2 For surface-water, preliminary dilution of the sample (using "static" head-space screening) is needed before the purging step in some cases. At the sub-ng/L level, the precision for surface-water is < 50% in most cases. At the 1-5 µg/L level, a precision of < 20% can be reached, falling to < 10% at the 5-10 µg/L level.

12.3 Analysis in the sub-µg/L range can only take place in clean laboratory air, which often requires purification with activated carbon. Working with analytes in the laboratory must be avoided as far as possible.

12.4 To avoid break-through of cartridges the safe sampling gas volume must be known. In general, a gas flow-rate of 10-150 mL/min, a cartridge temperature of less than 20°C and a sampling time of 20-40 min will not give rise to break-through, except in some cases. When the retention volume of an organic compound at 20°C under the conditions obtaining is known experimentally, the safe sampling volume is 50% of this value.

12.5 If compounds with a wide boiling range must be analysed, a combination of Tenax-GC and Ambersorb must be used. The cartridge is filled with 130 mg Tenax-GC and 70 mg Ambersorb on top of it. The purge gas must first pass the Tenax-GC. (For desorption, the direction is reversed.)

12.6 For most compounds there is a good correlation between retention volume and boiling point. Some safe sampling volumes are given in the following Table:

Organic compound	Boiling point (°C)	Safe sampling volume[a] (litres per g of Tenax-GC)
Chloromethane	- 24	Not applicable[b]
Dichloromethane	40	1.5
Trichloromethane	62	9.3
Tetrachloromethane	76	3.1
1,1-Dichloroethane	57	7.0
1,2-Dichloroethane	84	27
1,1,1-Trichloroethane	74	1.9
1,1,2-Trichloroethane	114	70
1,1,2,2-Tetrachloroethane	146	850
Vinyl chloride	- 14	Not applicable[b]
1,1-Dichloroethene	32	1.1
1,2-Dichoroethene	55	5.4
Trichloroethene	87	28
Tetrachloroethene	121	240
3-Chloro-1-propene	45	5
Bromomethane	4	Not applicable[b]
Trichlorofluoromethane	24	Not applicable[b]
2-chloro-2-bromo-1,1,1-trifluoroethane	50	0.8

[a] R.H. Brown & Purnell, C.J. (1979) Collection and analyses of trace organic vapour pollutants in ambient atmosphere. J. Chromatogr., 178, 79-90

[b] Ambersorb or Porapak N must be used to trap these compounds.

12.7 Optimum purging conditions can be reached at pH 7.1, while sodium sulfate (3.35 mol/L) is added to improve recoveries of analytes. Note that exposure of the volatile compounds to direct sunlight must be avoided.

13. SCHEMATIC REPRESENTATION OF PROCEDURE

Collect water sample and prepare blank sample
and calibration sample
Transport and store under identical conditions

↓

Check instrument performance by analysing calibration
sample (as below)
Check analytical system for memory effects
Check laboratory air for purity

↓

Fill purging vessel with 1 L of sample (4°C)
Heat to 80°C while maintaining N_2 flow (40 mL/min)
Two Tenax cartridges are placed in series on top of condenser

↓

After 20-30 min, cap cartridges and store if necessary

↓

Uncap cartridges and flash-desorb (250°C) in injection system
desorption unit
Trap analytes at -30°C

↓

Flash-desorb analytes from -30°C trap
onto GC column
Record detector responses

↓

Repeat purging + analysis for blank samples
and calibration samples
Record detector responses

↓

After every ten samples, analyse a blank cartridge

↓

Calculate analyte concentration in water sample
by comparison of detector responses for water sample
and calibration sample for a given analyte (corrected for blank response)

14. ORIGIN OF THE METHOD

National Institute of Public Health and Environmental Hygiene
P.O. Box 1
3720 BA Bilthoven, The Netherlands

Contact points: G.J. Piet
W.C.M.M. Luijten
P.C.M. van Noort

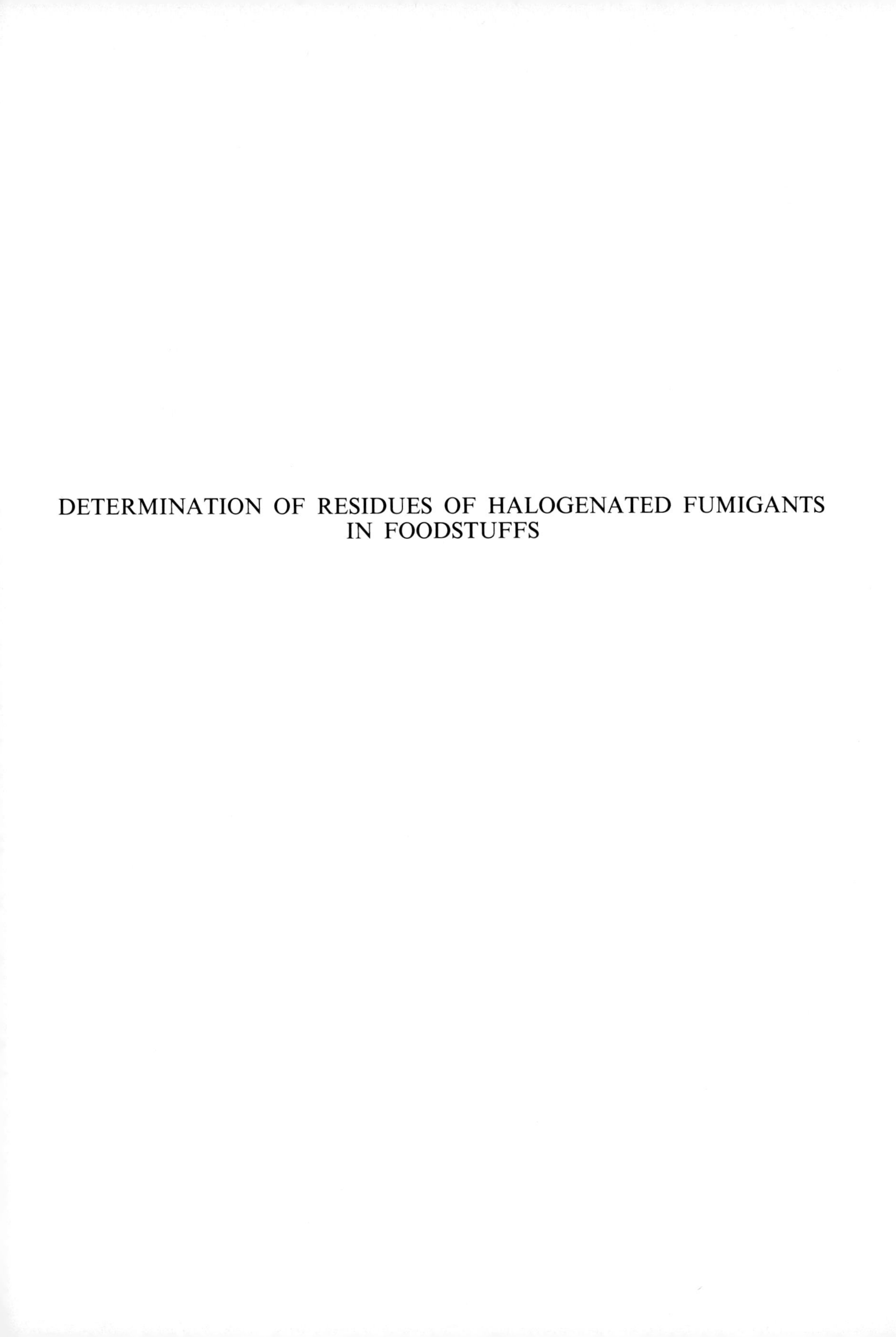

DETERMINATION OF RESIDUES OF HALOGENATED FUMIGANTS IN FOODSTUFFS

DETERMINATION OF RESIDUES OF HALOGENATED FUMIGANTS IN FOODSTUFFS

INTRODUCTION

K.A. Scudamore & A. MacKenzie Peers

1. FUMIGANTS AND FUMIGATION

Fumigation is used both as an emergency action when infestation is found in a food commodity and also to guarantee that a consignment is insect-free. It confers, however, no lasting protection, such as could be obtained by using less-volatile insecticides.

Volatile chemical compounds have been widely used as fumigants to disinfest food commodities containing insects and mites. A number of the halogenated aliphatic alkanes and alkenes have appropriate properties. They are volatile, toxic to insects, relatively cheap and readily available. Some of them, such as bromomethane (methyl bromide) and tetrachloromethane (carbon tetrachloride), are also nonflammable.

The use of gases or volatile liquids is necessary to permeate bulk commodities. As insects and mites are killed by the action of the compounds in the vapour state, high concentrations are necessarily present during treatment. Sorption of these compounds by a food commodity may thus be considerable and, although the amount of sorbed vapour may fall rapidly after the end of treatment (both by volatilization and chemical reaction), considerable amounts can remain associated with the foodstuff for a long period of time. The persistence of a volatile compound will depend on its chemical and physical properties, the nature and condition of the foodstuff and the conditions of storage. Because of the toxic nature of the compounds used, methods of determination of residual fumigants in foodstuffs are necessary and have been developed in a number of countries.

2. ANALYTICAL METHODS

2.1 Selection of methods

The analytical procedures included in this Manual for the determination of halogenated fumigant residues have been selected from those published during the last 20 years. Criteria for the inclusion of a method are that it has been (a) collaboratively tested, or used by the Author or personally recommended to him by other workers in the field and (b) found to be simple and reliable. Omission of a method does not imply that it is unsuitable or unreliable, but only that it does not meet the criteria applied above.

2.2 Scope of the methods

Methods are required both for the determination of residues of a specific compound and for the components of a mixture of several compounds. In addition, they should be applicable to routine screening of foodstuffs. To meet these requirements, methods have been developed for the determination of as many compounds as possible, using procedures in which parameters can be optimized for each individual compound if required.

2.3 Analytical considerations

2.3.1 Sampling

The following recommendations are taken from Bates, J.A.R., ed. (1982) IUPAC Reports on Pesticides (16). Pure & Appl. Chem., 54, 1400-1414. The full report (ibid., 1361-1450) should be consulted for more complete information.

Obtaining representative samples from bulk containers is difficult and, if possible, the sample should be taken at frequent, regular intervals from the stream during transfer to another container. A probe sample may be acceptable if it is possible to reach every part of the container and a large number of samples are mixed. Sampling of smaller lots, samples stored in bags or those composed of a large number of individually packaged items are discussed in the above report.

Volatility of fumigant residues require that handling and exposure of samples should be kept to a minimum. Samples must be placed immediately in clean inert containers. Nylon bags or glass bottles with screw caps, fitted with aluminium or teflon liners, are suitable. Avoid other plastic containers (e.g., PVC) which may interfere with the analytical method or be permeable to fumigant vapours. If cans are used, they must be checked for presence of oil films, etc., which could interfere with analysis. Glass containers should be thoroughly cleaned and rinsed with pesticide-free solvents, then dried before use.

Samples should be analysed as soon as possible, but if storage is required the sample should be stored whole, at or below -20°C. It is always advisable to protect samples and/or extracts from needless exposure to light.

2.3.2 Volatility

Fumigants are gases or volatile liquids and the use of methods that involve grinding could result in low or inconsistent recoveries. In the methods described here, the sample is taken in an undivided state and added directly to the extraction medium.

2.3.3 Reactivity

Some of the compounds used as fumigants are reactive at elevated temperatures and it is thus desirable to use methods that do not involve heating or refluxing. Most of the methods described are carried out at room temperature.

2.3.4 Concentration of extracts

The similarity between the boiling points of the solvents used for extraction and the compounds used as fumigants makes it impracticable to evaporate the solvent as a means of increasing the sensitivity of the method.

2.3.5 Chromatography of extracts

The solvent and its impurities will tend to elute from gas chromatographic columns with retention times similar to those of the compounds being determined. Use of multi-component solvents increases the complexity of gas chromatographic separations, particularly in multi-residue screening procedures. Careful checking of chromatographic purity of solvents is thus an important consideration.

2.3.6 Extractibility

The volatility of compounds used as fumigants facilitates diffusion into the individual grains or food granules. However, strong interactions with the food matrix often occur and the degree of interaction of the residue with the foodstuff is reflected in the rate at which it is released into the extraction solvent. Determination of extraction efficiencies is thus difficult and not often attempted. Some published methods for fumigant residues give data showing high recoveries of fumigant which has been added directly to the commodity-solvent matrix. This does no more than indicate losses or concentration effects during the analytical procedure. Typically, recovery of fumigants from unground commodities left to soak in the solvent increases with time and reaches a maximum value in 24 to 72 h, depending on the residue and the foodstuff. However, as the time elapsed since the original treatment increases, the residues remaining are progressively those more firmly bound and thus more difficult to remove, leading to slower and less efficient extraction. In using any of the following methods, this should always be considered.

2.3.7 Confirmation of identity

Coincidence of retention times of an unknown compound with that of a known standard is not proof of chemical identity. Identification may be made more certain by checking retention times on different chromatographic columns, but positive identification requires the use of other analytical methods.

METHOD 17

MULTI-RESIDUE GAS CHROMATOGRAPHIC METHOD FOR DETERMINATION OF FUMIGANT RESIDUES IN CEREAL GRAINS AND OTHER FOODS

K.A. Scudamore

Adapted in major part from Official methods of analysis of the Association of Official Analytical Chemists (13th Ed., 1980) sections 29.056-29.057 and from 'The Determination of Residues of Volatile fumigants in Grain" - a report by the Panel on fumigant residues in grain of the Committee for analytical methods for residues of pesticides and veterinary products in foodstuffs of the U.K., Ministry of Agriculture, Fisheries and Food (1974). Analyst, 99, 575-576

1. SCOPE AND FIELD OF APPLICATION

This method has been collaboratively tested for the determination of carbon tetrachloride, chloroform, trichloroethylene and 1,2-dibromoethane in fumigated wheat and maize, and has been found to be applicable to the following compounds:

Bromochloromethane	
Bromomethane	(Methyl bromide)
1,2-Dibromoethane	(Ethylene dibromide)
1,2-Dichloroethane	(Ethylene dichloride)
Iodomethane	(Methyl iodide)
Nitrotrichloromethane	(Chloropicrin)
Tetrachloroethene	(Perchloroethylene)
Tetrachloromethane	(Carbon tetrachloride)
1,1,1-Trichloroethane	(Methyl chloroform)
Trichloroethene	(Trichloroethylene)
Trichloromethane	(Chloroform)

This method has also been used successfully for the determination of residues of individual fumigants in a wide range of foodstuffs, including barley, oats, rice, sorghum, wheatflour, bread, groundnuts, sunflower seeds, rape seed, soya meal, lentils, chick peas, dried fruit and cocoa beans. Approximate limits of detection for each compound are given in Table 1.

Table 1. Retention times and sensitivity guide

Compound	Retention time		Sensitivity[a] (Apiezon-L) (pg)	Limit of detection[b] (mg/kg)
	Apiezon L 110°C (min)	PPG[c] 120°C (min)		
Methyl bromide	2.2	1.8	30 (70°)	0.15
Acetonitrile	2.2	2.5	-	-
Acetone	2.3	2.0	-	-
Methyl iodide	4.0	2.4	0.5	0.002
Bromochloromethane	5.8	3.8	3	0.01
Chloroform	5.9	3.8	6	0.02
Methyl chloroform	7.0	3.3	2.5	0.01
Ethylene dichloride	7.0	4.1	1 000	4
Carbon tetrachloride	8.7	3.4	0.5	0.002
Trichloroethylene	11.0	4.4	8	0.03
Chloropicrin[d]	15.5	6.4	1.5	0.005
Ethylene dibromide	23.5	9.7	14	0.05
Perchloroethylene	25.5	6.3	5	0.02

[a] 2% Full-scale deflection at a noise level of 0.5%

[b] Limit with commodities free from chromatographic interference

[c] Polypropylene glycol

[d] Use acetonitrile (reacts with acetone).
The residue is extracted from the whole commodity by soaking in solvent for 48 h. The number of samples which can be processed in this time is limited by the throughput of the gas chromatograph.

2. REFERENCES

Goodship, G. & Scudamore, K.A. (1982) disappearance of 1,1,1-trichloroethane from fumigated cereal grains and other foodstuffs during storage and processing. Pestic. Sci., 13, 139-148

Heuser, S.G. & Scudamore, K.A. (1967) Determination of ethylene chlorohydrin, ethylene dibromide and other volatile fumigant residues in flour and whole wheat. Chem. Ind., 1557-1560

Heuser, S.G. & Scudamore, K.A. (1968) Determination of residual acrylonitrile and ethylene dichloride in cereals after fumigation. Chem. Ind., 1154-1157

Heuser, S.G. & Scudamore, K.A. (1968) fumigant residues in wheat and flour: solvent extraction and gas chromatographic determination of free methyl bromide and ethylene oxide. Analyst, 93, 252-258

Heuser, S.G. & Scudamore, K.A. (1969) Determination of fumigant residues in cereal and other foodstuffs: a multi-detection scheme for gas chromatogrpahy of solvent extracts. J. Sci. Food Agric., 20, 566-572

Heuser, S.G. & Scudamore, K.A. (1970) Selective determination of ionized bromide and organic bromides in foodstuffs by gas-liquid chromatogrpahy with special reference to fumigant residues. Pestic. Sci., 1, 244-249

Jagielski, J., Scudamore, K.A. & Heuser, S.G. (1978) Residues of carbon tetrachloride and 1,2-dibromoethane in cereals and processed foods after liquid fumigant grain treatment for pest control. Pestic. Sci., 9, 117-126

Ministry of Agriculture, Fisheries & Food (U.K.) (1974) Determination of residues of volatile fumigants in grain. Analyst, 99, 570-576

Scudamore, K.A. & Heuser, S.G. (1970) Residual free methyl bromide in fumigated commodities. Pestic. Sci., 1, 14-17

Scudamore, K.A. & Heuser, S.G. (1973) Determination of carbon tetrachloride in fumigated cereal grain during storage. Pestic. Sci., 4, 1-12

3. DEFINITIONS

Not applicable

4. PRINCIPLE

Residues are extracted from an unground sample of the commodity by soaking in acetone:water or in acetonitrile:water. Water is removed from the organic solvent and the dried solution injected into a gas chromatograph, fitted with an electron-capture detector.

5. HAZARDS

Acetonitrile and fumigants are toxic solvents and should be handled in a fume hood. Methyl bromide is an odourless, toxic gas at room temperature. Preparation of standards from ice-cold liquid methyl bromide should be performed in a fume hood.

6. REAGENTS[1]

All reagents should be of analytical grade. The solvents cited below must be kept isolated from other halogenated solvents.

Acetone
Acetonitrile
Distilled water
Sodium chloride
Calcium chloride (anhydrous, 3-8 mesh)
Carbon tetrachloride
Chloroform
Chloropicrin
Ethylene dibromide
Ethylene dichloride
Methyl bromide (supplied in cylinders)
Methyl chloroform
Methyl iodide
Perchloroethylene
Trichloroethylene
Nitrogen (oxygen-free for gas chromatography)
Acetone:water (5:1, v/v)
Acetonitrile:water (5:1, v/v)

Standard solutions of each fumigant in acetone or acetonitrile are prepared singly, or as a multicomponent mixture, by serial dilution of weighed amounts.

---

[1] Reference to a company and/or product is for the purpose of information and identification only and does not imply approval or recommendation of the company and/or product by the International Agency for Research on Cancer, to the exclusion of others which may also be suitable.

7. APPARATUS[1]

7.1 Laboratory glassware

Conical flasks	250 mL, with 24/29 ground-glass sockets and glass stoppers
Graduated cylinders	25 mL and 10 mL, with glass stoppers
Micro syringes	1 µL or 5 µL capacity

7.2 Gas chromatography

Any suitable gas chromatograph fitted with an electron-capture detector and a recorder or integrator for measurement of detector response.

GC columns:

i) 4 m x 2.2 mm i.d. glass column, packed with 15% (w/w) polypropylene glycol (LB 550X, Ucon fluid) on Chromosorb W, 80-100 mesh.

(ii) 4 m x 2.2 mm i.d. glass column, packed with 15% (w/w) Apeizon L on Chromosorb P, 80-100 mesh.

8. SAMPLING

Obtain a representative sample (see INTRODUCTION, p. 348, section 2.3.1). If this cannot be analysed immediately, store in a vapour-tight container in a freezer, isolated from other volatile halogenated compounds.

9. PROCEDURE

9.1. Blank tests

9.1.1 Inject an aliquot of the extraction solvent into the gas chromatograph to check for presence of peaks which might interfere.

9.1.2 If an uncontaminated sample of foodstuff is available, extract and process as in sections 9.4-9.6 to check for interference from the commodity constituents.

---

[1] Reference to a company and/or product is for the purpose of information and identification only and does not imply approval or recommendation of the company and/or product by the International Agency for Research on Cancer, to the exclusion of others which may also be suitable.

9.1.3 Inject an aliquot of each standard solution to check its purity.

9.2 Check test

Not applicable

9.3 Test portion

50 g of well-mixed sample

9.4 Sample extraction and removal of water

9.4.1 Weigh a 50 g portion of the well-mixed sample and quickly immerse in 150 mL of acetone:water (5:1, v/v) or acetonitrile:water (5:1, v/v) in a 250 mL conical flask and insert the stopper.

9.4.2 Allow the flask to stand for 48 h in the dark at room temperature (20 to 25°C), with occasional shaking.

9.4.3 Pour a 20 mL portion of the supernatant liquid into a 25 mL graduated cylinder and add 2 g sodium chloride.

9.4.4 Stopper and shake the cylinder vigourously for 2 min, then allow to stand until the two layers separate (about 30 min).

9.4.5 Pour a 10 mL portion of the clear upper layer into a 10 mL graduated cylinder and add 1 g calcium chloride.

9.4.6 Stopper, shake vigourously for 2 min and allow to stand for 30 min with occasional shaking.

9.5 Gas chromatography conditions

Carrier gas	Nitrogen; flow-rate 30-40 mL/min
Injection temperature	150°C
Detector temperature	250°C
Oven temperature	70-140°C (isothermal)

NOTE: The above parameters will depend on the gas chromatograph and column used. Retention times for each compound at a compromise temperature are given in Table 1 (section 1). An oven temperature of 70°C is appropriate for methyl bromide, while for the more strongly-retained compounds, temperatures up to 140°C may be used. It is not recommended that the PPG column be used above 130°C.

9.6 Determination of residue in sample

9.6.1 Inject aliquots of volume equal to that employed in 9.7, using the clear upper layer of the dried sample extract (9.4.6).

9.6.2 If the response is greater than that of the highest standard, dilute as necessary with dry acetone or acetonitrile and re-inject.

9.7 Preparation of a standard curve

Inject a fixed volume (usually in the range 0.5-5 µL) of each standard solution into the injection block of the gas chromatograph (GC) using a suitable microlitre syringe.

Construct a calibration curve of response against mass for each compound injected.

9.8 Recovery

9.8.1 Add 140 mL of a mixture of acetone:water (5:1, v/v) or acetonitrile:water (5:1, v/v) to 50 g of sample, as in 9.4.1.

9.8.2 spike by adding 10 mL of acetone:water or acetonitrile:water containing a known amount of the compound of interest. Proceed as described in sections 9.4-9.6 and calculate the recovery.

NOTE: This provides a recovery figure for the analytical procedure only. It is very difficult to determine experimentally the extraction efficiency for volatile compounds absorbed in the samples.

10. METHOD OF CALCULATION

The mass fraction, w, of fumigant residue in the sample is given by,

$$w = \frac{mVF}{vM} \quad \text{(mg/kg)}$$

where

m = mass of analyte injected in 9.6.1, calculated from GC response and standard curve, 9.7 (ng)

V = total volume of organic solvent in 9.4.1 (mL)

F = dilution factor (see 9.6.2)

v = volume of extract injected into GC in 9.6.1 (µL)

M = mass of sample taken in 9.4.1 (g)

NOTE: The method assumes that water is removed from the organic solvent without loss of the organic phase. If 150 mL of 5:1 (v/v) organic:water extraction solution is used initially, V is taken as 125 mL.

11. REPEATABILITY AND REPRODUCIBILITY

The method has been collaboratively tested for residues of carbon tetrachloride, chloroform, dibromoethane and trichloroethane and the results published (Ministry of Agriculture, Fisheries and Food (U.K.), 1974).

12. NOTES ON PROCEDURE

Not applicable

13. SCHEMATIC REPRESENTATION OF PROCEDURE

Add 150 ml of acetone:water (5:1, v/v)
or acetonitrile:water (5:1, v/v)
to 50 g of food commodity

↓

Leave for 48 h at room temperature in the dark

↓

Decant 20 mL, shake with 2 g sodium chloride
and leave to stand for 30 min

↓

Decant 10 mL of upper layer
and shake vigourously with 1 g calcium chloride.
Stand for 30 min

↓

Inject upper layer directly or after dilution into GC
and compare peak height or area obtained
with those of standard solutions

14. ORIGIN OF METHOD

Pest Control Chemistry Department
Ministry of Agriculture, Fisheries and Food
Slough Laboratory
London Road
Slough, Berkshire, UK

Contact point: Mr K.A. Scudamore

METHOD 18

GAS CHROMATOGRAPHIC DETERMINATION OF CHLOROFORM, CARBON TETRACHLORIDE, ETHYLENE DIBROMIDE AND TRICHLOROETHYLENE IN CEREAL GRAINS AFTER DISTILLATION

K.A. Scudamore

1. SCOPE AND FIELD OF APPLICATION

This method has been used in the author's laboratory for the determination of the title compounds and also for 1,2-dichloroethane. In samples containing carbon tetrachloride, the apparent formation of small amounts of chloroform during extraction and distillation has been reported. Limits of detection are stated as being in the mg/kg range and in the author's experience are comparable to those obtainable using Method 17.

A result can be obtained in 2 to 3 hours. However the nature of the equipment necessary limits the number of samples which can be processed. (This method would have advantages where a quick result for a few samples is required.)

2. REFERENCE

Bielorai, R. & Alumot, E. (1966) Determination of residues of a fumigant mixture in cereal grain by electron-capture gas chromatography. J. Agric. Food Chem., 14, 622-625

3. DEFINITIONS

Not applicable

4. PRINCIPLE

The sample of unground cereal grains is extracted by refluxing with water. The fumigant residues are removed from the aqueous extract by steam distillation and trapped in a small volume of toluene in a continuous distillation apparatus. After drying, the compounds collected are determined by gas chromatography (GC).

5. HAZARDS

Fumigants and toluene are toxic solvents. Care should be taken to prevent exposure of operators to vapours of these compounds.

6. REAGENTS[1]

All should be of analytical grade. Toluene should be kept isolated from chlorinated solvents.

Toluene
Distilled water
Sodium sulfate (anhydrous)
Ethylene dibromide
Carbon tetrachloride
Trichloroethylene
Chloroform
Nitrogen (oxygen-free for GC)
Standard solutions of each fumigant in toluene are prepared by serial dilution.

7. APPARATUS[1]

7.1 Laboratory glassware

Flasks, round-bottomed	500 mL, with 50/50 ground-glass joint
Flasks, volumetric	25 mL, with glass stoppers
Microsyringes	1 to 5 μL capacity

Extraction and distillation apparatus shown in Figure 1.

7.2 Gas chromatography

Any suitable gas chromatograph fitted with an electron-capture detector and a recorder or integrator for measurement of detector response.

A 2 m, all-glass column, packed with 10% (w/w) silicone oil DC-710 on 80-100 mesh Chromosorb W HMDS (or other suitable column).

---

[1] Reference to a company and/or product is for the purpose of information and identification only and does not imply approval or recommendation of the company and/or product by the International Agency for Research on Cancer, to the exclusion of others which may also be suitable.

FIG. 1.

EXTRACTION AND DISTILLATION APPARATUS

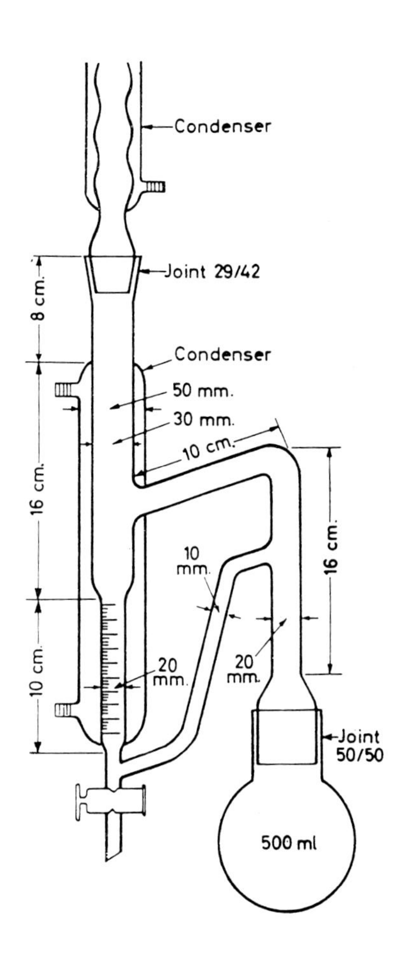

8. SAMPLING

Obtain a representative sample (see INTRODUCTION, p. 348, section 2.3.1). If this cannot be analysed immediately, store in a vapour-tight container in a freezer, isolated from other volatile halogenated compounds.

9. PROCEDURE

9.1 Blank tests

9.1.1 Inject an aliquot of pure dry toluene into the GC to check for presence of peaks which might interfere.

9.1.2 If an uncontaminated sample of foodstuff is available, extract and process as in sections 9.4-9.6 to check for interference from the commodity constituents.

9.1.3 Inject an aliquot of each standard solution to check its purity.

9.2 Check test

Not applicable

9.3 Test portion

10 to 50 g of well-mixed sample.

9.4 Sample extraction and distillation of residues

9.4.1 Weigh a 10-50 g portion of the well-mixed sample and add to 250 mL of water in a 500 mL round-bottom flask.

9.4.2 Place the flask in a heating mantle and add 5 mL of toluene.

9.4.3 Connect the distillation apparatus and condenser and fill the collection tube with water.

9.4.4 Run ice-cold water through the condenser and bring the contents of the distillation flask to the boil.

9.4.5 Distil for 30 min and allow the apparatus to cool.

9.4.6 Run off the water layer from the collection tube and run the toluene layer into a 25 mL conical flask, fitted with a glass stopper and containing 2 g of anhydrous sodium sulfate.

9.4.7 Rinse the collection tube several times with small aliquots of toluene and add to the volumetric flask.

9.4.8 Leave to dry for about 1 h and transfer to a 25 mL volumetric flask with washings from the sodium sulfate. Make up to volume.

9.5 Gas chromatography

Nitrogen carrier gas, flow-rate 20 mL/min
Column oven temperature, 60°C
Injection temperature, 150°C
Detector temperature, 250°C

NOTE: These parameters will depend on the GC and column used.

9.6 Determination of residues in sample

9.6.1 Inject aliquots of volume equal to that employed in 9.7, using the dried toluene solution from step 9.4.8.

9.6.2 If the response is greater than that from the most concentrated standard, dilute with dried toluene and re-inject.

9.7 Preparation of a standard curve

Inject a fixed volume (usually in the range 0.5-5 µL) of each standard solution into the injection block of the gas chromatograph, using a suitable microlitre syringe.
Construct a calibration curve of response against mass for each compound injected.

9.8 Recovery

9.8.1 Add 5 mL of toluene containing known amounts of fumigants to 50 g of commodity and 250 mL of water, contained in a 500 mL round-bottom flask.

9.8.2 Proceed as described in sections 9.4-9.6 and calculate the recovery.

NOTE: This provides a recovery figure for the analytical procedure only. It is very difficult to determine experimentally the extraction efficiency for volatile compounds absorbed in the samples.

10. METHOD OF CALCULATION

The mass fraction, w, of fumigant residue in the sample is given by,

$$w = \frac{mVF}{vM} \text{ (mg/kg)}$$

where

m = mass of analyte injected in 9.6.1, calculated from GC response and standard curve, 9.7 (ng)

V = final volume of toluene in 9.4.8 (mL)

F = dilution factor (see 9.6.2)

v = volume of extract injected into the GC in 9.6.1 (µL)

M = mass of sample taken in 9.4.1 (g)

11. REPEATABILITY AND REPRODUCIBILITY

No information is available

12. NOTES ON PROCEDURE

Not applicable

13. SCHEMATIC REPRESENTATION OF PROCEDURE

Add 10-50 g of cereal grain to 250 mL of water

↓

Add 5 mL toluene, bring to the boil and distill

↓

After 30 min, cool distillate and run off water layer

↓

Run the toluene into a 25 mL conical flask containing 2 g anhydrous sodium sulfate with washings from the collection tube

↓

Leave to dry for 1 h and transfer to a 25 mL volumetric with washings and make up to volume

↓

Inject into GC and compare peak height or area observed with those obtained with standard solutions

14. ORIGIN OF THE METHOD

Department of Animal Nutrition
Agricultural Research Organisation
The Volcani Center
P.O. Box 6
Bet-Dagan, Israel

Contact point: Dr R. Bielorai

METHOD 19

GAS CHROMATOGRAPHIC DETERMINATION OF CHLOROFORM, CARBON TETRACHLORIDE, TRICHLOROETHYLENE AND ETHYLENE DIBROMIDE IN CEREAL GRAINS AFTER DIRECT EXTRACTION

K.A. Scudamore

1. SCOPE AND FIELD OF APPLICATION

This method is suitable for the determination of the title compounds in cereal grains. The limits of detection are similar to those attainable with Method 17.

The residue is extracted from the whole commodity by soaking in the appropriate solvent for 48 h. The number of samples which can be processed in this time is limited by the throughput of the gas chromatograph.

2. REFERENCE

Greve, P.A. & Hogendoorn, E.A. (1979) Determination of fumigant residues in grain. Med. Fac. Landbouww. Rijksuniv. Gent, 44/2, 877-884

3. DEFINITIONS

Not applicable

4. PRINCIPLE

A mixture of hydrochloric acid and petroleum ether is used to extract the fumigant compounds. The whole grains swell due to uptake of water and the compounds released to the petroleum ether layer are determined by gas chromatography.

5. HAZARDS

Fumigants are toxic solvents. Care should be taken to prevent exposure of operators to vapours of these compounds.

6. REAGENTS[1]

All reagents should be of analytical grade if possible. Extraction solvents must be stored away from halogenated solvents.

Hydrochloric acid
Petroleum ether (40°-60°C boiling range)
Distilled water
Ethylene dibromide
Carbon tetrachloride
Trichloroethylene
Chloroform
Nitrogen (oxygen-free for gas chromatography (GC))
Prepare standards of each fumigant in petroleum ether by serial dilution.

NOTE: Petroleum ether is a mixture of compounds. If problems occur due to interference (especially with the use of alternative GC columns), *n*-heptane, *n*-hexane or *iso*-octane may be useful substitute solvents.

7. APPARATUS[1]

7.1 Laboratory glassware

Conical flasks	Capacity 150 mL, with 24/29 glass sockets and glass stoppers
Micro syringes	1 to 5 μL capacity

7.5 Gas chromatography

Any suitable gas chromatograph fitted with an electron-capture detector and a recorder or integrator for measurement of detector response.

A 1.2 m x 2 mm i.d. column, packed with 5% Carbowax-20M on Chromosorb 101, 80-100 mesh.

---

[1] Reference to a company and/or product is for the purpose of information and identification only and does not imply approval or recommendation of the company and/or product by the International Agency for Research on Cancer, to the exclusion of others which may also be suitable.

NOTE: 1,2-Dichloroethane, which has been commonly used for fumigation in mixtures with carbon tetrachloride, has the same retention time as the latter on the column indicated above. However, the electron-capture detector is several orders of magnitude more sensitive to carbon tetrachloride then to 1,2-dichloroethane. If doubt arises, an alternative column, such as 20% SP-2100: 0.1% Carbowax - 100 on Supelcoport, 100-120 mesh, will separate the two compounds.

8. SAMPLING

Obtain a representative sample (see INTRODUCTION, p. 348, section 2.3.1). If this cannot be analysed immediately, store in a vapour-tight container in a freezer, isolated from other volatile halogenated compounds.

9. PROCEDURE

9.1 Blank tests

9.1.1 Inject an aliquot of petroleum ether (or alternative solvent) into the gas chromatograph to check for the presence of peaks which might interfere.

9.1.2 If an unfumigated sample of the grain is available, process as in sections 9.4-9.6 to check for interference on the gas chromatogram.

9.1.3 Inject on aliquot of each standard to check its purity.

9.2 Check test

Not applicable

9.3 Test portion

50 g of well-mixed sample

9.4 Sample extraction

9.4.1 Weigh 50 g of a well-mixed sample of grain and add to a 150 mL conical flask containing 60 mL of 0.1 mol/L hydrochloric acid and 40 mL of petroleum ether (or alternative solvent).

9.4.2 Stopper the conical flask and allow to stand for at least 48 h at room temperature (20-25°C), with occasional shaking.

9.5 Gas chromatography conditions

Column temperature, 150°C
Other conditions to be optimized by the analyst.

9.6 Determination of residue in sample

9.6.1 Inject aliquots of volume equal to that employed in 9.7, using the petroleum ether layer from 9.4.2.

9.6.2 If the response is greater than that of the most concentrated standard, dilute as necessary with petroleum ether and re-inject.

9.8 Recovery

9.8.1 Spike 50 g samples of grain by adding known amounts of fumigant in petroleum ether.

9.8.2 Proceed as in sections 9.4-9.6 and calculate the recovery of each compound.

10. METHOD OF CALCULATION

The mass fraction, w, of fumigant residue in the sample is given by,

$$w = \frac{mVF}{vM} \text{ (mg/kg)}$$

where,

m = mass of analyte injected in 9.6.1, calculated from GC response and standard curve 9.7 (ng)

V = volume of petroleum ether in 9.4.1 (mL)

F = dilution factor (see 9.6.2)

v = volume of extract injected into the GC in 9.6.1 (µL)

M = mass of sample taken in 9.4.1 (g)

11. REPEATABILITY AND REPRODUCIBILITY

In a series of spiking experiments, Greve and Hogendoorn (1979) found recoveries of 91 ± 2% for chloroform, 99 ± 3% for carbon tetrachloride, 99 ± 4% for trichloroethylene and 93 ± 3% for 1,2-dibromoethane. Spiking levels corresponded to the range 0.01-40 mg/kg.

12. NOTES ON PROCEDURE

Not applicable

13. SCHEMATIC REPRESENTATION OF PROCEDURE

Add 50 g of cereal grain to 60 mL of 0.1 mol/L hydrochloric acid and 40 mL of petroleum ether

↓

Leave to stand for 48 h in the dark

↓

Inject upper layer directly, or after dilution, into the gas chromatograph

↓

Compare response with those from standard solutions

14. ORIGIN OF METHOD

Laboratory of Organic Chemistry and Environmental Hygiene
National Institute of Public Health
P.O. Box 1
Bilthoven, The Netherlands

Contact point: Dr P.A. Greve

METHOD 20

DETERMINATION OF METHYL BROMIDE IN GRAIN USING HEAD-SPACE ANALYSIS

K.A. Scudamore

1. SCOPE AND FIELD OF APPLICATION

This method has been used for the determination of methyl bromide residues in cereal grains, with a limit of detection of 10 µg/kg.

The residue is extracted from the whole commodity by soaking in the appropriate solvent for 24 h. The number of samples which can be processed in this time is limited by the throughput of the gas chromatograph.

2. REFERENCES

Greve, P.A. & Hogendoorn, E.A. (1979) Determination of fumigant residues in grain. Med. Fac. Landbouww. Rijksuniv. Gent, 44/2, 877-884

3. DEFINITIONS

Not applicable

4. PRINCIPLE

Methyl bromide is extracted from whole cereal grains by soaking in aqueous acetone at room temperature. A portion of the solvent extract is placed in a glass bottle closed with a screw cap fitted with a septum. After equilibration, an aliquot of the head-space vapour is injected into a gas chromatograph.

5. HAZARDS

Methyl bromide is an odourless, toxic gas at room temperature. Preparation of standards from ice-cold liquid methyl bromide should be performed in a fume cupboard.

6. REAGENTS[1]

All reagents should be of analytical grade. Extraction solvents must be isolated from halogenated solvents

Acetone	
Distilled water	
Methyl bromide	(supplied in gas cylinders)
Nitrogen	(oxygen-free, for gas chromatography)
Acetone:water	(9:1, v/v)
Standard solutions of methyl bromide	Prepare standard solutions in acetone: water (9:1, v/v), using liquid, ice-cold methylbromide. Prepare solutions weekly and store in glass bottles, closed with screw caps fitted with Teflon-faced liners. Fill each bottle to the brim.

7. APPARATUS[1]

7.1 Laboratory glassware

Glass bottles	300 mL capacity, equipped with screw cap
Glass bottles	Nominal 15 mL capacity, equipped with screw cap and rubber septum. The volumes of the 15 mL bottles must be within 1% of each other
Gas-tight syringes	250, 500 μL capacity

7.2 Gas chromatography

Any suitable gas chromatograph (GC), fitted with an electron-capture detector and a recorder or integrator for measurement of detector response.

A 1.2 m × 2 mm i.d. column, packed with 5% Carbowax - 20 m on Chromosorb 101, 80-100 mesh

---

[1] Reference to a company and/or product is for the purpose of information and identification only and does not imply approval or recommendation of the company and/or product by the International Agency for Research on Cancer, to the exclusion of others which may also be suitable.

8. SAMPLING

Obtain a representative sample (see INTRODUCTION, p. 348, section 2.3.1). If this cannot be analysed immediately, store in a vapour-tight container in a freezer, isolated from other halogenated volatiles.

9. PROCEDURE

9.1 Blank tests

9.1.1 Add 10 mL of freshly prepared acetone:water (9:1, v/v) to a 15 mL glass bottle. After equilibration, inject a suitable aliquot of the head-space vapour into the GC, using a gas-tight syringe (see 9.7.2) and check for possible interference with methyl bromide peak.

9.1.2 If an unfumigated sample of the foodstuff is available, proceed as in sections 9.4-9.6 and check for possible interference.

9.2 Check test

Not applicable

9.3 Test portion

See 9.4.2.

9.4 Sample extraction

9.4.1 Add 175 mL of acetone:water (9:1, v/v) to a 300 mL glass bottle equipped with screw cap. Weigh.

9.4.2 Fill the bottle with grain until the liquid reaches the brim. Close and weigh again. Calculate the weight of grain taken.

9.4.3 Leave at room temperature for at least 24 h.

9.4.4 Transfer 10 mL of extract from 9.4.3 to a 15 mL glass bottle, equipped with screw cap and rubber septum.

9.4.5 Close the bottle and let stand for 5 min at room temperature.

9.5 Gas chromatography conditions

Column temperature, 80°C
Other conditions to be optimized by the analyst.

9.6 Determination of residue in sample

9.6.1 From the head space above the sample extract (9.4.5), inject aliquots of volume equal to that employed in 9.7.2.

9.6.2 If the response is greater than that from the most concentrated standard, dilute the extract solution and re-inject.

9.7 Preparation of a calibration curve

9.7.1 Place 10 mL aliquots of standard solutions of methyl bromide in acetone:water (9:1) in 15 mL glass bottles, as in 9.4.4.

9.7.2 Inject a fixed volume (usually in the range 250-500 μL) of the head-space vapour into the GC by means of a gas-tight syringe.

9.7.3 Construct a calibration curve of response against mass of methyl bromide in the 10 mL aliquots.

9.8 Recovery

Place 175 mL of standard solutions of methyl bromide in acetone:water (9:1, v/v) in a 300 mL bottle and fill with grain as in 9.4.2. Proceed as described above for normal sample and calculate the recovery of methyl bromide.

NOTE: This provides a recovery figure for the analytical procedure only. It is very difficult to determine experimentally the extraction efficiency for volatile compounds absorbed in the samples.

10. METHOD OF CALCULATION

The mass fraction, w, of fumigant residue in the sample is given by,

$$w = \frac{mVF}{vM} \text{ (mg/kg)}$$

where

m = mass of analyte in 10 mL aliquot (9.4.4), calculated from GC response (9.6.1) and standard curve, 9.7 (µg)

V = volume of acetone:water in 9.4.1 (mL)

F = dilution factor (see 9.6.2)

v = volume of extract transferred to 15 mL bottle in 9.4.4 (mL)

M = mass of sample taken in 9.4.2 (g)

11. REPEATABILITY AND REPRODUCIBILITY

In a series of 19 recovery experiments over the spiking range corresponding to 0.93-10.4 mg/kg, the recovery obtained was 91 ± 3%.

12. NOTES ON PROCEDURE

Not applicable

13. SCHEMATIC REPRESENTATION OF PROCEDURE

Extract grain with 175 mL of acetone:water (9:1, v/v) by soaking for 24 h

↓

Transfer 10 mL extract to a 15 mL bottle

↓

Stand 5 min and inject head-space vapour into gas chromatograph

↓

Convert response to residue content using standard curve

14. ORIGIN OF THE METHOD

Unit for Residue Analysis
National Institute of Public Health
P.O. Box 1
Bilthoven, Netherlands

Contact point: Dr P.A. Greve

METHOD 21

DETERMINATION OF METHYL BROMIDE IN FOOD COMMODITIES USING DERIVATIVE GAS CHROMATOGRAPHY

K.A. Scudamore

1. SCOPE AND FIELD OF APPLICATION

This method is suitable for the determination of methyl bromide in a range of food commodities, including wheat, barley, maize, oats, wheat flour, rapeseed, groundnuts, cocoa beans, rice and dried milk. The limits of detection are below 10 µg/kg.

The residue is extracted from the whole commodity by soaking in the appropriate solvent for 24 h. The number of samples which can be processed in this time is limited by the throughput of the gas chromatograph.

2. REFERENCE

Fairall, R.J. & Scudamore, K.A. (1980) Determination of residual methyl bromide in fumigated commodities using derivative gas-liquid chromatography. Analyst, 105, 251-256

3. DEFINITIONS

Not applicable

4. PRINCIPLE

Methyl bromide is extracted from an unground sample of commodity by soaking in an acetone:water mixture. The methyl bromide is then converted to methyl iodide by reaction with sodium iodide in acetone:water solution. Methyl iodide is extracted with *n*-pentane and is determined by gas chromatography (GC).

5. HAZARDS

Methyl bromide is an odourless, toxic gas at room temperature. Preparation of standards from ice-cold, liquid methyl bromide should be performed in a fume cupboard.

6. REAGENTS

Analytical grade reagents should be used where possible. Extraction solvents should be kept isolated from chlorinated solvents or methyl iodide solutions.

Acetone	
Methyl bromide	supplied in gas cylinders
Methyl iodide	
_n_-Pentane	99.5%
Nitrogen	oxygen-free, for GC
Sodium iodide	
De-ionized or distilled water	
Acetone:water	5:1 (v/v)
Standard solutions of methyl iodide	Prepare standard solutions of methyl iodide in _n_-pentane
Standard solutions of methyl bromide	Prepare standard solutions of methyl bromide in acetone:water (5:1, v/v)

7. APPARATUS[1]

7.1 Laboratory glassware

Conical flasks	100 mL capacity, with 24/29 ground glass sockets and glass stopppers

---

[1] Reference to a company and/or product is for the purpose of information and identification only and does not imply approval or recommendation of the company and/or product by the International Agency for Research on Cancer, to the exclusion of others which may also be suitable.

Conical flasks	50 mL capacity, with 19/26 ground glass sockets and glass stoppers
Graduated cylinders	25 mL, with ground glass stoppers

7.2 Gas chromatography

Any suitable gas chromatograph, fitted with an electron-capture detector and a recorder or integrator for measurement of detector response.

A 4 m x 2 mm i.d. glass column, packed with 15% Apiezon L on Chromosorb P.

8. SAMPLING

Obtain a representative sample (see INTRODUCTION, p. 348, section 2.3.1). If this cannot be analysed immediately, store in a vapour-tight container in a freezer, isolated from other volatile halogenated compounds.

9. PROCEDURE

9.1 Blank tests

9.1.1 Inject an aliquot of pentane into the GC to check for presence of peaks which might interfere.

9.1.2 Add 20 mL acetone:water (5:1, v/v) to a 50 mL conical flask, then add 0.5 g sodium iodide. Proceed as in 9.5.2 to 9.5.7 inclusive. Inject an aliquot of the final solution to check for presence of interfering peaks.

9.1.3 Process a sample of the food commodity as specified in 9.4 and 9.5, omitting the addition of sodium iodide. Inject an aliquot of the final pentane extract (9.5.7) to check for interference.

9.1.4 If a sample of the unfumigated foodstuff is available, proceed as in 9.4 and 9.5. Inject an aliquot of the final pentane extract to check for interference.

9.1.5 Inject an aliquot of methyl iodide standard in pentane to check for purity.

NOTE: The checking procedure in 9.1.3 will confirm that no compound with a retention time similar to methyl iodide is extracted from the commodity. That described in 9.1.4 will confirm that no peak with a retention time corresponding to that of methyl iodide is produced during the derivatization step.

9.2 Check test

Not applicable

9.3 Test portion

10 to 20 g of well-mixed sample

9.4 Sample extraction

9.4.1 Weigh a 10-20 g portion of a well-mixed sample and quickly immerse in 60 mL of acetone:water (5:1, v/v) in a 100 mL conical flask and insert the glass stopper.

9.4.2 Allow to stand for 24 h in the dark at room temperature (20-25°C) with occasional shaking.

9.5 Derivatization (see Notes on Procedure, 12.2)

9.5.1 Transfer 20 mL of the extract to a 50 mL conical flask and add 0.5g sodium iodide.

9.5.2 Stopper and allow to stand for 1 h at room temperature.

9.5.3 Shake and transfer 3 mL of the reacted extract to a 25 mL graduated cylinder containing 5 mL pentane.

9.5.4 Add 15 mL of water, mix thoroughly and allow layers to separate.

9.5.5 Draw off or decant the upper layer as completely as possible and transfer to a 25 mL volumetric flask.

9.5.6 Wash the remaining acetone:water layer with two further 5 mL volumes of pentane and transfer the washings to the 25 mL flask.

9.5.7 Make the volume up to the mark with pentane.

9.6 Gas chromatography conditions

9.6.1 GC conditions

Column temperature, 100°C
Injection port temperature, 100°C
Detector temperature, 200°C
Carrier gas, nitrogen; flow-rate 25 mL/min

9.7 Determination of fumigant residue in sample

9.7.1 Inject aliquot of volume equal to that employed in 9.8, using the pentane solution 9.5.7 (but see Notes on Procedure, 12.3).

9.7.2 If response is greater than that obtained with most concentrated standard, dilute as necessary with pentane and re-inject.

9.8 Preparation of a standard curve

Inject a fixed volume (usually in the range of 0.5-5 μL) of each standard methyl iodide solution into the injection block of the GC. Construct a calibration curve of response against mass of methyl iodide injected.

9.9 Recovery

9.9.1 Take 20 mL aliquots of standard solutions of methyl bromide in acetone:water (5:1, v/v) and proceed as in 9.5.1 to 9.7.1, inclusive. Calculate the recovery of methyl bromide as methyl iodide.

9.9.2 Add standard solutions of methyl bromide in acetone:water (5:1, v/v) to 10-20 grams of commodity and proceed as in sections 9.4.1 to 9.7.1, inclusive. Calculate recovery of methyl bromide as methyl iodide in presence of commodity coextractives.

NOTE: This provides a recovery figure for the analytical procedure only. It is very difficult to determine experimentally the extraction efficiency for volatile compounds absorbed in the samples.

10. METHOD OF CALCULATION

The mass fraction, w, of methyl bromide residue in the sample is given by,

$$w = \frac{0.669\ mV_3V_1F}{V_4MV_2} \quad \text{(mg/kg)}$$

where,

0.669= ratio of molecucular weights of methyl bromide and methyl iodide
m = mass of methyl iodide injected in 9.7.1, calculated from GC response and standard curve, 9.8 (ng)
V_1 = volume of acetone:water employed in 9.4.1 (mL)
V_2 = volume of reacted extract used in 9.5.3 (mL)
V_3 = final volume of pentane solution, 9.5.7 (mL)
V_4 = volume of pentane solution injected into GC in 9.7.1 (μL)
M = mass of sample portion in 9.4.1 (g)
F = dilution factor (see 9.7.2)

NOTE: For the volumes specified in sections 9.4-9.5, (V_3V_1/V_2) = 500 mL

11. REPEATABILITY AND REPRODUCIBILITY

Recoveries of methyl bromide from solutions over the concentration range 2 to 4 000 ng/mL have been shown to be close to 100%.

12. NOTES ON PROCEDURE

12.1 Additional assurance concerning the identity of the residue can be obtained by measuring GC retention time on an alternative column, such as polypropylene glycol (see Method 17, section 7.2).

12.2 Chromatography of non-derivatized samples will give a peak for methyl bromide but will be useful only for confirming the presence of methyl bromide at higher concentrations (see Method 17, Table 1).

12.3 The sensitivity of the method may be increased five-fold by direct injection of the upper solvent layer obtained in 9.5.4.

13. SCHEMATIC REPRESENTATION OF PROCEDURE

Weigh out 10 to 20 g of commodity and add 60 mL of acetone:water (5:1, v/v)

↓

Allow to stand for 24 h at room temperature in the dark

↓

Transfer 20 mL, add 0.5 g sodium iodide and allow to stand for 20 min at room temperature

↓

With draw 3 mL and shake with 15 mL of water and 5 mL pentane

↓

Draw off the top layer and wash the aqueous acetone layer with two further 5 mL portions of pentane

↓

Combine the three 5 mL portions of pentane and make up to 25 mL

↓

Inject aliquots of the pentane solution into the GC

Convert response to residue content using standard curve

14. ORIGIN OF THE METHOD

Pest control Chemistry Department
Ministry of Agriculture, Fisheries and Food
Slough Laboratory
London Road
Slough, Berkshire, UK

Contact point: Mr K.A. Scudamore

METHOD 22

DETERMINATION OF ETHYLENE DIBROMIDE RESIDUES IN BISCUITS AND COMMERCIAL FLOUR BY GAS CHROMATOGRAPHY

D.M. Rains

1. SCOPE AND FIELD OF APPLICATION

This method is suitable for the determination of ethylene dibromide (EDB) residues in biscuits and flour. Two different extraction techniques are used, solvent soak and hexane steam distillation. The limit of detection is 5 μg/kg for flour and 1 μg/kg for biscuits. The analysis time is typically about 4 days for the solvent soak method and 4 hours for the hexane steam-distillation method. The 4-day soak does not involve substantial analyst time. The methods are subject to interferences from the solvents, particularly at the low μg/kg range. Care must be taken to screen all solvents by gas chromatography (GC) prior to use.

2. REFERENCE

Rains, D.M. & Holder, J.W. (1981) Ethylene dibromide residues in biscuits and commercial flour. J. Assoc. Off. Anal. Chem., 64, 1252-1254

3. DEFINITIONS

Not applicable

4. PRINCIPLE

Flour is extracted by the solvent soak method. The weighed sample is placed in a Teflon or foil-lined screw-cap vial. A measured amount of hexane is added, the samples are vigorously shaken by hand and are stored at room temperature in the dark. Once a day for 3 days, the samples are manually shaken for approximately 30 s. On the fourth day, the supernatant of the settled sample is injected directly onto the GC. Biscuits are analysed by the hexane steam-distillation method. The weighed sample is added to a distillation flask with water and hexane and assembled as shown in Figure 1. The sample is heated and the hexane and EDB are collected in the Barrett trap. The hexane extract is injected onto the GC.

FIG. 1. EDB (ETHYLENE DIBROMIDE) EXTRACTION APPARATUS

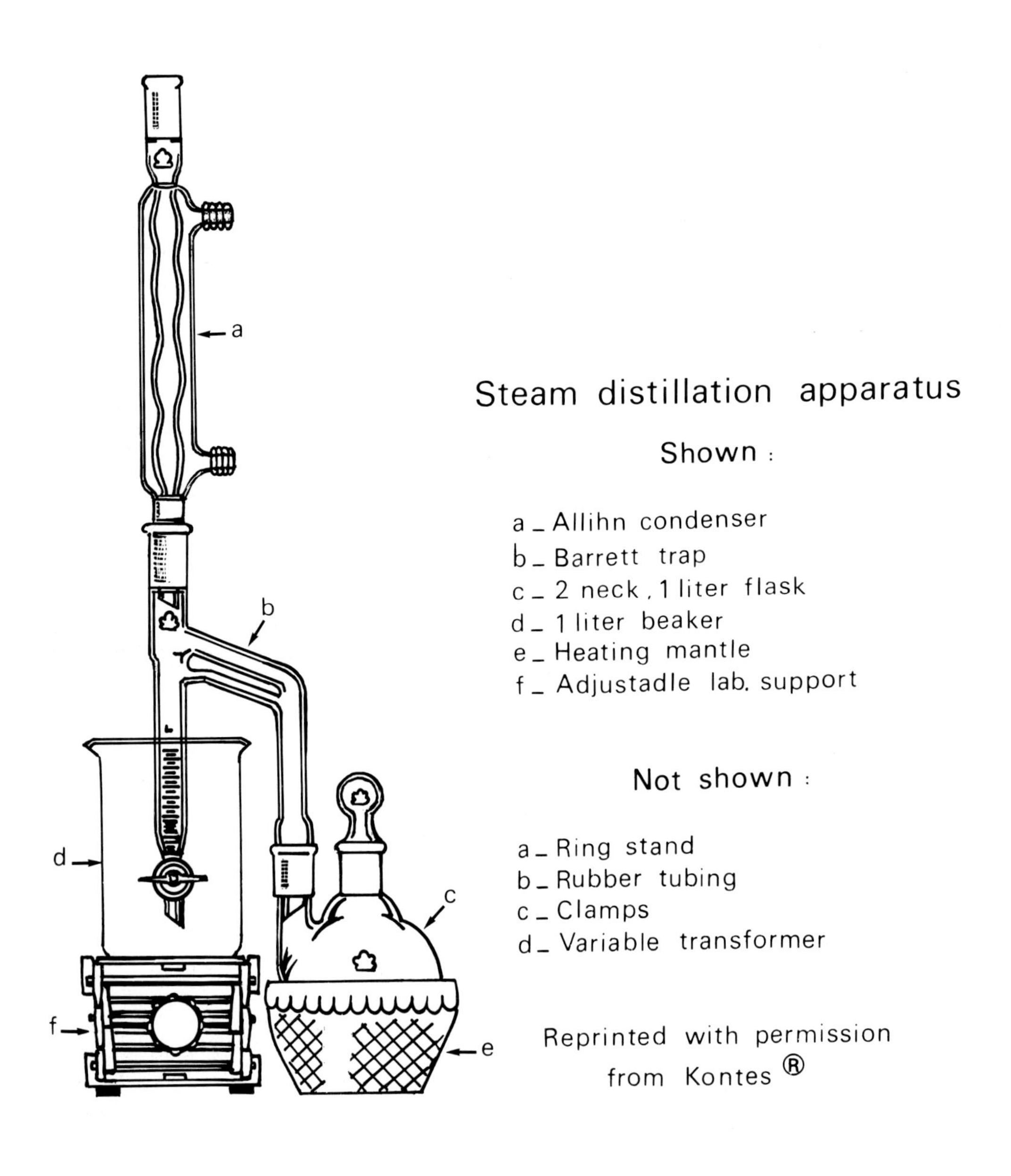

5. HAZARDS

EDB is an irritating liquid and a suspected carcinogen. Care should be taken when handling this chemical and the use of gloves and a fume hood are recommended.

6. REAGENTS[1]

Distilled water	Pesticide-free
_n_-Hexane	Pesticide-residue analysis quality, or equivalent. Check for interferences by injecting 6 µL onto GC (see 9.6).
Ethylene dibromide	Boiling range 130.5°-132.0°C, freezing point 9.3°C.
Standard solution	In a pre-weighed, stoppered 100-mL volumetric flask containing approximately 80 mL hexane, add 85 µL of EDB, using a microliter syringe with the needle below the surface of the hexane. Re-weigh the stoppered flask to calculate the amount of EDB added.
Working standard solutions	Prepare by serial dilutions of the stock solution to the 1-20 µg/L range, using 1-mL and 10-mL volumetric pipettes and 100-mL volumetric flasks.

7. APPARATUS[1]

Scintillation vials	20-mL capacity, glass, with Teflon or foil-lined screw-caps, or equivalent.
Volumetric flasks	Glass-stoppered, class A.

---

[1] Reference to a company and/or product is for the purpose of information and identification only and does not imply approval or recommendation of the company and/or product by the International Agency for Research on Cancer, to the exclusion of others which may also be suitable.

Steam distillation apparatus:

Distilling flask	1-litre, two-neck (both vertical), round-bottom
Barret trap	20-mL capacity, or equivalent
Heating mantle	soft-sided, 1 litre
Allihn condenser	300 mm, or equivalent
Glass stoppers	Ṩ 34/45 and Ṩ 24/40
Beaker	1000 mL, for cooling Barrett trap
Support platform	Adjustable, for raising and lowering beaker
Graduated centrifuge tubes	13-mL, glass-stoppered, 0.1 mL/division
Gas chromatograph	Equipped with ^{63}Ni electron-capture detector. Varian Model 3700, Hewlett Packard Model 5840, or equivalent
Syringes for GC	10-100 µL

8. SAMPLING

8.1 Mix flour samples manually by shaking the flour in a large polyethylene bag. Transfer to a foil-lined screw-cap jar and store in a freezer until required.

8.2 Cut biscuits into small pieces using a sharp knife, then store in a foil-lined glass jar in a freezer until required.

9. PROCEDURE

9.1 Blank tests

Water and hexane are tested for interferences by performing the hexane steam-distillation procedure without sample present. Use the procedure described in section 9.5.

9.2 Check test

Not applicable

9.3 Test portions

1.00-2.00 g flour; 20.0 g biscuits

9.4 Determination of EDB in flour

9.4.1 Weigh 1.00-2.00 g of sample into a scintillation vial and add 10 mL *n*-hexane.

9.4.2 Shake vigorously by hand for 30 s, then store at room temperature in the dark.

9.4.3 Once a day for three days, shake samples manually for 30 s.

9.4.4 On the fourth day, inject the settled supernatant directly onto GC (see 9.6).

9.5 Determination of EDB in biscuits

9.5.1 Weigh 20.0 g of biscuits into a 1-litre two-neck round-bottom flask and add 200 mL of distilled, deionized water and 10 mL of *n*-hexane.

9.5.2 Stopper the flask and shake vigorously, allowing the flask to be vented several times by removing the stopper for a few seconds.

9.5.3 Assemble the apparatus as shown in Figure 1.

9.5.4 Bring the water to a boil in about 40 min and remove heating mantle when 0.5-1.0 mL of water collects at the bottom of the Barrett trap.

9.5.5 Transfer the hexane and water to a graduated centrifuge tube and shake vigorously, then allow the water to settle. (A centrifugation step may be added to decrease the time for the separation of the water and hexane layers. No drying step is needed.)

9.5.6 Record the volume of water collected and subtract from total volume of solvent collected to determine the total volume of hexane recovered (typically, 8.5-9.8 mL of hexane are collected).

9.5.7 Inject aliquots of the hexane extract directly onto the GC (see 9.6).

9.6 GC operating conditions

(a) Column	Stainless steel, 1.83 m x 2.0 mm i.d., packed with 15% OV-17 on 80/100 mesh Chromosorb W.
detector temperature	260°C
column temperature	90°C
injector temperature	160°C
carrier gas (5% methane in argon) flow rate	120 mL/min

injection volume	150 pg EDB gave half-scale deflection on 1 mV recorder scale
(b) Alternative column	Glass, 1.83 m × 2.0 mm i.d., packed with 10% SP-1000 on 80/100 mesh Supelcoport
carrier gas (5% methane in argon) flow rate	40 mL/min

Retention time of EDB on column (a) was 4.2 min, retention time on column (b) with same operating temperatures was 3.1 min.

9.7 GC calibration

Using the same injection volume employed in 9.4.4 (or 9.5.7), choose a working standard solution of concentration such that the area of the peak obtained is comparable with that obtained in 9.4.4 (or 9.5.7). Inject and record peak area.

10. METHOD OF CALCULATION

The mass fraction, w, of EDB in the sample is given by,

$$w = \frac{\rho A_e V}{A_s m} \; (\mu g/kg)$$

where

ρ = concentration of working standard solution in 9.7 (µg/L)
A_s = peak area obtained with standard solution in 9.7
A_e = peak area obtained with sample extract (9.4.4 or 9.5.7)
V = total volume of hexane extract in 9.4.4 or 9.5.6 (mL)
m = mass of sample in 9.4.1 or 9.5.1 (g)

11. REPEATABILITY AND REPRODUCIBILITY

No data are available

12. NOTES ON PROCEDURE

Not applicable

13. SCHEMATIC REPRESENTATION OF PROCEDURE

FLOUR

Weigh flour into 20-mL screw-capped vial

↓

Add 10.0 mL of _n_-hexane

↓

Close vial and shake manually

↓

Store at room temperature in the dark and shake once daily

↓

Inject an aliquot of hexane extract onto the GC on the fourth day

↓

Inject same volume of a standard solution of EDB in hexane

BISCUITS

Weigh 20 g of biscuit into extraction flask

↓

Add 200 mL water and 10 mL hexane and stopper flask

↓

Mix by shaking, allowing pressure to equilibrate by briefly removing stopper

↓

Assemble steam-distillation apparatus

↓

Heat sample until 0.5-1.0 mL water collects at the bottom of the Barrett trap

↓

Decant the hexane and water into a glass-stoppered centrifuge tube. Mix well by shaking

↓

Allow the water to settle out and calculate the volume of hexane recovered by subtracting the amount of water from the total volume of solvent

↓

Inject hexane extract onto gas chromatograph

↓

Inject same volume of standard solution of EDB in hexane

14. ORIGIN OF THE METHOD

U.S. Environmental Protection Agency
ARC-EAST Bldg. 306, Room 125
Beltsville, Maryland 20705, USA

Contact point: Diane M. Rains

BIOLOGICAL MONITORING

BREATH ANALYSIS

METHOD 23

BREATH SAMPLING

E.D. Pellizzari, R.A. Zweidinger & L.S. Sheldon

1. SCOPE AND FIELD OF APPLICATION

This technique has been designed to collect breath samples from people exposed to volatile pollutants. The breath samples are collected at (or near) the subjects residence or work-place. The technique was therefore developed for sampling in a mobile unit (van).

2. REFERENCES

Not applicable

3. DEFINITIONS

Not applicable

4. PRINCIPLE

A Tedlar bag is filled with purified, humidified air which the subject inhales using a special mouthpiece. The subject then exhales into the mouthpiece, filling a second Tedlar bag with breath. The breath from the filled (exhale) bag is then drawn through a Tenax cartridge by a Nutech pump (two cartridges and two pumps in parallel are used for duplicate collections). The Tenax cartridge is then dried over calcium sulfate and stored for analysis.

5. HAZARDS

Methanol, used to rinse part of the equipment is flammable and toxic. It should never be used or stored in the van.

6. REAGENTS[1]

All reagents used should be analytical reagent grade

Compressed air[a]	0.1 THC grade
Compressed helium	99,9% grade
Distilled water[a]	
Charcoal filters[a]	
Calcium sulfate[a]	Drierite, indicating Drierite, non-indicating, cleaned by heating in a muffle-furnace at 400°C for 2 hours
n-Pentane	
Methanol	Distilled in glass
Tenax GC	2,6-Diphenyl-p-phenyleneoxide polymer (Applied Science, State College, Pa)
Glass wool	Virgin Tenax (or Tenax to be recycled) must be extracted in a Soxhlet apparatus for at least 24 h with methanol, followed by 24 h with n-pentane. Dry the extracted Tenax in a vacuum oven at 100°C for 24 h at 28 inches of water (7.0 kPa), then purge in a nitrogen box for 24h. Sieve the Tenax under nitrogen to obtain 40/60 particle size range (all sieving and cartridge preparation must be conducted in a "clean" room).

7. APPARATUS[1]

Spirometer

The spirometer is a device for the collection of breath samples and is shown in Figure 1. An all-Teflon mouthpiece with Tedlar flap valves and a stainless-steel ball valve is used. A bubbler filled with distilled-deionized water is placed in-line with the air tank to humidify the air for subject comfort. Each Tenax cartridge is connected to a separate Nutech 221 sampler equipped with a dry gas meter, so that the amount of air drawn through each cartridge may be accurately measured. The mouthpiece is mounted on an adjustable bar located on the front of the spirometer body. The glass Y, to which the Tenax cartridges are attached, is secured to the frame of the spirometer and the floor.

---

[a] These reagents are necessary for sample collection and should be carried in the van at all times.

[1] Reference to a company and/or product is for the purpose of information and identification only and does not imply approval or recommendation of the company and/or product by the International Agency for Research on Cancer, to the exclusion of others which may also be suitable.

Tedlar bags	40-L (Nutech Inc.)
Teflon tubing	1/4 inch o.d.
Air regulator[a]	(plus one spare air regulator)
Cajon Ultra Torr Union 316[a]	Stainless-steel, with viton O-rings
Straight unions	Stainless steel, 1/4 inch
Teflon ferrules	1/4 inch
Charcoal filters	See Figure 1

Teflon mouthpiece

Teflon plug valves	1/4 inch (Cole Palmer-6392-20)
Teflon tubing	1/4 inch o.d.
Ball valve with Teflon seats	Stainless steel, 1 inch
Nacom Teflon unions	1 inch
Viton O-rings[a]	See 8.5.3
Mouthpiece union, with Tedlar flap valves	
Tedlar flap valves[a]	
Teflon mouthbits[a]	
Solid Teflon mouthbit plug[a]	

Sampling lines

Latex tubing	1/4 inch i.d.
Male Quick-Connect	Single-end shut-off (connects sampling line to pump)

---

[a] These materials should be carried in the van at all times. They include materials for sample collection, as well as spare parts for the spirometer and Nutech samplers.

Plastic in-line filters (Drierite)

Cajon Ultra Torr Unions 316[a]

FIG. 1. SPIROMETER APPARATUS

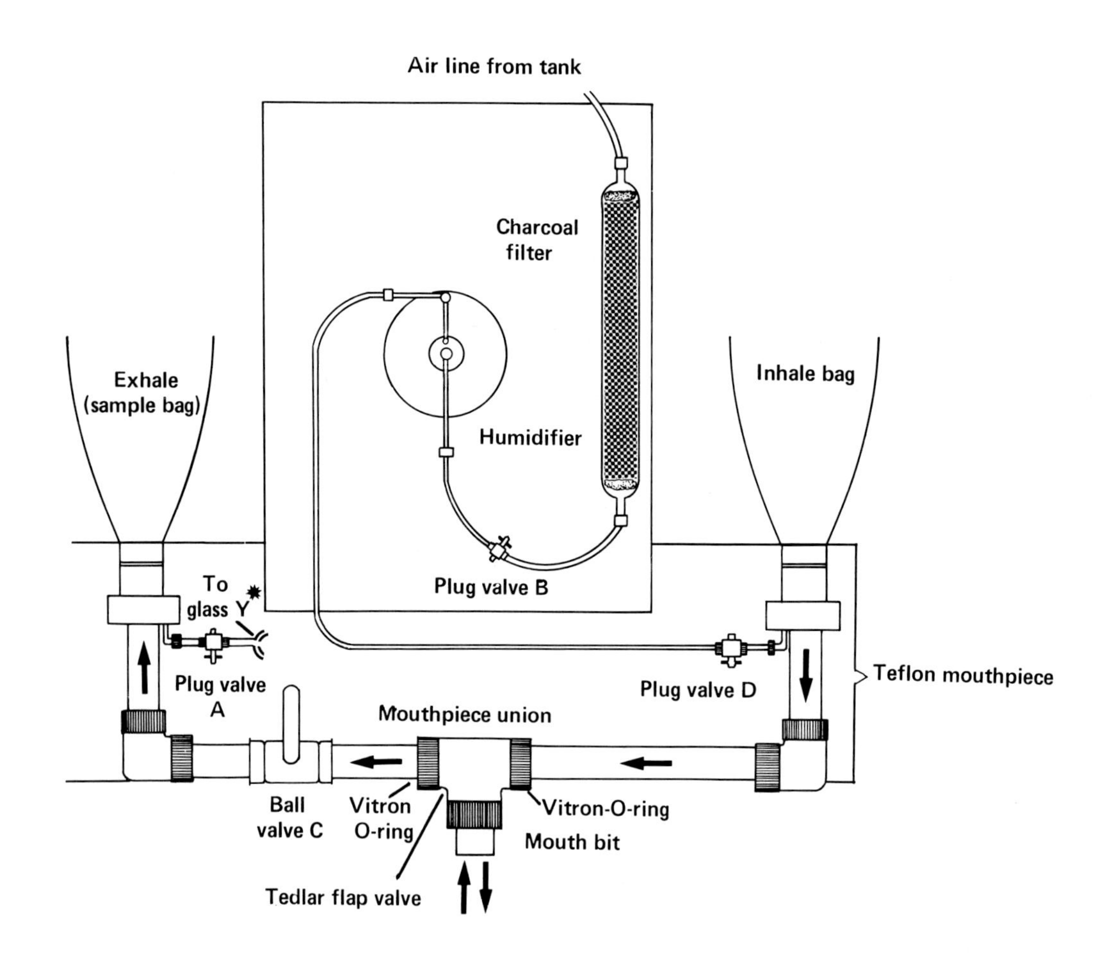

* See Figure 2.

[a] These materials are necessary for sample collection and should be carried in the van at all times.

Stainless steel needles	1.25 inch, 21 gauge[a]
Thermogreen septum	Supelco 2-0668
Glass Y's[a]	
Nutech Model 221 Gas Samplers	High-volume gas pump
Auto batteries	12 volts

<u>Humidifier[a]</u>

Flat-bottom boiling flask	500 mL, 24/40 joint
Midget impinger stopper	12/5 ball joint, 24/40 stopper (Lab Glass-6891)
Ground socket joint	12/5 joint (Lab Glass-1041)
Pinch clamps	size 12 (Lab Glass-1045)

<u>Miscellaneous equipment</u>

Large forceps[a]	
Small forceps	
Empty culture tubes, Kimax	With Teflon liner and Teflon-lined screw-caps. 2.5 × 15 cm
Glass tubes	10 × 1.5 cm i.d.
Plywood bag press[a]	
Small stool	
Copper wire	
Three-prong clamps	Various sizes
Clamp holders	
Small plastic bags	
Glass beaker[a]	10 mL
Dummy cartridges[a]	See NOTE, section 8.7.3
Stopwatch/calculator[a]	
Noseclip[a]	
Mercury thermometer[a]	
Kimwipes[a]	8 × 5 inches
Aluminum foil	
Binder clips	2-inch (to seal Tedlar bags, see 8.2.3)

---

[a] These materials are necessary for sample collection and should be carried in the van at all times.

NOTE: Wash all glassware in "Isoclean"/water, rinse with deionized-distilled water, acetone and air-dry. Heat glassware to 450-500°C for 2 h to insure that all organic material has been removed prior to use.

Sonicate Teflon liners in methanol, then pentane, for approximately 15 min each, then dry in a vacuum oven.

Rinse the Teflon mouthbits in methanol and air-dry for at least 12h before use.

Methanol is flammable and toxic and should at no time be used or stored in the van. The mouthbits must be cleaned only in a designated area of the workroom. Never allow the mouthbits to soak in methanol longer than 5 min. Air-dry and store in an upright position on a Kimwipe or clean towel.

8. SAMPLING

8.1 Preparation of sampling cartridges

8.1.1 Prepare the sampling cartridges by packing a glass tube (10 cm x 1.5 cm i.d.) with 6 cm of 40/60 mesh Tenax GC, using a glass wool plug.

8.1.2 Condition prepared cartridges at 270°C with a purified helium flow of 30 mL/min for >120 min. (Prior to entering the Tenax GC cartridge, the helium is purified by passing through a liquid N_2-cooled trap).

8.1.3 Transfer the conditioned cartridges to Kimax® (2.5 cm x 15 cm) culture tubes.

8.1.4 Seal immediately, using Teflon-lined screw-caps, and allow to cool. Store the cartridges in a sealed field collection can.

8.2 Preparation of Tedlar bags

NOTE: The 40-L Tedlar bags used are designed to fit the Teflon mouthpiece of the spirometer. Tedlar is a brittle material and should never be folded or creased, since this may cause it to crack or split. "Inhale" and "exhale" bags, although identical, should always remain segregated in order to minimize cross contamination.

8.2.1 Label each bag either "inhale" or "exhale" (one inhale bag is usually required for three exhale bags).

8.2.2 Attach the in-line charcoal filter to the helium tank and fill the bags with helium from the charcoal filter (do not over-inflate the bags).

8.2.3 Seal the mouth with a two-inch binder-clip (do not fold the mouth of the bag when sealing with the binder clip), then set aside the helium-filled bags for at least two hours.

8.2.4 Remove the binder clip, place the bag in the wooden bag press and allow the helium to flow out under the weight of the press.

8.2.5 Repeat the bag purging (8.2.2-8.2.4) at least two more times (six times in the case of new Tedlar bags).

8.2.6 At least 30 min before leaving for sample collection, refill each bag with helium and seal the mouth with a two-inch binder clip. Carry the inflated bags to the van and hang on the hooks provided.

NOTE: The Tedlar bags are easily damaged by grit, sand or sharp objects. They must be inspected for small holes or cracks. If the bag has been used frequently and the mouth is very wrinkled or cracked, it may be trimmed back, provided it will still fit the mouthpiece.

8.3 Purging of humidifier water

NOTE: This operation may be performed at any time prior to filling the first inhale bag with humidified air. (Freshly-purged water may be used for up to one week before being replaced, if it is kept sealed in the humidifier and purged each day before use.) Purge the humidifier water for at least 15 min each time it is changed, or when additional water is added. Proceed as follows:

8.3.1 Install the in-line glass charcoal filter (Fig. 1).

8.3.2 Attach the air regulator to the gas cylinder and assemble the humidifier apparatus as shown in Figure 1.

8.3.3 Half-fill the humidifier flask with distilled water.

8.3.4 Refer to Figure 1 and open plug valves B and D (the Tedlar bags are not attached).

8.3.5 Close the shut-off valve on the regulator, then slowly open the compressed air cylinder.

8.3.6 Gradually open the shut-off valve until a steady bubbling through the water is observed. Adjust the pressure (regulator diaphragm) to 21 kPa (3 psi).

8.3.7 After 15 min, turn off the air supply at the regulator and close plug valves B and D (Plug valve B must be closed whenever water is stored in the humidifier. If left open, water from the humidifier will saturate the charcoal filter and render it inactive.)

8.3.8 Close the air cylinder valve, remove the regulator, and cap the tank.

8.4 Equipment assemblage

When the van reaches its destination, the equipment should be made ready for the collection of the breath sample. Proceed as follows:

8.4.1 Assemble the mouthpiece as shown in Figure 1 and place the Teflon mouthbit in the mouthpiece.

8.4.2 Check the two Tedlar flap valves (in the mouthpiece union) by breathing in and out through the mouthpiece. Replace if damaged (i.e., if air flow is obstructed). The flap valves operate in only one direction. They must be installed so that air will flow from the inhale to the exhale bag only.

8.4.3 Attach all of the Teflon lines (using stainless steel unions and Teflon ferrules), including the line from the exhale side of the mouthpiece to the glass Y.

8.4.4 Secure the glass Y to a support rod, approximately two feet above the gas cylinder, using a small three-prong clamp and copper wire.

8.4.5 Attach two Cajon Ultra Torr unions to the glass Y as depicted in Figure 2.

8.4.6 Place the spirometer upright and secure it with the compressed air cylinder and attach the gas inlet line to the air regulator.

NOTE: The van should not be driven with the regulator attached to the gas cylinder. Sudden movements may fracture regulator stem or cylinder valve which could lead to serious injury. The cylinder valve should be protected by the cap at all times when not in use.

8.5 Inhale bag filling procedure

8.5.1 Using the bag press, expel all of the helium from the inhale bag.

FIG. 2. TENAX CARTRIDGE AND GLASS Y ASSEMBLY

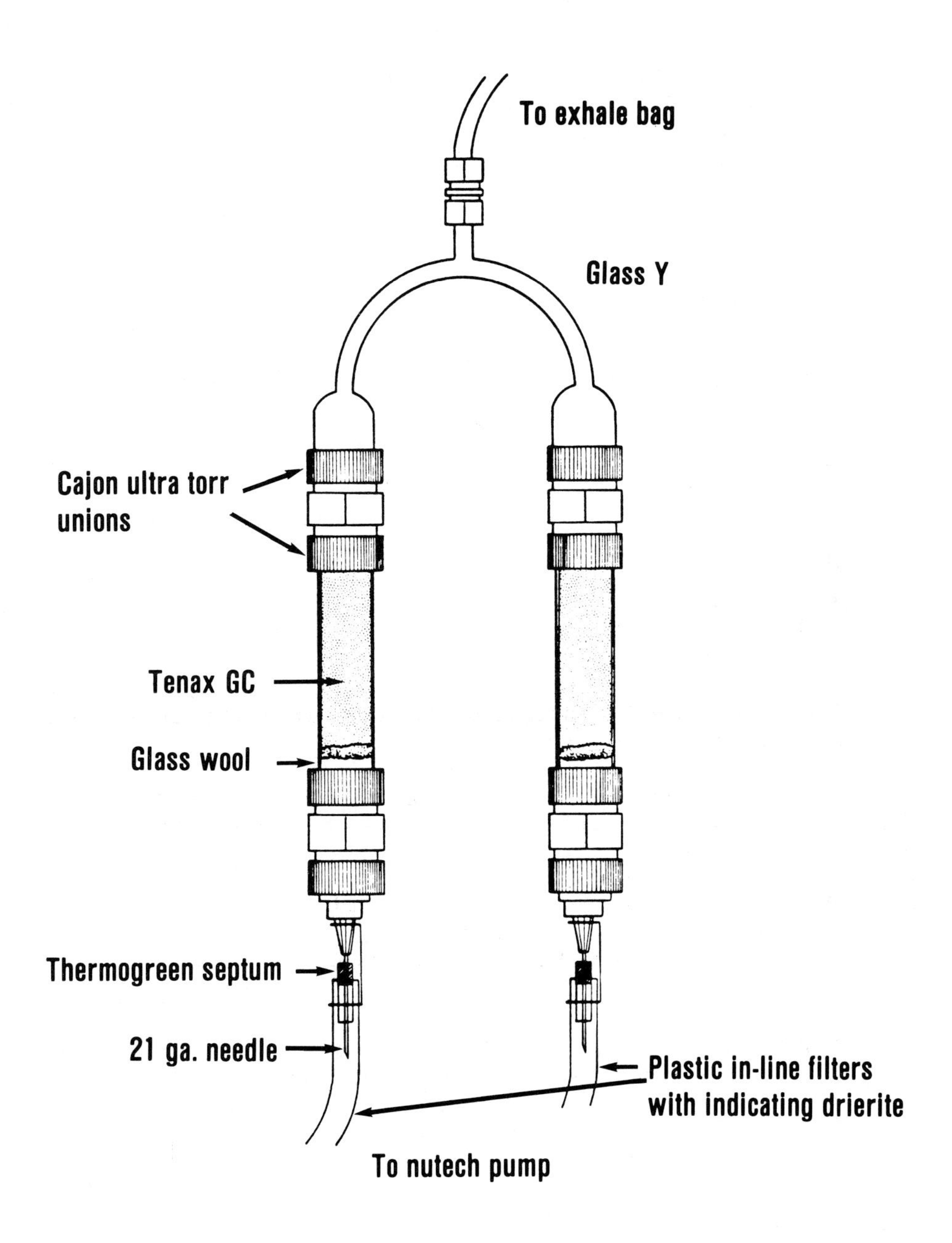

8.5.2 Immediately install the inhale bag on the right side of the spirometer mouthpiece (hang the bag by the grommet from the hook on the spirometer body).

8.5.3 Carefully roll the two large O-rings over the mouth of the bag, thus sealing the bag against the Teflon mouthpiece.

8.5.4 Place the solid Teflon mouthbit plug in the mouthpiece, open valves B and D and close valves A and C.

8.5.5 Attach the air regulator to the air cylinder and carefully bleed air through the humidifier into the inhale bag. (Never open the air cylinder before closing the shut-off valve on the regulator. A sudden surge of air may destroy the humidifier. Keep the back pressure relatively low, controlling the air flow with the shut-off valve).

8.5.6 Once the inhale bag is full, turn off the air supply and close valve D, then close the air cylinder valve and the air regulator shut-off valve.

8.5.7 Close plug valve B only after turning off the air supply (otherwise, the in-line back pressure may force the stopper from the humidifier, contaminating the air within and possibly destroying the humidifier).

8.6 Preparation of the Nutech 221 Samplers

NOTE: Two samplers are operated simultaneously, splitting the collected breath sample into two equal portions and drawing it through two parallel Tenax cartridges. Each sampler is powered by a standard 12-volt automotive battery.

8.6.1 While the inhale bag is being filled, connect the Nutech samplers to the batteries and determine if they are both operational (make sure the sampler number and correction factor are displayed on the front panel).

8.6.2 Attach a sampling line to each Nutech sampler, using the male Quick-Connect fitting, and fill the in-line filter of the sampling line with indicating Drierite.

8.6.3 Inspect the sampling line 21 ga. needles daily for blockage.

8.7 Tenax cartridge installation

8.7.1 When the participant has arrived, remove the top pad of glass wool (using forceps) from a culture tube containing a Tenax cartridge and lay it on a clean Kimwipe®.

8.7.2 Using a clean Kimwipe®, remove the Tenax cartridge and connect the void end to the Cajon fitting of the sampling line leading to the Nutech 221 sampler.

8.7.3 Carefully connect the other end of the Tenax cartridge to the fitting on the glass Y. Label the cartridge.

NOTE: A standard breath sample consists of only one Tenax cartridge; therefore, an unused or "dummy" cartridge is placed in line with the second Nutech sampler. If duplicate samples are to be collected for a participant, a clean Tenax cartridge is used in place of the dummy Tenax cartridge. If a dummy Tenax cartridge is used, place a distinctive marking on it so that it cannot be mistaken for the breath sample.

8.8 Breath collection

NOTE: The subjects must be cautioned to breath at a normal rate, otherwise they may hyperventilate.

8.8.1 Place an exhale bag on the left side of the spirometer by repeating steps 8.4.1 to 8.4.3, inclusive.

8.8.2 Instruct the participant to breath only through the mouth, to keep their lips sealed around the mouthbit, and not to stop before the exhale bag is full.

8.8.3 Have the participant put on the noseclip so that no air can pass through the nose.

8.8.4 Remove the solid Teflon plug from the mouthpiece and install a clean mouthbit.

8.8.5 Begin the breath collection by having the participant exhale, then place their lips on the mouthbit.

8.8.6 Open ball valve C and instruct the participant to begin breathing. (Start timer to determine duration of breath collection.)

NOTE: Observe the participant for a moment and remind them to breathe at a normal rate and to maintain a good seal on the mouthbit with their lips.

8.8.7 Using large forceps, remove the bottom pad of glass wool from the culture tube and place it on a clean Kimwipe®.

8.8.8 Fill a 10-mL beaker to the 5-mL mark with non-indicating Drierite, pour the Drierite into the culture tube and replace the bottom pad of glass wool. Cap the tube and retain for step 8.9.6.

8.8.9 When the exhale bag is full, close valve C and tell the participant to stop breathing into the mouthbit. (Stop timer and record duration of exhalation).

8.8.10 Remove the mouthbit and place the solid Teflon plug in the mouthpiece.

8.9 Cartridge loading procedure

8.9.1 Open plug valve A

8.9.2 Record temperature and the serial number, correction factor and meter reading for the Nutech sampler which will be used to collect the sample (assuming duplicate samples are not required), then begin pumping the breath sample through each Tenax cartridge.

8.9.3 Using the needle valve on the Nutech sampler, adjust the flow rates so that a back pressure of 150 to 175 torr is observed on each sampler (it is imperative that the pressure drop does not exceed 175 torr (0.23 atm), or loss of target compounds may result). A normal sample collection (30 L divided evenly between two cartridges) should take approximately 15 min.

8.9.4 Using the needle valves on the samplers, adjust the flow rates (1 L/min) to be as equal as possible while maintaining approximately equal vacuum readings.

8.9.5 When a total of 30 L have been pumped, turn off both samplers and record the pumping time. Record the final dry gas meter reading from the Nutech sampler and close plug valve A.

8.9.6 Using a clean Kimwipe®, carefully remove the Tenax cartridge from the Cajon fittings and place it in the culture tube, then add the top pad of glass wool (using forceps) on the Tenax cartridge and cap the culture tube (do not allow excess glass wool to lie over the top edge of the culture tube, preventing the cap from forming a proper seal).

8.9.7 Return the culture tube to the field collection can and seal the can.

NOTE: Previous experiments have shown that the organic vapors collected on Tenax GC are stable and can be quantitatively recovered up to at least 4 weeks after sampling, when the cartridges are tightly closed in culture tubes and placed in a second sealed container, protected from light and stored at -20°C.

8.9.8 Remove the exhale bag from the spirometer, place it in the wooden bag press and allow any remaining breath sample to flow out under the weight of the press. Hang the empty exhale bag on the hooks in the van. If additional breath samples are to be collected, leave the inhale bag in place.

METHOD 24

GC/MS DETERMINATION OF VOLATILE HYDROCARBONS IN BREATH SAMPLES

E.D. Pellizzari, R.A. Zweidinger & L.S. Sheldon

1. SCOPE AND FIELD OF APPLICATION

this method is suitable for the analysis of the halocarbons listed in Table 1 (a description of breath sampling is given in Method 23). It is not suitable for methyl chloride, methyl bromide, vinyl chloride, or methylene chloride.

Table 1. Some halocarbons for which the method is suitable

Chloroform	Bis-(chloromethyl)-ether
Carbon tetrachloride	Chloromethyl methyl ether
1,1,1-Trichloroethane	Haloethane
Hexachloroethane	Dibromochloropropane
1,2-Dichloroethane	1,2-Dibromoethane
Trichloroethylene	Bromoform
Tetrachloroethylene	Bromodichloromethane
Epichlorohydrin	Dibromochloromethane
Allyl chloride	Chlorobenzene
Trichlorobenzenes	Dichlorobenzenes
1,1,2-Trichloroethane	1,1,2,2-Tetrachloroethane

The linear range for the analysis of a volatile organic compound depends mainly on the breakthrough volume (Table 2) of the compound on the Tenax GC sampling cartridge and on the sensitivity of the mass spectrometer. The linear range for quantification using glass capillaries on a gas chromatograph/mass spectrometer/computer (GC/MS/COMP) is generally three orders of magnitude (5 - 5 000 ng). Table 3 lists measured detection limits for some volatile organic compounds. No interference has been observed. The analyses of a single breath sample requires 1.5 h.

Table 2. Tenax GC breakthrough volumes for some halocarbons[a]

Compound	b.p. (°C)	Temperature (°C)					
		10	15.6	21	26.7	32.2	37.8
Chloroform	61	56	41	32	24	17	13
Bromodichloromethane	87	82	61	45	34	25	-
Carbon tetrachloride	77	45	36	28	21	17	13
1,2-Dichloroethane	83	71	55	41	31	24	19
1,1,1-Trichloroethane	75	31	24	20	16	12	9
1,1,2-Trichloroethane	112	302	212	155	112	92	58
Tetrachloroethylene	121	481	356	261	192	141	104
Trichloroethylene	87	120	89	67	51	37	28
Chlorobenzene	132	1 989	871	631	459	332	241
m-Dichlorobenzene	173	2 393	1 758	1 291	948	697	510

[a] For a Tenax GC bed of 8.0 cm × 1.5 cm i.d. Volumes are in litres.

2. REFERENCES

Annual Book of ASTM Standards, Part 11.03 Atmospheric Analysis, American Society for Testing and Material, Philadelphia, Pennsylvania

Eight Peak Index of Mass Spectra (1970) Vol. 1 (Tables 1 & 2) and II (Table 3) Mass spectrometry Data Centre, AERE, Aldermaston, Reading, RF74PR, UK

Krost, K.J., Pellizzari, E.D., Walburn, S.G. & Hubard, S.A. (1982) Collection and analysis of hazardous organic emissions. Anal. Chem., 54, 810-817

Pellizzari, E.D. (1977) Analysis of Organic Air Pollutants by Gas Chromatography and Mass Spectroscopy (EPA-600/2-77-100), U.S. Environmental Protection Agency, Cincinnati, OH, 114 pp.

Pellizzari, E.D. (1980) Evaluation of the Basic GC/MS Computer Analysis Technique for Pollutant Analysis (EPA Contract No. 68-02-2998), U.S. Environmental Protection Agency, Cincinnati, OH

Pellizzari, E.D., Bunch, J.E., Berkley, R.E. & McRae, J. (1976) Determination of trace hazardous organic vapor pollutants in ambient atmosphere by gas chromatography/mass spectrometry/computer. Anal. Chem., 48, 803-806

Pellizzari, E.D., Bunch, J.E., Berkley, R.E. & McRae, J. (1976) Collection and analysis of trace organic vapor pollutants in ambient atmospheres. Anal. Lett., 9, 45

Table 3. Approximage measured limits of detection and quantification limits for selected organic compounds in breath

Compound	m/z	LOD[a] (µg/m³)	QL (µg/m³)
Chloroform	83/ 85	0.11	0.55
1,2-Dichloroethane	98/ 62	0.16	0.82
1,1,1-Trichloroethane	97/ 99	0.22	1.10
Vinylidene chloride	96/ 98	0.16	0.82
Trichloroethylene	130/132	0.22	1.10
Tetrachloroethylene	164/166	0.33	1.65
Bromodichloromethane	127/ 83	0.33	1.65
Chlorobenzene	112/114	0.22	1.10
1,1,2-Trichloroethane	97/ 99	0.22	1.10
m-Dichlorobenzene	146/148	0.27	1.37

[a] The limit of detection (LOD) is defined as S/N - 4 for the ion selected for quantification. The quantification limit (QL) is defined as 5 × LOD. Limits are based on a collection volume of 20L or breakthrough volume (21°C), whichever is smaller, for 8.0 cm × 1.5 i.d. Tenax GC bed.

3. DEFINITIONS

Not applicable

4. PRINCIPLE

The breath sample is collected on a Tenax GC cartridge, dried over calcium sulfate and analysed by thermal desorption of volatile components into a gas chromatograph/mass spectrometer (GC/MS) (Fig. 1).

FIG. 1. GC/MS ANALYTICAL SYSTEM FOR ANALYSIS OF ORGANIC VAPORS TRAPPED FROM AIR ONTO CARTRIDGES

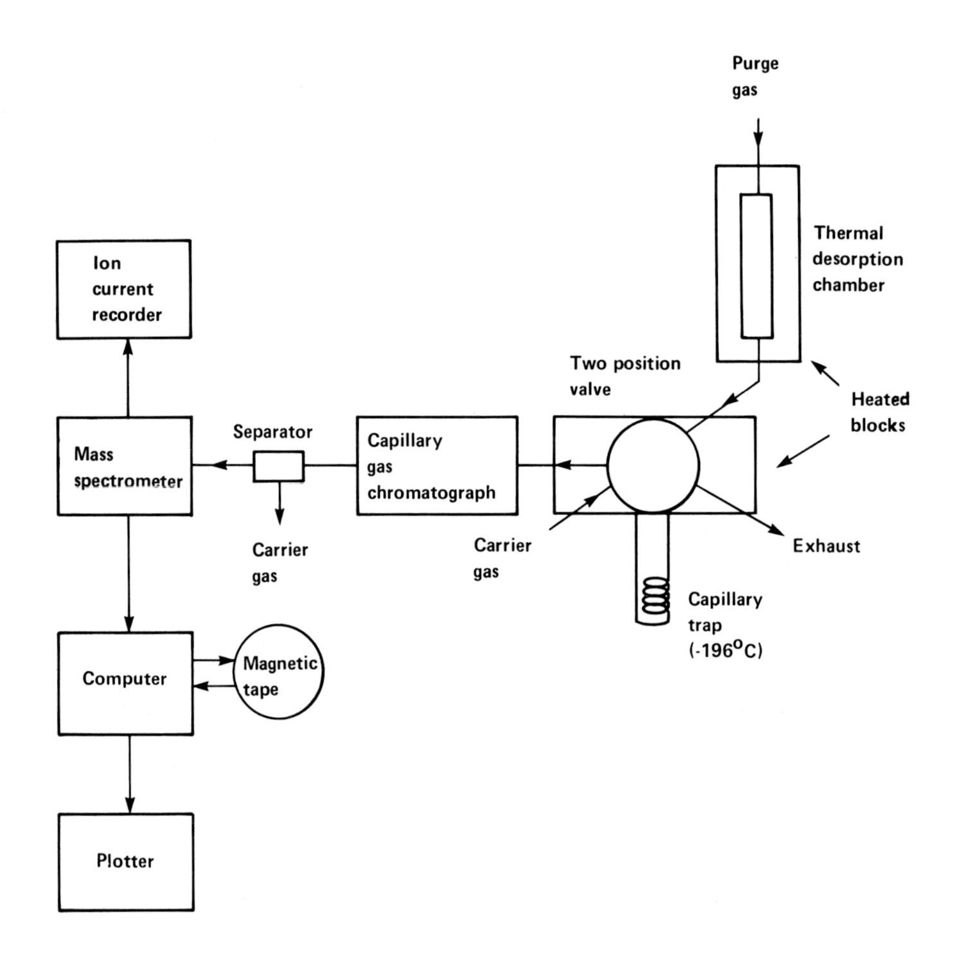

5. HAZARDS

Some of the volatile halocarbons are known carcinogens, while others are suspected carcinogens. Because of their volatility, care must be exercised to avoid their inhalation as well as dermal exposure. They should be handled cautiously in well-ventilated fume hoods and operators should wear protective face masks, clothing and gloves which do not readily absorb the substances.

6. REAGENTS[1]

Compressed helium, 99.9% grade

Compressed nitrogen

Calibration standards:
- perfluorotoluene
- perfluorotributylamine (low mass)
- trisperfluoro-(heptyl)S-triazine (TRIS) (high mass for Finnigan 3300 MS)
- perfluorotributylamine (FC 43) for Finnigan 4021 MS

Pure analytes (see Table 1) in reagent bottles or permeation tubes (section 7)

Deuterated analytes for detecting breakthrough

7. APPARATUS[1]

Inlet manifold	Desorption chamber, valve and capillary trap interfaced to GC/MS system for thermal recovery of vapours trapped on Tenax sampling cartridges (see Fig. 1)
Gas chromatography/mass spectrometer/computer	A Finnigan 9500 or Finnigan 9610 GC with a fused-silica capillary column, which is directly coupled to the ion source of the Finnigan 3300 or Finnigan 4021 MS systems, respectively. A mass-flow controller (Tylan) is used to regulate the flow of carrier gas. Such an analytical system is shown in Figure 1. The characteristics of the GC/Mass Spectrometer/Computer systems are specified in Table 4.
Permeation system (Fig. 2)	
Permeation tubes	Sealed plastic tubes with permeable walls, containing perfluorotoluene and analytes of interest (Metronics Corp., santa Clara, CA, USA)

---

[1] Reference to a company and/or product is for the purpose of information and identification only and does not imply approval or recommendation of the company and/or product by the International Agency for Research on Cancer, to the exclusion of others which may also be suitable.

Table 4. GC/MS specifications[a]

	Finnigan 3300	Finnigan 4021
1) Type	Low-resolution quadrupole	Low-resolution quadrupole
2) Resolution	Mass range to 1000 with unit resolution	Mass range to 1000 with unit resolution
3) Scan speed	1 s-10 min over entire range	0.1 s-10 min over entire range
4) Routine mass calibration standards	Perfluorotributylamine (low mass); trisper-fluoroheptyl) S-triazine (high mass)	FC-43 (perfluorotributyl-amine)
5) Mode	Electron impact; chemical ionization (CH_4, NH_3, isobutane)	electron impact; chemical ionization, positive and negative
6) GC	Finnigan 9500	Finnigan 9610
7) GC columns	Glass capillaries (SCOT); packed; fused silica capillaries	Glass capillaries (WCOT, SCOT); fused silica capillaries; packed
8) GC injection	Glass capillaries thermal desorption	Thermal desorption; splitter; Grob type; liquid
9) GC-MS	Single stage glass jet; capillary direct coupling	Direct coupling or glass jet
10) Sample introduction	GC; direct probe interchangeable with molecular leak heated inlet	GC, interchangeable with direct probe
11) Computer	Data General NOVA 3	Data General NOVA 3

	Finnigan 3300	Finnigan 4021
12) Computer hardware	- 32K central processor with Tektronix 4010-1 graphic terminal keyboard - Versatec electrostatic printer/plotter - 4 Perkin-Elmer disk drives (5 megaword double density disks) - Wangco Model 1045 9-track 800 BPI, 45 IPS, industry-compatible magnetic tape - external interface	- 32K central processor with Tektronix 4010-1 graphic terminal and keyboard - Versatec electrostatic printer/plotter - 4 Perkin-Elmer disk drives (5 megaword double density disks) - Wangco Model 1045 9-track 800 BPI, 45 IPS, industry-compatible magnetic tape - external interface
13) Computer interface	- simultaneous dual mass spectrometer interface	- simultaneous dual mass spectrometer interface
14) Software capabilities	- control full scan mass spectral acquisition - acquire multiple ion detection data for up to 25 ions - reconstruct gas chromatograms - subtract background - reconstruct mass chromatograms - calculate peak area from mass chromatograms - plot normalized or maximum intensity mass spectra - library search (EPA/NIH library) - reverse library search	- control full scan mass spectral acquisition - acquire multiple ion detection for up to 25 ions - reconstruct gas chromatograms - subtract background - reconstruct mass chromatograms - calculate peak area from mass chromatograms - plot normalized or maximum intensity mass spectra - library search (EPA/NIH library) - reverse library search

[a] This equipment is employed by the authors.

FIG.2. PERMEATION SYSTEM FOR GENERATING AND LOADING AIR VAPOUR MIXTURES[a]

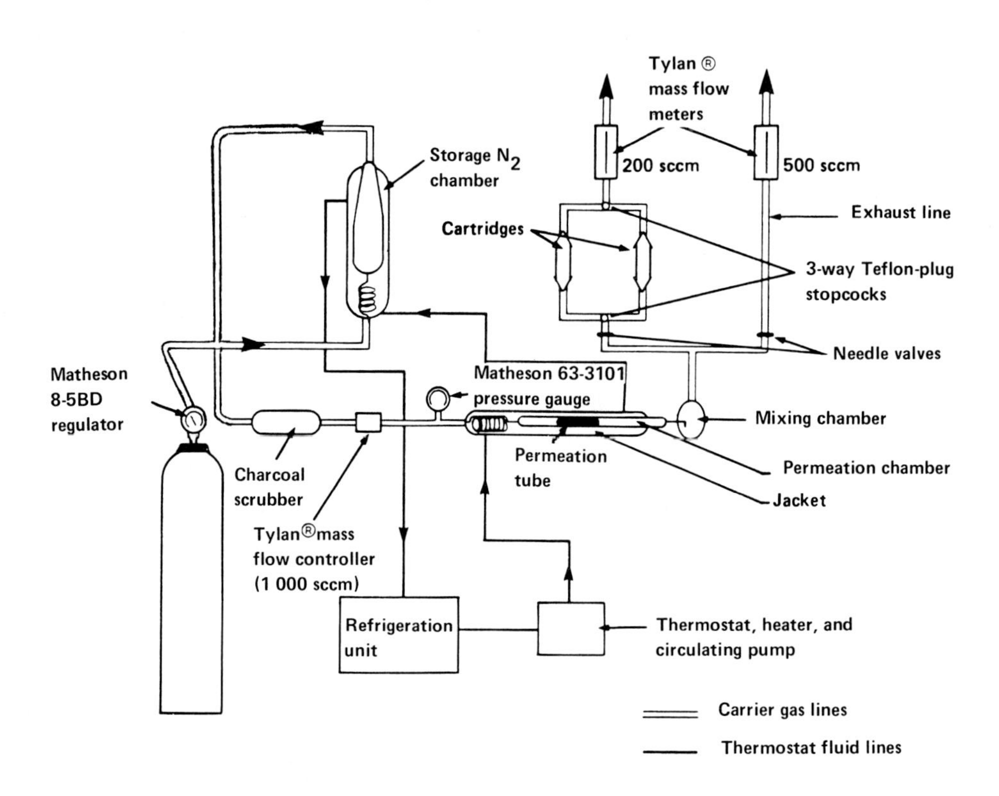

[a] The detailed operation of the permeation system is described by Krost *et al.* (1982).

8. SAMPLING

See Method 23 for detailed sampling procedure.

8.1 For detection of cartridge breakthrough, load ~200 mg of deuterated analogues of each of the analytes of interest onto one in ten of the cartridges to be used for breath sampling (see 9.4.2 for procedure). Alternatively, employ only d_5-1,2-dichloroethane, d_3-trichloroethane and d_5-chlorobenzene to cover most of the breakthrough spectrum.

8.2 For each group of 5 breath samples, set aside one Tenax cartridge from the same batch to serve as a blank (field control). Blank cartridges travel to the field site and are returned to the laboratory unused, to be stored with the field samples at -20°C until required.

8.3 From the same batch of cartridges (8.2) set aside two others for the preparation of perfomance and calibration standards (9.4). The eight cartridges concerned (8.2, 8.3) should be analysed as a group (9.5) and three consecutive groups should employ cartridges from the same uniform batch (see 10.1).

9. PROCEDURE

9.1 Blank test

Each field blank (8.2) is analysed with the breath samples from the same set, according to steps 9.4.3 and 9.5.4.

9.2 Check tests

Check mass calibration periodically, using perfluorotributylamine (the mass calibration is stable for months). On a daily basis, compare the spectrum obtained for perfluorotoluene in a system performance standard (9.4.1) with the data given in Table 5.

Table 5. Mass (m/z) and relative ion abundances from perfluorotoluene acceptable for quantification with quadrupole instruments[a]

Perfluorotoluene (Finnigan 3300)		
m/z	Relative abundance	
	Mean	Range
69	18	17-39
79	8	4-12
93	16	9-23
117	46	34-58
167	16	11-19
186	66	55-77
217	100	100
236	56	47-65

[a] To be achieved in the chromatography mode

9.3 Test portion

Not applicable

9.4 Preparation of standards

9.4.1 System performance standards:

Load all of the compounds in Table 6 onto a single Tenax GC sampling cartridge (8.3), using the flash evaporation system (Fig. 3) and the following procedure[1]. Prepare standard solutions of the compounds in methanol (75 or 150 mg/L, see Table 6) and inject a 2-μL aliquot of each through the septum of the heated (250°C) loading tube. Carry the vapour onto the Tenax cartridge with a stream of purified helium (60 mL/min) for 15 min. Store at -20°C until required. The system performance standards are employed to determine the sensitivity and performance of the GC/MS/Computer system on a daily basis (see also, 12.1).

[1] The "void" (exit) end of all loaded cartridges should be marked and desorption (9.5) should take place with the carrier gas flowing in the opposite direction to that obtaining during loading (see also Method 23, step 8.7.2).

Table 6. GC/MS/COMP system performance and quantification standards

Compound	Quantity (ng)
Perfluorotoluene	150
Ethylbenzene	300
o-Xylene	300
n-Octane	300
n-Decane	300
1-Octanol	300
5-Nonanone	300
Acetophenone	300
2,6-Dimethylaniline	300
2,6-Dimethylphenol	300

FIG. 3. SCHEME OF VAPORIZATION UNIT FOR LOADING ORGANICS DISSOLVED IN METHANOL ONTO TENAX GC CARTRIDGES

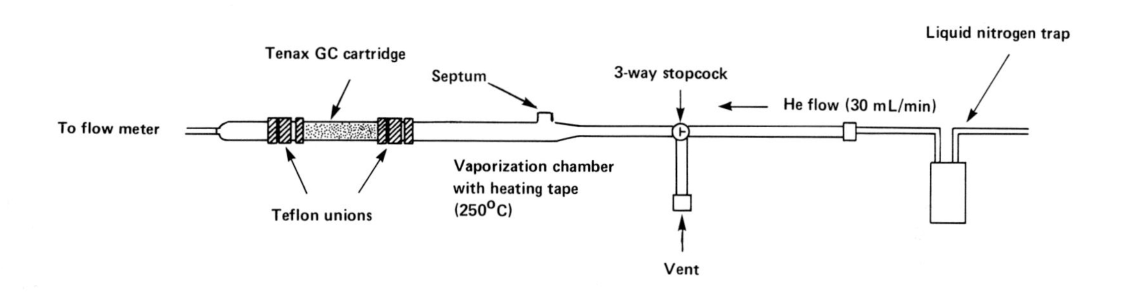

9.4.2 Calibration (relative response factor) standards:

Using the permeation system (Fig. 2) or the flash-evaporator (see 9.4.1), load 250-450 ng of each of the analytes of interest, plus ~150 ng of perfluorotoluene (quantification standard), onto a single Tenax cartridge (8.3). If the permeation system is used, prepare a nitrogen/vapour mixture containing 1-4 ng analyte/mL and pass a known volume (~200 mL) of the mixture through the Tenax cartridge. The amounts of the compounds loaded must be accurately known. Store at -20°C until required.

9.4.3 Using the permeation system, as in 9.4.2, add a known amount (~150 ng) of perfluorotoluene to the blank and to the five breath samples (8.2).

9.5 Analyte determination

NOTE: The operations described in 9.5 must be carried out in the order given.

9.5.1 Using the GC/MS operating conditions specified in Table 7, place the system performance standard (9.4.1) in the pre-heated desorption chamber and pass helium through the cartridge to carry the vapours into the capillary cold trap (Fig. 1).

9.5.2 When desorption is complete (8 min), rotate the inlet value and raise the temperature of the capillary trap rapidly. (>100°C/min), whereupon the carrier gas introduces the sample onto the GC column. When all analytes have eluted into MS, cool the column to ambient temperature.

9.5.3 Repeat 9.5.1 and 9.5.2 with the calibration standard (9.4.2).

9.5.4 Repeat 9.5.1 and 9.5.2 with the blank cartridge (8.2).

9.5.5 Repeat 9.5.1 and 9.5.2 with the five breath samples (8.2).

9.5.6 Repeat the cycle 9.5.1-9.5.5 until all breath samples have been analysed. If a cycle cannot be completed in a working day, each day must nevertheless begin with the analysis of a system performance standard (9.5.1-9.5.2), before completing the cycle.

Table 7. Operating parameters for GC/MS system

Parameter	Setting
Inlet manifold	
Desorption chamber and valve	270°C
Capillary trap - minimum	-195°C (cooled with nitrogen)
- maximum	240°C
Thermal desorption time	8 min
He purge flow	15 mL/min
GC	
60 m DB-1 wide-bore fused silica	40°C (hold 5 min) → 240°C, 4°C/min
Carrier (He) flow	1.0 mL/min
Separator oven	240°C
MS	
Finnigan 3300	
scan range	m/z 35 → 350
scan cycle, automatic	1.9 s/cycle
filament current	0.5 mA
electron multiplier	1 600 volts
analyzer vacuum	18 mTorr
ion source vacuum	18 mTorr
inlet vacuum	25 mTorr
hold time	0.1 s

9.6 Data interpretation

9.6.1 Qualitative analysis:

For qualitative analysis, an ion chromatogram is constructed from the mass spectra. This will generally indicate whether the run is suiable for further processing, since it provides some idea of the number of unknown compounds in the sample and the resolution obtained using the particular GC column and conditions. See section 12.2 for compound identification procedures.

9.6.2 Quantitative analysis:

The time-dependent characteristic-ion spectra are employed to obtain chromatograms of breath samples and calibration standards (9.4.2). Both samples and standards contain known amounts of the quantification standard, perfluorotoluene. The ratios, peak area/mass loaded, for analyte and for perfluorotoluene, obtained with the calibration standard, are employed to calculate the relative response factor, F, which permits quantification of that analyte on the sample cartridge (see section 10).

10. METHOD OF CALCULATION

10.1 Determination of relative response factor (F)

The relative response factor, F_a, for a given analyte is obtained from,

$$F_a = A_a m_s / A_s m_a$$

where,

A_a = area of analyte peak on calibration standard chromatogram.

A_s = area of quantification standard peak on calibration standard chromatogram.

m_a = mass of analyte on calibration standard cartridge (μg).

m_s = mass of quantification standard on calibration standard cartridge (μg).

10.2 Determination of analyte mass in sample

Using the symbols employed in 10.1, the mass of the given analyte on the sample cartridge is obtained from,

$$m_a = A_a m_s / F_a A_s \text{ (μg)}$$

where the values of A_a, A_s and m_s are those obtained from the sample cartridge.

NOTE: F_a is an average value, determined from at least three independant analyses carried out during analysis of a set of breath samples.

10.3 The mass concentration of the given analyte in the breath sample, ρ_a, is obtained from,

$$\rho_a = 10^3(m_a - m_b)/V \ (\mu g/m^3)$$

where,

m_b = mass of analyte on blank cartridge (µg)

V = volume of breath sample (L)

and m_a is defined in 10.2.

11. REPEATABILITY AND REPRODUCIBILITY

The reproducibility of this method has been determined to range from ±10 to ±30% (relative standard deviation) for different substances when replicate sampling cartridges are examined.

The accuracy of analysis is generally ±10 to ±30%, but depends on the chemical and physical nature of the compound.

12. NOTES ON PROCEDURE

12.1 Assessment of chromatographic performance

The quality of the chromatography is of the utmost importance for the accuracy and precision of qualitative and quantitative analysis. Glass capillary columns are evaluated according to the following criteria:

(1) percent peak asymmetry factor (PAF)

$$\% \ PAF = 100 \ B/F$$

where

B = the area of the back half of a chromatographic peak

F = area of the front half of the chromatographic peak both measured 10% above baseline

(2) Height equivalent to an effective theoretical plate ($HETP_{eff}$)

$$HETP_{eff} = \frac{L}{5.54\ (X/Y)^2}$$

where

X = the retention distance (corrected for sweep time) of the compound,

Y = chromatographic peak width at 1/2 peak height,

L = column length (mm)

(3) separation number (SN)

$$SN = \frac{D}{(Y_1 + Y_2)} - 1$$

where

D = the distance between two peaks,

Y_1, Y_2 = widths at 1/2 height

(4) resolution (R)

$$R = \frac{2\ \Delta\ t}{W_1 + W_2}$$

where

Δt = distance between peak tops
W = peak width at base

(5) $$\text{Acidity} = \frac{\text{weak base (peak area or height)}}{\text{acetophenone (peak area or height)}}$$

$$\text{Basicity} = \frac{\text{weak acid (peak area or height)}}{\text{acetophenone (peak area or height)}}$$

The use of the compounds listed in Table 8 provides information regarding the degree of adsorption and the type of adsorption mechanism. The peak assymetry of 1-octanol and 5-nonanone serves to determine the extent of deactivation of the glass surface (PAF). The acidity and basicity of the glass capillary column are assessed respectively by the adsorption of weak bases (e.g., 2,6-dimethylaniline) and acids (e.g., 2,6-dimethylphenol).

The resolution and separation number are determined for the compound pairs ethylbenzene:p-xylene and octane:decane, respectively. Table 8 lists the minimum performance specifications acceptable for breath analysis.

Table 8. Minimum performance specifications for glass capillary columns

Parameter	Test compounds	Value
Resolution	Ethylbenzene: -xylene	> 1.0
Separation No.	Octane:decane	< 40
% Peak asymmetry factor	1-Octanol	< 250
	Nonanone	< 160
	Acetophenone	< 300
Acidity	2,6-Dimethylaniline: acetophenone	0.7-1.3
Basicity	2,6-Dimethylphenol: acetophenone	0.7-1.3

12.2 The computer automatically assigns masses during data acquisition by the use of the mass calibration table obtained for perfluorotributylamine. After the spectra are obtained in mass-converted form, processing proceeds either manually or by computer comparison with a library. Compound identification can involve several levels of certainty.

Level 1.- The raw data generated from the analysis of samples are subjected only to computerised deconvolution/library search. Compound identification made using this approach has the lowest level of confidence. In general, it is reserved for only those cases where compound verification is the primary intent of the qualitative analysis.

Level 2.- The plotted mass spectra are manually interpreted by a skilled interpretor and compared to spectra compiled in a data compendium. In general, a minimum of five masses and intensities (±5%) should match between the unknown and library spectrum.

NOTE: This level does not utilise any further information, such as retention time.

Level 3.- The mass spectra are manually interpreted (as in level 2) and spectra and retention times are compared with those of the authentic compounds, using identical operating conditions.

13. SCHEMATIC REPRESENTATION OF PROCEDURE

Breath samples and blank cartridges (one breath-sample cartridge in ten is spiked with deuterated analytes for breakthrough control)

↓

Add ~150 mg quantification standard to each cartridge

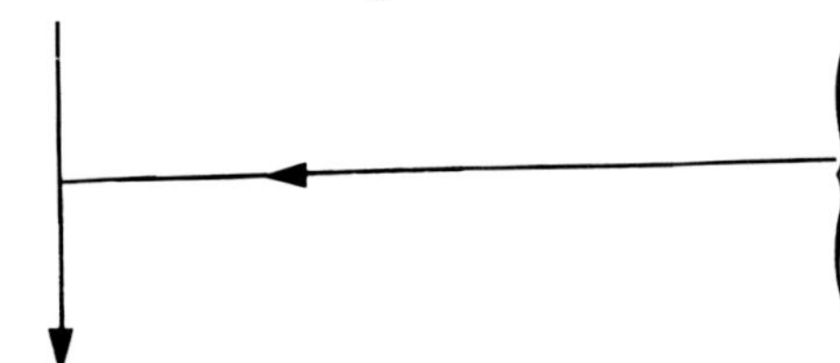

Prepare the following spiked cartridges:
1. system performance standards
2. calibration (relative response factor) standards

Analyse system performance standard, calibration standard, blank and breath samples, in that order, by GC/MS/Computer

↓

Calculate relative response factor from calibration standard chromatogram.
Calculate analyte concentration in sample using sample chromatogram and relative response factor.

14. ORIGIN OF THE METHOD

Research Triangle Institute
Research Triangle Park, North Carolina 27709, USA

Contact point: Dr D. Pellizzari, Vice-President, Analytical and Chemical Sciences

BIOLOGICAL MONITORING

BLOOD AND TISSUES

METHOD 25

GC/MS DETERMINATION OF VOLATILE HALOCARBONS IN BLOOD AND TISSUE

E.D. Pellizzari, L.S. Sheldon & J.T. Bursey

1. SCOPE AND FIELD OF APPLICATION

This method is suitable for the analysis of the halocarbons listed in Table 1 in blood and tissues. For a 10 mL blood sample, the limit of detection is about 3 ng/mL. Detection limits of about 6 ng/g are typical for 5 g tissue samples. Upper limits for these samples are ~10^4 x lower limits.

Table 1. Some halocarbons for which the method is suitable

Chloroform	Bis-(chloromethyl)-ether
Carbon tetrachloride	Chloromethyl methyl ether
1,1,1-Trichloroethane	Haloethane
Hexachloroethane	Dibromochloropropane
1,2-Dichloroethane	1,2-Dibromoethane
Trichloroethylene	Bromoform
Tetrachloroethylene	Bromodichloromethane
Epichlorohydrin	Dibromochloromethane
Allyl chloride	Chlorobenzene
Trichlorobenzenes	Dichlorobenzenes
1,1,2-Trichloroethane	1,1,2,2-Tetrachloroethane

2. REFERENCES

Pellizzari, E.D. (1974) Development of Method for Carcinogenic vapor Analysis in Ambient Atmospheres (EPA Contract No. 68-02-1228), U.S. Environmental Protection Agency, cincinnati, OH

Krost, K.J., Pellizzari, E.D., Walburn, S.G. & Hubbard, S.A. (1982) Collection and analysis of hazardous organic emissions. Anal. Chem., 54, 810-817

Michael, L.C., Erickson, M.D., Parks, S.P. & Pellizzari, E.D. (1980) Volatile environmental pollutants in biological matrices with a headspace purge technique. Anal. Chem., 52, 1836-1841

Pellizzari, E.D., Hartwell, T.D., Harris, B.S.H., III, Waddel, R.D., Whitaker, D.A. & Erickson, M.D. (1982) Purgeable organic compounds in mother's milk. Bull. Environ. Contam. & Toxicol., 28, 322-328

3. DEFINITIONS

Not applicable

4. PRINCIPLE

Volatile halocarbons are recovered from a blood sample by warming the sample and passing an inert gas over the warm sample. Tissues are first macerated in water, then treated in the same manner as blood. The halocarbon vapours are trapped on a Tenax GC® cartridge, then recovered by thermal desorption and analysed by gas chromatography/mass spectrometry.

5. HAZARDS

Some of the volatile halocarbons are known carcinogens, while others are suspected carcinogens. Because of their volatility, care must be exercised to avoid inhalation or skin exposure. They should be handled cautiously in well-ventilated fume hoods and operators should wear protective face masks, lab coats and glvoes which do not readily absorb the substances.

6. REAGENTS[1]

All reagents used should be analytical Reagent grade

Pure analytes	See Method 24, section 6.
Distilled water	Organic-free
Tenax GC®(60/80 mesh)	See Method 23, section 6 for cleaning and sieving instructions.

---

[1] Reference to a company and/or product is for the purpose of information and identification only and does not imply approval or recommendation of the company and/or product by the International Agency for Research on Cancer, to the exclusion of others which may also be suitable.

Compressed helium	99.9999% grade
Calibration standards	See Method 24, section 6
Drierite	Non-indicating, baked at 400°C for 2 h

7. APPARATUS[1]

Glass cartridges	10 cm long × 1.5 cm i.d.
Glass wool	
Soxhlet apparatus with condenser	
Vacuum oven	
Stainless steel mesh screens	For 60/80 fraction
Kimax® culture tubes	2.5 cm × 15 cm with Teflon®-lined screw-caps
Vacutainer tubes 10-mL	Venoject L 428, Kimble
Glass syringes	10-mL
Shell vials	10-mL with Teflon®-lined screw-caps
Magnetic stirring bar	Teflon®-coated
Heating mantle	For 100-mL round-bottom flask
Magnetic stirrer	
Mercury thermometers	
One-gallon paint cans	With press-fit lids
Ice bath	
Disposable Pasteur pipettes and bulbs	
Virtis tissue homogenizer	

---

[1] Reference to a company and/or product is for the purpose of information and identification only and does not imply approval or recommendation of the company and/or product by the International Agency for Research on Cancer, to the exclusion of others which may also be suitable.

Purge apparatus

The purge apparatus is shown in Figure 1.

Tenax GC® cartridges

Sampling tubes are prepared by packing a 10 cm long × 1.5 cm i.d. glass tube containing 6.0 cm of 60/80 mesh Tenax GC (~1.6 g), using glass wool in the ends to provide support. See Method 23, section 6, for preliminary extraction of Tenax and section 8.1 (Method 23) for cartridge preparation and conditioning procedure.

NOTE: Cleaning procedures for glassware and Teflon liners are described in Method 23, at the end of section 7. Store cleaned glass cartridge tubes and culture tubes in sealed, one-gallon paint cans.

Gas chromatography/mass spectrometer/computer See Method 23, section 7

8. SAMPLING

8.1 Collection of blood samples

8.1.1 In the field, collect 10 mL blood samples by brachial venipuncture, using 10 mL vacutainer tubes (See Notes on Procedure, 12.1).

8.1.2 As soon as possible, chill the blood sample to 4°C and transfer it to a clean shell vial with a Teflon-lined screw-cap. Seal using Teflon tape.

8.2 Collection of tissue samples

Collect samples with minimum exposure to plastic or rubber and store in cleaned, oven-treated, glass jars sealed with either Teflon or foil-lined caps.

Collect tissue samples from cadavers a short time following death and freeze immediately in a cleaned glass container with as small a "head-space" as possible.

FIG. 1. HEAD-SPACE PURGE APPARATUS FOR BLOOD AND TISSUE SAMPLES

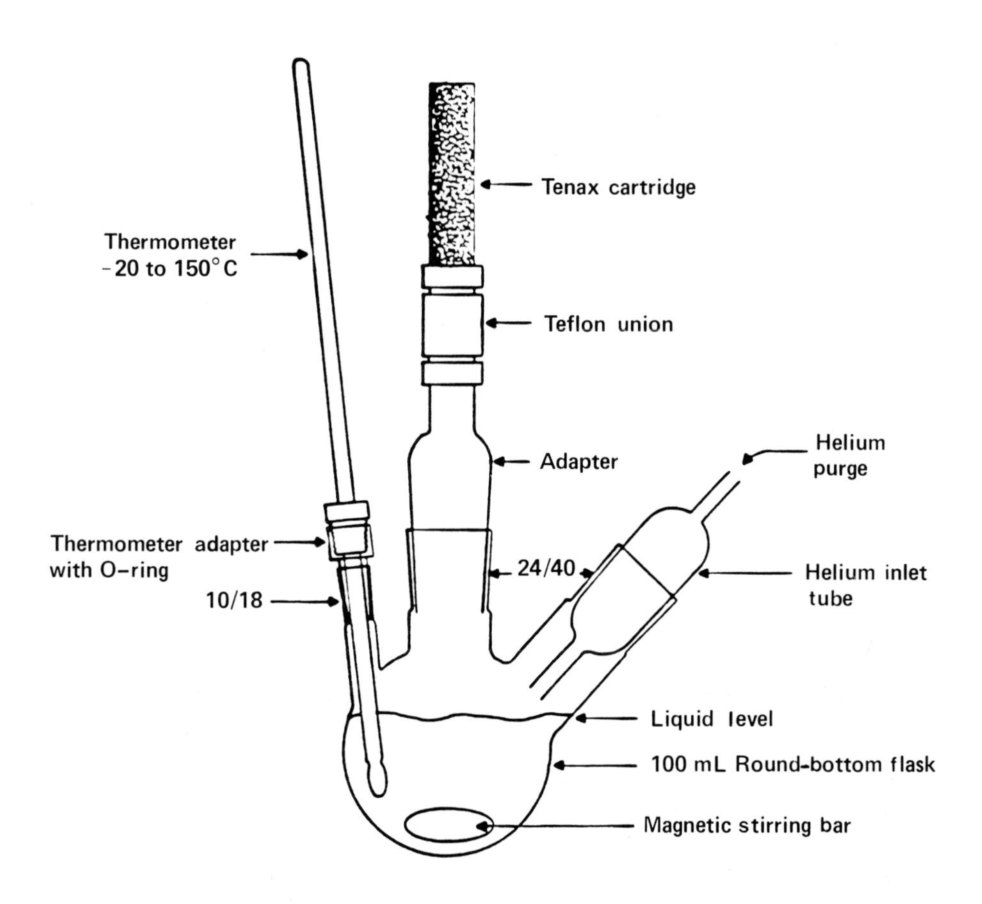

9. PROCEDURE

9.1 Blank test

Not applicable

9.2 Check test

See Method 24, section 9.2.

9.3 Test portions

1-10 mL whole blood; 5 g frozen tissue

9.4 Preparation of standards

See Method 24, sections 9.4.1 and 9.4.2.

9.5 Purging and trapping of analytes from blood sample

9.5.1 Measure an aliquot of whole blood, chilled to 4°C into the purge flask (Fig. 1).

9.5.2 Dilute to ~50 mL with purged, distilled, organic-free water and add a stirring bar. Assemble the apparatus as in Figure 1, begin stirring and raise the temperature to 50°C, with a helium flow-rate of 25 mL/min.

9.5.3 After 90 min, terminate the purge and transfer the Tenax cartridge to a Kimax culture tube containing 5 mL non-indicating Drierite, covered with a pad of glass wool. Cap the tube and store in a freezer until required for analysis.

9.6 Purging and trapping of analytes from tissue samples

9.6.1 Section an aliquot (5 g) of frozen tissue and transfer it to the purge flask (Fig. 1).

9.6.2 Dilute the aliquot to ~50 mL with purged, distilled, organic-free water and macerate the mixture in an ice bath, using a Virtis tissue homogenizer.

9.6.3 Immediately assemble the purging apparatus as in Figure 1, begin stirring and raise the temperature to 50°C, with a helium flow of 25 mL/min. After 30 min, terminate the purge and store the Tenax cartridge in a Kimax culture tube containing calcium sulfate desiccant, as described in 9.5.3.

9.7 Analyte determination

See Method 24, sections 9.4.3, 9.5 and 9.6.

10. METHOD OF CALCULATION

10.1 The mass concentration, ρ_a, of a given analyte in a blood sample is obtained from

$$\rho_a = m_a/V \ (\mu g/L)$$

where

m_a = mass of analyte recovered from Tenax cartridge (ng)

V = volume of aliquot in 9.5.1 (mL)

The value of m_a is obtained as in Method 24, sections 10.1 and 10.2.

10.2 The mass fraction, w_a, of the analyte in a tissue sample is given by

$$w_a = m_a/m_t \ (\mu g/kg)$$

where m_a is defined in 10.1 and m_t = wet weight of tissue aliquot in 9.6.1 (g).

11. REPEATABILITY AND REPRODUCIBILITY

11.1 Blood

The purge-and-trap technique was validated using both ^{14}C-labeled model compounds and "cold" model compounds. Results of these recovery studies are presented in Tables 2 and 3. Based on these data, expected recoveries of purgeable halogenated organic compounds from blood are about 80% or better.

11.2 Tissue

The purge-and-trap technique was validated using "cold" model compounds. Results of these recovery studies are presented in Table 4. Expected recoveries of purgeable halogenated organics from tissue are about 50%. See Notes on Procedure, 12.2.

12. NOTES ON PROCEDURE

12.1 Glass syringes represent the optimal collection device, since no polymeric material which may contaminate the sample comes in contact with the blood. However, sterilization of large numbers of glass syringes in the field is not practical.

Possible contamination by permeation through the rubber septum caps of the vacutainers is a cause for concern. Teflon-lined vacutainers are not available, but the manufacturers recommended special vacutainers "suitable for GC" (Venoject L 428, Kimble). Validation experiments have found the background of these tubes to be acceptable.

Leakage of the vacutainer caps has been observed, however, and permeation through the cap material is suspected. Accordingly, these containers are not suitable for storage.

Table 2. Percent recovery of ^{14}C-labelled compound from blood[a]

Compound	Mass loading (µg)	dpm loaded	% Recovery	Average % recovery
Chloroform	1.46	91 755.1	94.2 93.1 46.5	93.7
Carbon tetrachloride	0.785	77 902.3	92.6 92.2 83.5	89.4
Chlorobenzene	1.12	89 538.1	43.5[b] 88.0 92.2	90.1
Bromobenzene	1.21	84 493.9	77.1 74.4 90.3	80.6

[a] Head-space purge of whole blood (25 mL) diluted 1:1 with distilled water and purged in 100-mL, 3-neck flasks at 50°C for 90 min with helium at 25 mL/min.

[b] Leaking desorption chamber cap; not included in average.

12.2 Difficulty may be encountered in quantitative introduction of a representative fortified sample into the container for analysis. Consequently variations in recovery may be attributed to losses during tissue maceration and transfer. Thus, the analysis of tissue samples for volatile halocarbons should be regarded as semi-quantitative.

Table 3. Recovery of halogenated hydrocarbons from human blood spiked from gas mixing bulb

Compound	Amount spiked (μg)	% Recovery				Average ± std. dev.
		1	2	3	4	
Methylene chloride	7.96	105	116	a	a	111 ± 7.8
Chloroform	8.16	84.6	145	139	a	123 ± 33.1
Bromodichloromethane	10.4	105	159	94.9	94.9	113 ± 30.7
Tetrachloroethylene	8.52	108	121	82.8	97.5	99.8 ± 17.9
Chlorobenzene	5.53	108	120	85.6	78.6	98.1 ± 19.3
m-Dichlorobenzene	6.44	85.1	101	74.6	84.5	86.3 ± 10.9
Mean						104.3 ± 22.7

a Not quantified because of background interferences.

Table 4. Recovery of halogenated hydrocarbon from human adipose tissue

Compound	% Recovery				Average + S.D.
	1	2	3	4	
Methylene chloride	a	78.2	69.8	92.8	80.3 ± 11.6
Chloroform	a	62.3	14.8	59.7	45.6 ± 26.7
Bromodichloromethane	64.4	15.5	13.4	49.4	35.7 ± 25.3
Tetrachloroethylene	39.4	15.2	112	40.6	51.8 ± 41.8
Chlorobenzene	21.5	6.8	7.3	17.9	13.4 ± 7.5
m-Dichlorobenzene	82.1	a	56.5	30.8	56.5 ± 25.7
Mean					45.3 ± 30.8

a Not quantified due to peak interference.

13. SCHEMATIC REPRESENTATION OF PROCEDURE

Disperse sample in water and purge head-space
with helium, trapping volatile vapours on Tenax GC cartridge

↓

Add quantification standard to each cartridge
and continue as in section 13, Method 24

14. ORIGIN OF THE METHOD

Research Triangle Institute
Research Triangle Park, North Carolina 27709, USA

Contact point: Dr D. Pellizzari, Vice-President
Analytical and Chemical Sciences

METHOD 26

DETERMINATION OF 1,1,1-TRICHLOROETHANE IN BLOOD

K. Pekari & A. Aitio

1. SCOPE AND FIELD OF APPLICATION

The method is suitable for the determination of 1,1,1-trichloroethane in blood specimens. the detection limit is $\leq$ 0,07 µmol/L. Twenty-five to thirty analyses can be performed in one day.

2. REFERENCES

Lindner J. & Angerer, J. (1978) Method No. 2, Trichloroethane, Band 2. In: Henschler, D., ed., Analysen in Biologischem Material, Verlag Chemie, GmBH, D-6940 Weinheim, pp. D1-D5

3. DEFINITIONS

Not applicable

4. PRINCIPLE

The blood specimen is extracted with n-hexane and the concentration of 1,1,1-trichloroethane in the organic phase is determined by gas chromatography (GC) with electron-capture detection.

5. HAZARDS

1,1,1-Trichloroethane is decomposed by heat and forms poisonous fosgene gas.

n-Hexane is readily flammable and may react vigorously in the presence of oxidizing compounds.

6. REAGENTS[1]

1,1,1-Trichloroethane	Puriss, Fluka AG, M.W. = 133.4, D_4^{20} = 1.366
_n_-Hexane	HPLC-grade, Ratburn Chemicals LTD (Should be chosen for the column to be used so that no peaks occur at the same retention time as 1,1,1-trichloroethane. With the capillary column SE-30, the Ratburn product has been suitable.)
Stock standard solutions	I 1 mL of 1,1,1-trichloroethane is made up to 100 mL with _n_-hexane II 1 mL of solution I is made up to 100 mL with _n_-hexane III 1 mL of solution II is made up to 100 mL with _n_-hexane
Working standard solutions	Working standards are prepared form stock solution III (10.02 µmol/L) as follows:

St.1	1 mL of solution III is made up to 100 mL with _n_-hexane	0.10 µmol/L
St.2	1 mL of solution III is made up to 50 mL with _n_-hexane	0.20 µmol/L
St.3	1 mL of solution III is made up to 25 mL with _n_-hexane	0.40 µmol/L
St.4	1 mL of solution III is made up to 20 mL with _n_-hexane	0.50 µmol/L
St.5	1 mL of solution III is made up to 10 mL with _n_-hexane	1.00 µmol/L
St.6	2 mL of solution III is made up to 10 mL with _n_-hexane	2.00 µmol/L

The standards are kept in sealed 2-mL glass vials with Teflon septa, ready for use in the autosampler of the gas chromatograph. They are stable at least six months when stored at + 5°C.

---

[1] Reference to a company and/or product is for the purpose of information and identification only and does not imply approval or recommendation of the company and/or product by the International Agency for Research on Cancer, to the exclusion of others which may also be suitable.

7. APPARATUS[1]

Gas chromatograph	with electron-capture (^{3}H) detector and an autosampler
Mechanical shaker	Mixer, Reax 2, with adjustable rotation speed and universal adapter for racks. (Heindolph-Electro KG, Werk Schwabach, 8420 Kelheim, West Germany)
Centrifuge	With 4 inserts for 6 tubes. (Sorvall, GLC 2B, General laboratory instruments, Du Pont Instruments). Mean radius, 90 mm; 2 000 rev/min
Sample tubes	15-mL, with ground-glass joint and plastic stopper
Syringes	Hamilton, 10-µL, used with an autosampler

8. SAMPLING

NOTE: 1,1,1-Trichloroethane enters the body mainly through the lungs and 98-99% of it is eliminated unchanged by the same route. As 1,1,1-trichloroethane accumulates to some extent in fat, from which its elimination is very slow, it can be found in the blood several days after exposure.

8.1 When the compound is used daily, take the blood samples before the exposure in the morning of the last working day of the week. (An estimate is thus obtained of the level of the exposure during the working week.) When the compound is not used continuously, take the specimens immediately after the working day.

8.2 The volume of blood sample required is 5 mL. For sampling, fill heparinized tubes (5 mL or 10 mL) to the rim with blood (to avoid loss of solvent to the air phase in the tube) and cap. Store the samples at + 5°C until analysed. If possible, carry out extraction immediately upon receipt in the laboratory.

---

[1] Reference to a company and/or product is for the purpose of information and identification only and does not imply approval or recommendation of the company and/or product by the International Agency for Research on Cancer, to the exclusion of others which may also be suitable.

9. PROCEDURE

9.1 Blank test

Analyse a solvent blank with every analytical series. No peak should be seen at the same retention time as 1,1,1-trichloroethane.

9.2 Check test

Not applicable

9.3 Test portion

2.00 ± 0.01 mL (use a pipette)

9.4 Sample extraction

Extract 2 mL of whole blood with 4 mL of n-hexane in a centrifuge tube for 15 min, using a mechanical shaker, then centrifuge, if necessary, for 5 min (2 000 rev/min). Retain for gas chromatography (GC).

9.5 GC conditions

Column	25 m × 0.2 mm (i.d.) vitreous silica capillary, coated with SE-30
Column temperature	60°C
Injector temperature	120°C
Detector temperature	220°C

9.6 GC determination of 1,1,1-trichloroethane

Inject 1 µL of the hexane phase (9.4) onto the GC and record the peak height.

9.7 Calibration curve

Using the working standard solutions in n-hexane, inject 1 µL of each solution onto the GC. (The calibration curve is determined with each series of samples to verify the linear response range of the detector.) Specimens with high concentrations of 1,1,1-trichloroethane must be diluted appropriately. Normally, the calibration curve is linear up to 1 µmol/L.

A linear relation between peak height (y) and concentration of 1,1,1-trichloroethane (C, µmol/L) is calculated by the method of least squares. The correlation coefficient of the equation $y = (aC + b)$ should be better than 0.99.

The C-intercept should be $< |0.05|$ µmol/L.

9.8 Recovery

The recovery, R, of the spiked specimens is 87.9 ± 5.7 %, as determined from additions of 0.33 - 0.99 µmol/L of 1,1,1-trichloroethane to blood with initial concentrations of trichloroethane of 0.24 - 0.81 µmol/L (n = 13).

10. METHOD OF CALCULATION

The concentration, C, of 1,1,1-trichloroethane in the specimen is calculated from

$$C = \frac{100d\ (y-b)}{Ra} \quad (\mu mol/L)$$

where

d = dilution factor (d = 2 in section 9.7)

R = recovery (%)

and

a and b are obtained from the equation of section 9.7.

11. REPEATABILITY AND REPRODUCIBILITY

The coefficient of variation calculated from duplicate determinations in the concentration range 0.3-0.8 µmol/L is 3% (n = 7).

12. NOTES ON PROCEDURE

Not applicable

13. SCHEMATIC REPRESENTATION OF PROCEDURE

5 mL blood sample

↓

Extract 2 mL blood with 4 mL *n*-hexane
(shake 15 min, centrifuge 5 min)

↓

Inject 1 µL hexane phase onto GC

↓

Calculate analyte concentration from
peak height and calibration curve

14. ORIGIN OF THE METHOD

Institute of Occupational Health
Laboratory of Biochemistry
Arinatie 3
SF-00370 Helsinki
Finland

Contact point: K. Pekari

METHOD 27

DETERMINATION OF TETRACHLOROETHYLENE IN BLOOD

K. Pekari & A. Aitio

1. SCOPE AND FIELD OF APPLICATION

The method is suitable for the determination of tetrachloroethylene in blood. The limit of detection is $\leq$ 0.03 µmol/L. Twenty five to thirty analyses can be completed in one day.

2. REFERENCES

Lindner, J. & Angerer, J. (1978) Method No. 2, Trichloroethane Band 2. In: Henschler, D., ed., Analysen in Biologischem Material, Verlag Chemie, GmBH, D-6940 Weinheim, pp. D1-D5

3. DEFINITIONS

Not applicable

4. PRINCIPLE

The specimen is extracted with *n*-hexane and the concentration of tetrachloroethylene in the organic phase is determined by gas chromatography, with electron-capture detection.

5. HAZARDS

Tetrachloroethylene is moderately toxic and irritant to eyes and skin. *n*-Hexane is readily flammable.

6. REAGENTS[1]

Tetrachloroethylene — p.a., Merck

Stock standard solutions:

I. 1 mL of tetrachloroethylene made up to 100 mL with *n*-hexane
II. 1 mL of solution I made up to 100 mL with *n*-hexane
III. 1 mL of solution II made up to 100 mL with *n*-hexane

Working standard solutions: The working standards are prepared from stock solution III (9.78 µmol/L) as follows:

St.1	1 mL of solution III made up to 100 mL with *n*-hexane	0.10 µmol/L
St.2	1 mL of solution III made up to 50 mL with *n*-hexane	0.20 µmol/L
St.3	1 mL of solution III made up to 25 mL with *n*-hexane	0.39 µmol/L
St.4	1 mL of solution III made up to 20 mL with *n*-hexane	0.49 µmol/L
St.5	1 mL of solution III made up to 10 mL with *n*-hexane	0.98 µmol/L

The standards are kept in 2-mL, sealed glass vials with Teflon septa, ready for use in the autosampler of the gas chromatograph. They are stable several months when stored at + 5°C.

n-Hexane — HPLC-grade, Ratburn chemicals Ltd. (Should be chosen for the column to be used so that no peaks occur at the same retention time as tetrachloroethylene. With the capillary column SE-30, the Ratburn product has been suitable).

7. APPARATUS[1]

Gas chromatograph — With an electron-capture (^3H) detector and an autosampler.

---

[1] Reference to a company and/or product is for the purpose of information and identification only and does not imply approval or recommendation of the company and/or product by the International Agency for Research on Cancer, to the exclusion of others which may also be suitable.

Mechanical shaker	Mixer, Reax 2, with adjustable rotation speed and universal adapter for racks (Heindolph-Electro KG, Werk Schwabach, 8420 Kelheim, West Germany)
Centrifuge	With 4 inserts for 6 tubes (Sorvall, GLC 2B, General laboratory instruments, Du Pont Instruments). Mean radius, 90 mm, 2 000 rev/min.
Sample tubes	15-mL, with ground-glass joint and a plastic stopper
Syringes	Hamilton 10-µL syringes, used with an auto-sampler

8. SAMPLING

When workers are exposed to tetrachloroethylene daily, take blood specimens before exposure in the morning of the last working day in the week (An estimate is thus obtained of the level of the exposure during the working week). When the compound is not used continuously, take the samples immediately after the working day.

The volume of blood required for analysis is 5 mL. Fill heparinized tubes (5 mL or 10 mL) to the rim with blood (to avoid loss of solvent to the air phase in the tube) and cap. Store the specimens at + 5°C until analysed, If possible, extract the specimens immediately upon arrival in the laboratory.

9. PROCEDURE

9.1 Blank test

Analyse a solvent blank with every analytical series. No peak should be seen at the same retention time as tetrachloroethylene.

9.2 Check test

Not applicable

9.3 Test portion

2.00 ± 0.01 mL (use a pipette)

9.4 Sample extraction

Extract 1 mL of whole blood with 5 mL of n-hexane in a centrifuge tube for 15 min, using a mechanical shaker, then centrifuge, if necessary, for 5 min (2 000 rev/min). Retain for gas chromatography (GC).

9.5 GC conditions

Column	25 m × 0.2 mm (i.d.) vitreous silica capillary, coated with SE-30
Column temperature	90°C
Injector temperature	120°C
Detector temperature	220°C

9.6 GC determination of tetrachloroethylene

Inject 1 µL of the hexane phase from 9.4 onto the GC. Record the peak height.

9.7 Calibration curve

Using the working standard solutions in n-hexane, inject 1 µL of each solution onto the GC (the calibration curve is determined with each series of samples to verify the linear concentration range of the detector). Specimens with high concentrations of tetrachloroethylene must be appropriately diluted. Normally, the calibration curve is linear up to 0.8 µmol/L.

The linear relation between peak height (y) and concentration of tetrachloroethylene (C, µmol/L) is calculated by the method of least squares. The correlation coefficient of the equation $y = aC + b$ should be better than 0.99. The C-intercept should be less than $|0.02|$ µmol/L.

9.8 Recovery

The recovery, R, from spiked specimens is 94.1 ± 4.5%, as determined from additions of 0.98 µmol/L of tetrachloroethylene to blood with initial concentrations of tetrachloroethylene of 0.36-1.57 µmol/L (n = 7).

10. METHOD OF CALCULATION

The concentration, C, of tetrachloroethylene in the specimen is calculated from

$$C = \frac{100d\ (y - b)}{Ra} \quad (\mu mol/L)$$

where

d = dilution factor (d = 5 in section 9.1)
R = recovery (%) and
a and b are obtained from the equation of section 9.7

11. REPEATABILITY AND REPRODUCIBILITY

The coefficient of variation calculated from duplicate determinations in the concentration range 0.4-0.6 µmol/L is 3% and in the concentration range 1.8-2.5 µmol/L is 2% (n = 5)

12. NOTES ON PROCEDURE

Not applicable

13. SCHEMATIC REPRESENTATION OF PROCEDURE

5 mL blood sample
↓
Extract 1 mL blood with 5 mL *n*-hexane
(shake 15 min, centrifuge 5 min)
↓
Inject 1 µL hexane phase on GC
↓
Calculate analyte concentration from peak height
and calibration curve

14. ORIGIN OF THE METHOD

Institute of Occupational Health
Laboratory of Biochemistry
Arinatie 3
SF-00370 Helsinki 37, Finland

Contact point: K. Pekari

METHOD 28

DETERMINATION OF HALOTHANE IN BLOOD BY GAS CHROMATOGRAPHY (GC)

D.J. Jones

1. SCOPE AND FIELD OF APPLICATION

This method may be used for the determination of halothane (2-bromo-2-chloro-1,1,1-trifluoroethane) and other halogenated volatile anaesthetics in blood. The method does not require extraction procedures and can be accomplished (with standard curve) within one hour. The limit of detection is 10 mg/L.

2. REFERENCE

Jones, D.J. (1978) rapid gas chromatographic assay for volatile anesthetics in blood. J. Pharmacol. Methods, 1, 155-160

3. DEFINITIONS

Not applicable

4. PRINCIPLE

After anaerobic sampling of blood and equilibration of the halothane with the head-space air in an airtight 5-mL serum vial (37°C), a sample of the head-space air is analysed by GC.

5. HAZARDS

Not applicable

6. REAGENTS[1]

Nitrogen	
Compressed air	Filtered to remove particulates
Hydrogen	
Standard solutions	Prepare calibration standards by adding 2.5 µL of halothane to 10 mL of heparinized blood or saline in a 15-mL serum bottle. Seal bottle with rubber and metal caps and agitate vigorously for 5 min. This "stock" solution has a halothane concentration of 469 mg/L (2.37 mmol/L). The calibration standards for halothane, made up to 1 mL with a volume of uncontaminated blood or saline in 5-mL vials, are noted in Table 1.

Table 1. Calibration standard dilutions and concentrations

Volume of "stock" solution (mL)	0.1	0.2	0.4	0.6	1.0
Concentration of halothane (mg/L)	A 46.8	B 43.6	C 187.2	D 280.2	E 468

7. APPARATUS[1]

Tuberculin syringes	1-mL, with air-tight plastic caps
Serum bottle	5-mL (Wheaton #223738, Wheaton Scientific, Millville, New Jersey, USA)
Rubber stoppers	Wheaton #224124

[1] Reference to a company and/or product is for the purpose of information and identification only and does not imply approval or recommendation of the company and/or product by the International Agency for Research on Cancer, to the exclusion of others which may also be suitable.

Aluminum 2 mm, one-piece seals	Wheaton #224183
Cap crimper	Wheaton #224303
Gas chromatograph	Fitted with flame-ionization detector
Shaking incubation bath	
Gas-tight liquid syringe	50-μL

8. SAMPLING

8.1 Draw accurate, 1-mL samples of either venous or arterial blood anaerobically into heparinized, disposable tuberculin syringes and tightly cap with air-tight plastic caps (the blood usually remains in the air-tight syringes for less than 2 min).

8.2 Quickly transfer sample into a 5-mL serum bottle, stopper and place an aluminum 2 mm one-piece seal on top of the rubber stopper. Crimp on the top to form an air-tight unit, using an appropriate-sized crimper.

9. PROCEDURE

9.1 Blank test

Prepare a 1.0-mL control standard which does not contain anaesthetic and follow instructions of sections 8 to 10, as for blood samples.

9.2 Check test

Not applicable

9.3 Test portion

Not applicable

9.4 GC conditions

GC column	Stainless steel columns, 3.05 m × 3.2 mm i.d., packed with 5% OV 101 (100/120) on chromosorb W (AW-DMCS) solid support. A 1.83 m × 3.2 mm column with i.e., 10% OV 101 (100/120) on Chromosorb G (AW-DMCS) solid support has also been used, with similar results.

Gas flow-rates — Nitrogen, 60 mL/min; hydrogen, 17-20 mL/min; compressed air, 38-40 mL/min

Temperatures — Injection port, 125°C; column, 100°C; detector, 100°C.

NOTE: Pyrex glass liners are placed in the injection port adjacent to a silicone rubber septum. The septums are usually changed every 50-100 assays (or when air leaks occur).

9.5 Analytical procedure

9.5.1 Place the samples and standards in a 37°C water bath to equilibrate for 15 min. During this period, gently swirle the samples to hasten equilibration of the anesthetic gas between head-space and blood.

9.5.2 At the time of assay, remove the center 0.5 cm aluminum seal to facilitate entry of the needle of the microliter syringe (the air-tight seal is still intact).

9.5.3 Insert the needle of a 50 µL Hamilton syringe through the rubber stopper and place the needle approximately 0.5 cm above the blood.

9.5.4 Move the plunger rapidly up and down at least 10 times, adjust volume to 10 µL, then rapidly inject into the gas chromatograph. Inject three such head-space samples for each blood sample.

9.6 Calibration curve

Obtain the calibration curve by plotting the observed peak height against the calculated mg of anaesthetic per litre of blood or saline standard solutions (see Fig. 1).

10. METHOD OF CALCULATION

Measure concentrations of analyte in samples by reading from the calibration curve at the point of the average peak height of the three sample injections (9.5.4). Subtract the concentration observed in the blank.

FIG. 1. HALOTHANE AND ENFLURANE STANDARD CURVES

Generated from data of 10 different standard curves. Each "X" represents the average peak height of triplicate samples. Computer-assisted linear regression was used to draw the line.

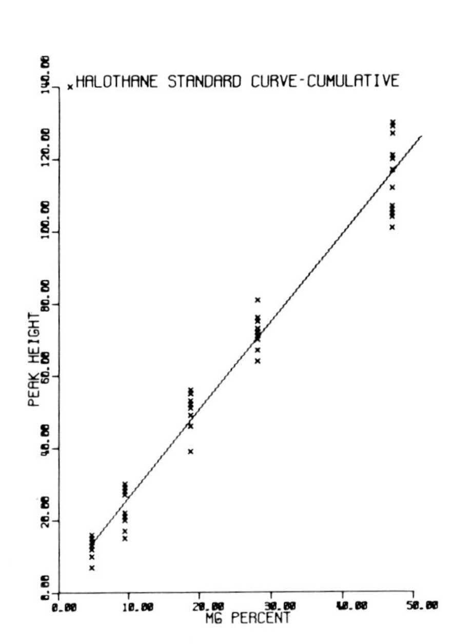

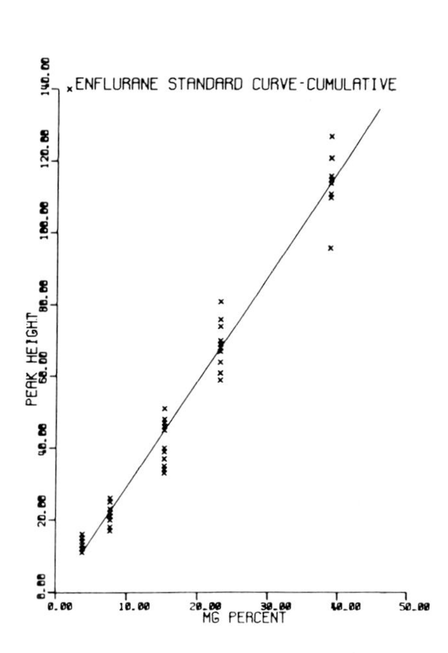

11. REPEATABILITY AND REPRODUCIBILITY

11.1. A print-out from GC analysis of three 10-µL standards of either halothane or enflurane is shown in Figure 2.

11.2 Plots of average values of 10 different standard curves of halothane and enflurane are shown in Figure 1. These curves were generated over a period of 5 weeks.

FIG. 2. TYPICAL CHROMATOGRAMS OF TRIPLICATE 10-μL SAMPLES FROM HALOTHANE AND ENFLURANE CALIBRATION STANDARDS

A-E for halothane correspond to concentrations noted in Table 1.

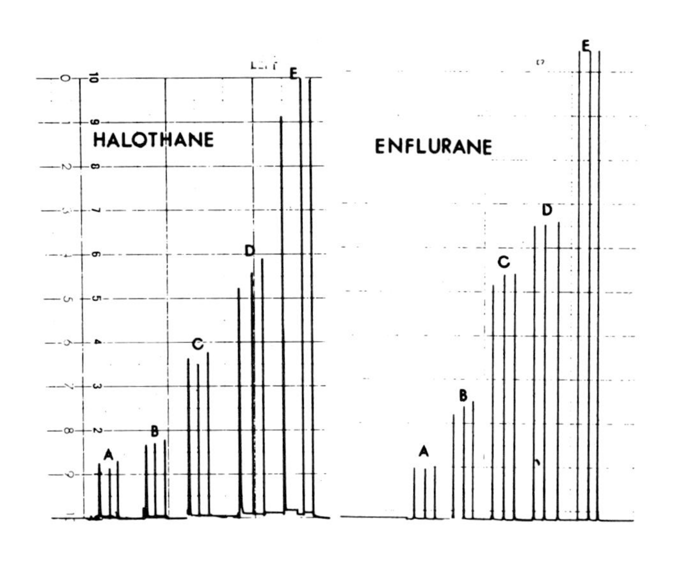

12. NOTES ON PROCEDURE

12.1 Rather than inject the blood into the vial through the rubber stopper, the vial is capped immediately following addition of the 1 mL of blood. This procedure was used so as to minimize any pressure changes which might lead to leakage or alterations in equilibration pressure within the vial.

12.2 The 10-μL samples of head-space air represent 0.2% of the total gas space. Up to six repeated samplings did not decrease the peak height.

12.3 If samples can not be analysed immediately, they can be stored at 2-4°C for up to 48 h without loss of sampling accuracy. Standards stored in this manner also do not exhibit a decrease in peak height or a change in linearity. Such a finding attests to the lack of leakage and solubility loss into the rubber stopper under these storage conditions.

12.4 As is evident from Figure 2, some variability in peak height occurs. This appears largely due to the injection technique. It is necessary to clean the syringe carefully between gas-phase injections from different blood samples and to adhere rigidly to the requirement of flushing the syringe with gas sample 10 times, prior to injection. Also one must insure that no air leaks occur at the time of piercing the GC septum, as this will give rise to a small peak prior to the halothane peak.

13. SCHEMATIC REPRESENTATION OF PROCEDURE

1 mL heparinized blood sample into 5-mL serum vial

↓

Cap and crimp vial

↓

Incubate at 37°C with gentle swirling for 15 min

↓

Remove center portion of aluminum
cap and insert needle of 50-μL syringe
to approximately 0.5 cm above blood

↓

Move plunger up and down at least 10 times,
withdraw a 10-μL aliquot and rapidly inject into GC port

↓

Repeat two more times

↓

Measure peak height and calculate halothane concentration
from standard curve

14. ORIGIN OF THE METHOD

Department of Anesthesiology
The University of Texas Health Science Center
7703 Floyd Curl Drive
San Antonio, TX 78284, USA

Contact point: David J. Jones, Ph.D.

BIOLOGICAL MONITORING

URINE

METHOD 29

DETERMINATION OF TRICHLOROACETIC ACID IN URINE

K. Pekari & A. Aitio

1. SCOPE AND FIELD OF APPLICATION

This method is suitable for the determination of trichloroacetic acid in urine for the biological monitoring of workers exposed to trichloroethylene. The limit of detection is $\leq$ 5 µmol/L. Twenty-five to thirty analyses can be completed in one day.

2. REFERENCES

Eben, A., Henschler, D. & Pilz, W. (1978) Method No. 1, Trichloroacetic acid, Band 2. In: Henschler, D., ed., Analysen in Biologischem Material, Verlag Chemie, GmbH, D-6940 Weinheim, pp. D1-D5

3. DEFINITIONS

Not applicable

4. PRINCIPLE

The method is an application of the Fujiwara reaction, where trichloroacetic acid is decarboxylated to chloroform and the absorbance of the coloured compound that is formed in pyridine is measured at 530 nm with a photometer.

5. HAZARDS

Sodium hydroxide is a strong, caustic base. Pyridine decomposes when heated, forming poisonous cyanide vapours. It also reacts strongly with oxidizing agents.

6. REAGENTS[1]

Trichloroacetic acid (TCA)	Merck, p.a.
Sodium hydroxide	Merck, p.a.
Sodium hydroxide	Merck, Titrisol^R, 0.01 mol/L, sealed ampoules
Sodium sulfate	Anhydrous, Merck, p.a.
Pyridine	Merck, p.a.
Standard solution of TCA	Dissolve 50 mg TCA in 1 L distilled water. Because TCA is hygrosocpic, the exact concentration of the solution (~ 300 µmol/L) is determined by titration with 0.002 mol/L sodium hydroxide with phenolphthalein as an indicator. Only sodium hydroxide prepared from freshly-opened ampoules may be used, as sodium hydroxide tends to take up carbon dioxide from the air.

7. APPARATUS

Test tubes	25-mL, with screw cap
Colorimeter	The spectral half-width of the coloured reaction product is 80 nm, so that a colorimeter with a filter with a band pass of 50-70 nm can be used. A colorimeter with round cuvettes is preferable, as bubble formation in tetragonal cuvettes may cause difficulties.
Mechanical shaker	Mixer, Reax 2, with adjustable rotation speed and universal adapter for racks (Heindolph-Electro KG, Werk Schwabach, 8320 Kelheim, West Germany)

[1] Reference to a company and/or product is for the purpose of information and identification only and does not imply approval or recommendation of the company and/or product by the International Agency for Research on Cancer, to the exclusion of others which may also be suitable.

Centrifuge	With 4 inserts for 6 tubes (Sorvall, GLC 2B, General laboratory instruments, Du Pont Instruments) Mean radius, 90 mm, 2 000 rev/min
Syringes	Hamilton 10-µL, used with an autosampler

8. SAMPLING

Trichloroethylene enters the body through the lungs and the skin. About seventy per cent of the absorbed trichloroethylene is metabolized to trichloroacetic acid and trichloroethanol, which are then excreted in urine.

The exposure of workers to trichloroethylene can be monitored by determining the concentration of trichloroacetic acid or trichloroethanol in the urine. The rapid metabolism and excretion of trichloroethanol makes the measurement of its concentration in urine suitable for detecting only the exposure during the last working hours before sampling. Due to the slow metabolism of trichloroethylene to trichloroacetic acid, the concentration of trichloroacetic acid in urine indicates the exposure level over several days before sampling.

For the analyses, only 2 mL of urine (single voiding in the morning following the exposure) is needed, but for representative sampling, $\geq$ 50 mL of urine is recommended. Store the specimens at + 5°C until analysed.

9. PROCEDURE

9.1 Blank test

Analyse 0.5 mL of distilled water by the method of sections 9.4-9.5.

9.2. Check test

Within each series, analyse two control specimens with known TCA levels. These are prepared form urine of workers exposed to trichloroethylene. Frozen control specimens are stable at least 12 months when kept at -18°C.

9.3 Test portion

0.5 ± 0.01 mL (use a pipette)

9.4 Conversion of TCA

9.4.1 Pipette 0.5 mL of urine sample into 25 mL test tube and add 10 mL of 25% sodium hydroxide.

9.4.2 With each group of urine samples, repeat 9.4.1 using 0.5 mL of standard TCA solution in place of urine.

9.4.3 Close tubes with screw caps and shake for one min in mechanical shaker.

9.4.4 Add 6 mL of pyridine to each tube and shake vigourously a few times by hand.

9.4.5 Loosen screw caps and place tubes in a water bath at 80.0 ± 0.5°C for 5 min.

9.4.6 Transfer the pyridine phase by pipette to a test tube containing 2g anhydrous sodium sulfate.

9.4.7 Close the tubes with screw caps and shake for 0.5 min, then centrifuge for 25 min (2 000 rev/min).

9.5 Colorimetric determination

Pour pyridine phase from 9.1.7 into colorimeter cuvette and measure absorbance (A) against that of pyridine.

9.6 Recovery

The recovery of the spiked specimens in the concentration range 5-300 µmol/L is 93.5 ± 1.7% (mean ± standard deviation, n = 8). The colour reaction follows Beer's law up to a concentration of at least 700 µmol/L.

10. METHOD OF CALCULATION

The concentration, C, of TCA in the specimen is calculated from the equation,

$$C = C_s \frac{(A - A_o)}{(A_s - A_o)} \quad (\mu mol/L)$$

where

C_s = concentration of TCA standard solution (µmol/L)

A_s = absorbance of the standard solution

A_o = absorbance of the blank

A = absorbance of the specimen

Because the standards are analysed together with the specimens, correction for recovery is not needed.

11. REPEATABILITY AND REPRODUCIBILITY

The coefficient of variation calculated from duplicate determinations in the concentration range 140-370 µmol/L is 3% (n = 10).

12. NOTES ON PROCEDURE

Not applicable

13. SCHEMATIC REPRESENTATION OF DETERMINATION

Add to 25 mL test tubes:
0.5 mL distilled water + 10 mL NaOH (25%)
0.5 mL standard TCA solution + 10 mL NaOH (25%)
0.5 mL urine + 10 mL NaOH (25%)

↓

Shake each tube for ~ 1 min

↓

Add 6 mL pyridine and shake

↓

Place in water bath (80°C) for 5 min

↓

Transfer pyridine to anhydrous sodium sulfate, shake and centrifuge for 25 min

↓

Measure absorbance of pyridine phase against pure pyridine

↓

Calculate urine TCA concentration from absorbance measurements

14. ORIGINE OF THE METHOD

Institute of Occupational Health
Laboratory of biochemistry
Arinatie 3,
SF-00370 Helsinki 37

Contact point: K. Pekari

METHOD 30

DETERMINATION OF 2,2,2-TRICHLOROETHANOL IN URINE

K. Pekari & A. Aitio

1. SCOPE AND FIELD OF APPLICATION

This method is suitable for the determination of the total (free + conjugated) 2,2,2-trichloroethanol in urine of workers exposed to trichloroethylene. The detection limit of the method is 0.5 µmol/L. One technician is able to analyse about 10 specimens in a working day.

2. REFERENCES

Vesterberg, O., Gorczak, J. & Krasts, M. (1975) Methods for measuring trichloroethanol and trichloroacetic acid in blood and urine after exposure to trichloroethylene. Scand. J. Work Environ. & Health, 1, 243-248

3. DEFINITIONS

Not applicable

4. PRINCIPLE

Conjugates of trichloroethanol in urine are hydrolyzed with acid, trichloroethanol is extracted into iso-octane and quantified by gas chromatography (GC), using electron-capture detection.

5. HAZARDS

Sulfuric acid is a strong, caustic acid. Iso-octane is easily flammable and may react vigorously in the presence of oxidizing compounds.

6. REAGENTS[1]

2,2,2-Trichloroethanol	96% Merck, p.a., ($d_{20°C}$ = 1.552)
Sulfuric acid	Merck, p.a., conc.
Iso-octane	Merck, p.a.
Stock standard solutions	I 30 µL of trichloroethanol made up to 100 mL with distilled water — 3000 µmol/L II 10 mL of the solution I made up to 100 mL with distilled urine — 300 µmol/L
Working standard solutions	Working standards are prepared from the stock solution II as follows:

St.0	trichloroethanol-free urine from non-exposed persons	
St.1	1 mL of stock solution II made up to 50 mL with St.0 urine	6.0 µmol/L
St.2	2 mL of the stock solution II made up to 50 mL with St.0 urine	12.0 µmol/L
St.3	3 mL of the stock solution II made up to 50 mL with St.0 urine	18.0 µmol/L
St.4	1 mL of the stock solution II made up to 10 mL with St.0 urine	30.0 µmol/L

7. APPARATUS[1]

Mechanical shaker	Mixer, Reax 2, with adjustable rotation speed and universal adapter for racks (Heindolph-Electro KG, Werk Schwabach, 8420 Kelheim, West Germany)

---

[1] Reference to a company and/or product is for the purpose of information and identification only and does not imply approval or recommendation of the company and/or product by the International Agency for Research on Cancer, to the exclusion of others which may also be suitable.

Centrifuge	With 4 inserts for 6 tubes (Sorvall, GLC 2B, General laboratory instruments, Du Pont Instruments) Mean radius, 90 mm; 2 000 rev/min
Sample tube	10-mL glass tube with a screw cap
Syringes	Hamilton 10 µL, used with an autosampler
Gas chromatograph	With an electron-capture (^3H) detector and an autosampler

8. SAMPLING

Trichloroethylene enters the body through the lungs and the skin. About seventy per cent of the absorbed trichloroethylene is metabolized to trichloroacetic acid and trichloroethanol, which are then excreted in urine.

Exposure of workers to trichloroethylene can be followed by determining the concentration of trichloroacetic acid or trichloroethanol in urine. The rapid metabolism and excretion of trichloroethanol makes the measurement of its concentration in urine suitable for detecting only the exposure during the last working hours before sampling.

For the analyses, only 2 mL of urine (single voiding after the working day) is needed, but for representative sampling, $\geq$ 50 mL of urine is recommended. Store the specimens at +5°C until analysed.

9. PROCEDURE

9.1 Blank test

Analyse a solvent blank (iso-octane) with each series of urine samples. No peak should be seen at the same retention time as 2,2,2-trichloroethanol.

9.2 Check test

Within each series, analyse two control specimens with known trichloroethanol levels. These are prepared form urine of workers exposed to trichloroethylene. Frozen control specimens are stable at least six months when kept at -18°C.

9.3 Test portion

0.500 ± 0.006 mL (use a micropipette, 200-1000 µL capacity).

9.4 Hydrolysis and extraction

9.4.1 Pipette 0.5 mL of urine and 0.25 mL of concentrated sulfuric acid into a 10 mL tube and seal with screw cap.

9.4.2 Place tubes in boiling water bath for 30 min.

9.4.3 Cool tubes, add 2 mL *iso*-octane and shake for 5 min, then centrifuge for 5 min at 2 000 rev/min. Proceed directly to 9.6.

9.5 GC conditions

Column	25 m × 0,2 mm (i.d.) vitreous silica capillary, coated with SE-30
Column temperature	110°C
Injector temperature	120°C
Detector temperature	220°C

9.6 GC determination of trichloroethanol

Inject 1 µL of the organic phase from 9.4.3 onto the GC.

9.7 Calibration curve

Prepare a 5-point calibration curve for each run using the 5 working standard solutions in order to verify the linear range of the detector. Inject 1 µL of each standard solution onto the GC. (Specimens with high concentrations of trichloroethanol must be appropriately diluted.) Normally, the calibration curve is linear up to 30 µmol/L.

An equation ($y = aC + b$) of the dependence of the peak heights on the 2,2,2-trichloroethanol concentration of the standards (C, µmol/L) is calculated using the least squares method. The correlation coefficient should be > 0.99 and the C-intercept < |0.3| µmol/L.

9.8 Recovery

The recovery of the method is 98.2 ± 6.0% (mean ± s.d., n = 6), when 5-30 µmol of trichloroethanol per litre urine is added to specimens with an initial concentration of 0.9 µmol/L of trichloroethanol.

10. METHOD OF CALCULATION

The concentration, C, of 2,2,2-trichloroethanol in the specimen is given by

$$C = (y - b)/a \ (\mu mol/L)$$

9.4 Hydrolysis and extraction

9.4.1 Pipette 0.5 mL of urine and 0.25 mL of concentrated sulfuric acid into a 10 mL tube and seal with screw cap.

9.4.2 Place tubes in boiling water bath for 30 min.

9.4.3 Cool tubes, add 2 mL iso-octane and shake for 5 min, then centrifuge for 5 min at 2 000 rev/min. Proceed directly to 9.6.

9.5 GC conditions

Column	25 m × 0,2 mm (i.d.) vitreous silica capillary, coated with SE-30
Column temperature	110°C
Injector temperature	120°C
Detector temperature	220°C

9.6 GC determination of trichloroethanol

Inject 1 µL of the organic phase from 9.4.3 onto the GC.

9.7 Calibration curve

Prepare a 5-point calibration curve for each run using the 5 working standard solutions in order to verify the linear range of the detector. Inject 1 µL of each standard solution onto the GC. (Specimens with high concentrations of trichloroethanol must be appropriately diluted.) Normally, the calibration curve is linear up to 30 µmol/L.

An equation ($y = aC + b$) of the dependence of the peak heights on the 2,2,2-trichloroethanol concentration of the standards (C, µmol/L) is calculated using the least squares method. The correlation coefficient should be > 0.99 and the C-intercept < $|0.3|$ µmol/L.

9.8 Recovery

The recovery of the method is 98.2 ± 6.0% (mean ± s.d., n = 6), when 5-30 µmol of trichloroethanol per litre urine is added to specimens with an initial concentration of 0.9 µmol/L of trichloroethanol.

10. METHOD OF CALCULATION

The concentration, C, of 2,2,2-trichloroethanol in the specimen is given by

$$C = (y - b)/a \ (\mu mol/L)$$

where y is the sample peak height and
a and b are obtained from the equation of section 9.7.

As the standards are treated like the specimens, no correction for recovery is needed.

11. REPEATABILITY AND REPRODUCIBILITY

The coefficient of variation calculated from double determinations in the concentration area 30-40 μmol/L is 0,01 (n = 5).

12. NOTES ON PROCEDURE

Not applicable

13. SCHEMATIC REPRESENTATION OF PROCEDURE

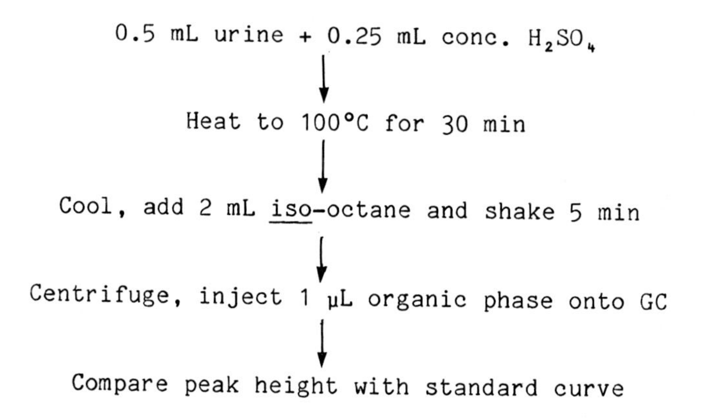

14. ORIGIN OF THE METHOD

Institute of Occupational Health
Laboratory of Biochemistry
Arinatie 3
SF-00370 Helsinki 37, Finland

Contact point: K. Pekari

INDEX OF AUTHORS

PUBLICATIONS OF THE INTERNATIONAL AGENCY FOR RESEARCH ON CANCER

SCIENTIFIC PUBLICATIONS SERIES

(Available from Oxford University Press)

No. 1 LIVER CANCER (1971)
176 pages; £10-

No. 2 ONCOGENESIS AND HERPES VIRUSES (1972)
Edited by P.M. Biggs, G. de Thé & L.N. Payne, 515 pages; £30.-

No. 3 *N*-NITROSO COMPOUNDS - ANALYSIS AND FORMATION (1972)
Edited by P. Bogovski, R. Preussmann & E.A. Walker, 140 pages; £8.50

No. 4 TRANSPLACENTAL CARCINOGENESIS (1973)
Edited by L. Tomatis & U. Mohr, 181 pages; £11.95

No. 5 PATHOLOGY OF TUMOURS IN LABORATORY ANIMALS. VOLUME 1. TUMOURS OF THE RAT. PART 1 (1973)
Editor-in-Chief V.S. Turusov, 214 pages; £17.50

No. 6 PATHOLOGY OF TUMOURS IN LABORATORY ANIMALS. VOLUME 1. TUMOURS OF THE RAT. PART 2 (1976)
Editor-in-Chief V.S. Turusov 319 pages; £17.50

No. 7 HOST ENVIRONMENT INTER-ACTIONS IN THE ETIOLOGY OF CANCER IN MAN (1973)
Edited by R. Doll & I. Vodopija, 464 pages; £30.-

No. 8 BIOLOGICAL EFFECTS OF ASBESTOS (1973)
Edited by P. Bogovski, J.C. Gilson, V. Timbrell & J.C. Wagner, 346 pages; £25.-

No. 9 *N*-NITROSO COMPOUNDS IN THE ENVIRONMENT (1974)
Edited by P. Bogovski & E.A. Walker 243 pages; £15.-

No. 10 CHEMICAL CARCINOGENESIS ESSAYS (1974)
Edited by R. Montesano & L. Tomatis, 230 pages; £15.-

No. 11 ONCOGENESIS AND HERPES-VIRUSES II (1975)
Edited by G. de-Thé, M.A. Epstein & H. zur Hausen
Part 1, 511 pages; £30.-
Part 2, 403 pages; £30.-

No. 12 SCREENING TESTS IN CHEMICAL CARCINOGENESIS (1976)
Edited by R. Montesano, H. Bartsch & L. Tomatis, 666 pages; £30.-

No. 13 ENVIRONMENTAL POLLUTION AND CARCINOGENIC RISKS (1976)
Edited by C. Rosenfeld & W. Davis 454 pages; £17.50

No. 14 ENVIRONMENTAL *N*-NITROSO COMPOUNDS - ANALYSIS AND FORMATION (1976)
Edited by E.A. Walker, P. Bogovski & L. Griciute, 512 pages; £35.-

No. 15 CANCER INCIDENCE IN FIVE CONTINENTS. VOL. III (1976)
Edited by J. Waterhouse, C.S. Muir, P. Correa & J. Powell, 584 pages; £35.-

No. 16 AIR POLLUTION AND CANCER IN MAN (1977)
Edited by U. Mohr, D. Schmahl & L. Tomatis, 331 pages; £30.-

No. 17 DIRECTORY OF ON-GOING RESEARCH IN CANCER EPI-DEMIOLOGY 1977 (1977)
Edited by C.S. Muir & G. Wagner, 599 pages; out of print

SCIENTIFIC PUBLICATIONS SERIES

No. 18 ENVIRONMENTAL CARCINOGENS - SELECTED METHODS OF ANALYSIS
Editor-in-Chief H. Egan
Vol. 1 - ANALYSIS OF VOLATILE NITROSAMINES IN FOOD (1978)
Edited by R. Preussmann, M. Castegnaro, E.A. Walker & A.E. Wassermann, 212 pages; £30.-

No. 19 ENVIRONMENTAL ASPECTS OF *N*-NITROSO COMPOUNDS (1978)
Edited by E.A. Walker, M. Castegnaro, L. Griciute & R.E. Lyle, 566 pages; £35.-

No. 20 NASOPHARYNGEAL CARCINOMA: ETIOLOGY AND CONTROL (1978)
Edited by G. de-Thé & Y. Ito, 610 pages; £35.-

No. 21 CANCER REGISTRATION AND ITS TECHNIQUES (1978)
Edited by R. MacLennan, C.S. Muir, R. Steinitz & A. Winkler, 235 pages; £11.95

No. 22 ENVIRONMENTAL CARCINOGENS - SELECTED METHODS OF ANALYSIS
Editor-in-Chief H. Egan
Vol. 2 - METHODS FOR THE MEASUREMENT OF VINYL CHLORIDE IN POLY(VINYL CHLORIDE), AIR, WATER AND FOODSTUFFS (1978)
Edited by D.C.M. Squirrell & W. Thain, 142 pages; £35.-

No. 23 PATHOLOGY OF TUMOURS IN LABORATORY ANIMALS. VOLUME II. TUMOURS OF THE MOUSE (1979)
Editor-in-Chief V.S. Turusov, 669 pages; £35.-

No. 24 ONCOGENESIS AND HERPESVIRUSES III (1978)
Edited by G. de-Thé, W. Henle & F. Rapp
Part 1, 580 pages; £20.-
Part 2, 522 pages; £20.-

No. 25 CARCINOGENIC RISKS - STRATEGIES FOR INTERVENTION (1979)
Edited by W. Davis & C. Rosenfeld, 283 pages; £20.-

No. 26 DIRECTORY OF ON-GOING RESEARCH IN CANCER EPIDEMIOLOGY 1978 (1978)
Edited by C.S. Muir & G. Wagner, 550 pages; out of print

No. 27 MOLECULAR AND CELLULAR ASPECTS OF CARCINOGEN SCREENING TESTS (1980)
Edited by R. Montesano, H. Bartsch & L. Tomatis, 371 pages; £20.-

No. 28 DIRECTORY OF ON-GOING RESEARCH IN CANCER EPIDEMIOLOGY 1979 (1979)
Edited by C.S. Muir & G. Wagner, 672 pages; out of print

No. 29 ENVIRONMENTAL CARCINOGENS - SELECTED METHODS OF ANALYSIS
Editor-in-Chief H. Egan
Vol. 3 - ANALYSIS OF POLYCYCLIC AROMATIC HYDROCARBONS IN ENVIRONMENTAL SAMPLES (1979)
Edited by M. Castegnaro, P. Bogovski, H. Kunte & E.A. Walker, 240 pages; £17.50

No. 30 BIOLOGICAL EFFECTS OF MINERAL FIBRES (1980)
Editor-in-Chief J.C. Wagner
Volume 1, 494 pages; £25.-
Volume 2, 513 pages; £25.-

No. 31 *N*-NITROSO COMPOUNDS: ANALYSIS, FORMATION AND OCCURRENCE (1980)
Edited by E.A. Walker, M. Castegnaro, L. Griciute & M. Börzsönyi, 841 pages; £30.-

No. 32 STATISTICAL METHODS IN CANCER RESEARCH
Vol. 1. THE ANALYSIS OF CASE-CONTROL STUDIES (1980)
By N.E. Breslow & N.E. Day, 338 pages; £17.50

No. 33 HANDLING CHEMICAL CARCINOGENS IN THE LABORATORY - PROBLEMS OF SAFETY (1979)
Edited by R. Montesano, H. Bartsch, E. Boyland, G. Della Porta, L. Fishbein, R.A. Griesemer, A.B. Swan & L. Tomatis, 32 pages £3.95

No. 34 PATHOLOGY OF TUMOURS IN LABORATORY ANIMALS. VOLUME III. TUMOURS OF THE HAMSTER (1982)
Editor-in-Chief V.S. Turusov,
461 pages; £30.-

No. 35 DIRECTORY OF ON-GOING RESEARCH IN CANCER EPIDEMIOLOGY 1980 (1980)
Edited by C.S. Muir & G. Wagner,
660 pages; out of print

No. 36 CANCER MORTALITY BY OCCUPATION AND SOCIAL CLASS 1851-1971 (1982)
By W.P.D. Logan, 253 pages £20.-

No. 37 LABORATORY DECONTAMINATION AND DESTRUCTION OF AFLATOXINS B_1, B_2, G_1, G_2 IN LABORATORY WASTES (1980)
Edited by M. Castegnaro, D.C. Hunt, E.B. Sansone, P.L. Schuller, M.G. Siriwardana, G.M. Telling, H.P. Van Egmond & E.A. Walker,
59 pages; £5.95

No. 38 DIRECTORY OF ON-GOING RESEARCH IN CANCER EPIDEMIOLOGY 1981 (1981)
Edited by C.S. Muir & G. Wagner,
696 pages; out of print

No. 39 HOST FACTORS IN HUMAN CARCINOGENESIS (1982)
Edited by H. Bartsch & B. Armstrong
583 pages; £35.-

No. 40 ENVIRONMENTAL CARCINOGENS. SELECTED METHODS OF ANALYSIS
Editor-in-Chief H. Egan
Vol. 4. SOME AROMATIC AMINES AND AZO DYES IN THE GENERAL AND INDUSTRIAL ENVIRONMENT (1981)
Edited by L. Fishbein, M. Castegnaro, I.K. O'Neill & H. Bartsch, 347 pages;
£20.-

No. 41 *N*-NITROSO COMPOUNDS: OCCURRENCE AND BIOLOGICAL EFFECTS (1982)
Edited by H. Bartsch, I.K. O'Neill, M. Castegnaro & M. Okada,
755 pages; £35.-

No. 42 CANCER INCIDENCE IN FIVE CONTINENTS. VOLUME IV (1982)
Edited by J. Waterhouse, C. Muir, K. Shanmugaratnam & J. Powell,
811 pages; £35.-

No. 43 LABORATORY DECONTAMINATION AND DESTRUCTION OF CARCINOGENS IN LABORATORY WASTES: SOME *N*-NITROSAMINES (1982) Edited by M. Castegnaro, G. Eisenbrand, G. Ellen, L. Keefer, D. Klein, E.B. Sansone, D. Spincer, G. Telling & K. Webb, 73 pages £6.50

No. 44 ENVIRONMENTAL CARCINOGENS. SELECTED METHODS OF ANALYSIS
Editor-in-Chief H. Egan
Vol. 5. SOME MYCOTOXINS (1983)
Edited by L. Stoloff, M. Castegnaro, P. Scott, I.K. O'Neill & H. Bartsch,
455 pages; £20.-

No. 45 ENVIRONMENTAL CARCINOGENS. SELECTED METHODS OF ANALYSIS
Editor-in-Chief H. Egan
Vol. 6: *N*-NITROSO COMPOUNDS (1983)
Edited by R. Preussmann, I.K. O'Neill, G. Eisenbrand, B. Spiegelhalder & H. Bartsch, 508 pages; £20.-

No. 46 DIRECTORY OF ON-GOING RESEARCH IN CANCER EPIDEMIOLOGY 1982 (1982)
Edited by C.S. Muir & G. Wagner,
722 pages; out of print

No. 47 CANCER INCIDENCE IN SINGAPORE (1982)
Edited by K. Shanmugaratnam, H.P. Lee & N.E. Day, 174 pages; £10.-

No. 48 CANCER INCIDENCE IN THE USSR (1983) Second Revised Edition
Edited by N.P. Napalkov, G.F. Tserkovny, V.M. Merabishvili, D.M. Parkin, M. Smans & C.S. Muir, 75 pages; £10.-

No. 49 LABORATORY DECONTAMINATION AND DESTRUCTION OF CARCINOGENS IN LABORATORY WASTES: SOME POLYCYCLIC AROMATIC HYDROCARBONS (1983)
Edited by M. Castegnaro, G. Grimmer, O. Hutzinger, W. Karcher, H. Kunte, M. Lafontaine, E.B. Sansone, G. Telling & S.P. Tucker, 81 pages; £7.95

No. 50 DIRECTORY OF ON-GOING RESEARCH IN CANCER EPIDEMIOLOGY 1983 (1983)
Edited by C.S. Muir & G. Wagner, 740 pages; out of print

No. 51 MODULATORS IN EXPERIMENTAL CARCINO GENESIS (1983)
Edited by R. Montesano & V.S. Turusov, 307 pages; £25.-

No. 52 SECOND CANCER IN RELATION TO RADIATION TREATMENT FOR CERVICAL CANCER: RESULTS OF A CANCER REGISTRY COLLABORATION (1983)
Edited by N.E. Day & J.C. Boice, Jr, 207 pages; £17.50

No. 53 NICKEL IN THE HUMAN ENVIRONMENT (1984)
Editor-in-Chief, F.W. Sunderman, Jr, 529 pages; £30.-

No. 54 LABORATORY DECONTAMINATION AND DESTRUCTION OF CARCINOGENS IN LABORATORY WASTES: SOME HYDRAZINES (1983)
Edited by M. Castegnaro, G. Ellen, M. Lafontaine, H.C. van der Plas, E.B. Sansone & S.P. Tucker, 87 pages; £6.95

No. 55 LABORATORY DECONTAMINATION AND DESTRUCTION OF CARCINOGENS IN LABORATORY WASTES: SOME *N*-NITROSAMIDES (1983)
Edited by M. Castegnaro, M. Benard, L.W. van Broekhoven, D. Fine, R. Massey, E.B. Sansone, P.L.R. Smith, B. Spiegelhalder, A. Stacchini, G. Telling & J.J. Vallon, 65 pages; £6.95

No. 56 MODELS, MECHANISMS AND ETIOLOGY OF TUMOUR PROMOTION (1984)
Edited by M. Börszönyi, N.E. Day, K. Lapis & H. Yamasaki, 532 pages, £30.-

No. 57 *N*-NITROSO COMPOUNDS: OCCURRENCE, BIOLOGICAL EFFECTS AND RELEVANCE TO HUMAN CANCER (1984)
Edited by I.K. O'Neill, R.C. von Borstel, C.T. Miller, J. Long & H. Bartsch, 1013 pages, £75.-

No. 58 AGE-RELATED FACTORS IN CARCINOGENESIS (1985)
Edited by A. Likhachev, V. Anisimov & R. Montesano (in press)

No. 59 MONITORING HUMAN EXPOSURE TO CARCINOGENIC AND MUTAGENIC AGENTS (1985)
Edited by A. Berlin, M. Draper, K. Hemminki & H. Vainio (in press)

No. 60 BURKITT'S LYMPHOMA: A HUMAN CANCER MODEL (1985)
Edited by G. Lenoir, G. O'Conor & C.L.M. Olweny (in press)

No. 61 LABORATORY DECONTAMINATION AND DESTRUCTION OF CARCINOGENS IN LABORATORY WASTES: SOME HALOETHERS (1985)
Edited by M. Castegnaro, M. Alvarez, M. Iovu, E.B. Sansone, G.M. Telling & D.T. Williams
55 pages, £5.95

No. 62 DIRECTORY OF ON-GOING RESEARCH IN CANCER EPIDEMIOLOGY 1984 (1984)
Edited by C.S. Muir & G.Wagner;
728 pages; £18.-

No. 63 VIRUS-ASSOCIATED CANCERS IN AFRICA (1984)
Edited by A.O. Williams, G.T. O'Conor, G.B. de-Thé & C.A. Johnson, 773 pages, £20.-

No. 64 LABORATORY DECONTAMINATION AND DESTRUCTION OF CARCINOGENS IN LABORATORY WASTES: SOME AROMATIC AMINES AND 4-NITROBIPHENYL (1985)
Edited by M. Castegnaro, J. Barek, J. Dennis, G. Ellen, M. Klibanov, M. Lafontaine, R. Mitchum, P. Van Roosmalen, E.B. Sansone, L.A. Sternson & M. Vahl (in press)

No. 65 INTERPRETATION OF NEGATIVE EPIDEMIOLOGICAL EVIDENCE FOR CARCINOGENICITY
Edited by N.J. Wald & R. Doll
(in press)

NON-SERIAL PUBLICATIONS

(Available from IARC)

ALCOOL ET CANCER (1978)
by A.J. Tuyns (in French only)
42 pages; Fr.fr. 35.-; Sw.fr. 14.-

CANCER MORBIDITY AND CAUSES OF DEATH AMONG DANISH BREWERY WORKERS (1980) By O.M. Jensen
145 pages; US$ 25.00; Sw.fr. 45.-

IARC MONOGRAPHS ON THE EVALUATION OF THE CARCINOGENIC RISK OF CHEMICALS TO HUMANS

(English editions only)

(Available from WHO Sales Agents)

Volume 1
Some inorganic substances, chlorinated hydrocarbons, aromatic amines, N-*nitroso compounds, and natural products* (1972)
184 pp.; out of print

Volume 2
Some inorganic and organometallic compounds (1973)
181 pp.; out of print

Volume 3
Certain polycyclic aromatic hydrocarbons and heterocyclic compounds (1973)
271 pp.; out of print

Volume 4
Some aromatic amines, hydrazine and related substances, N-*nitroso compounds and miscellaneous alkylating agents* (1974)
286 pp.; US$7.20; Sw.fr. 18.-

Volume 5
Some organochlorine pesticides (1974)
241 pp.; out of print

Volume 6
Sex hormones (1974)
243 pp.; US$7.20; Sw.fr. 18.-

Volume 7
Some anti-thyroid and related substances, nitrofurans and industrial chemicals (1974)
326 pp.; US$12.80; Sw.fr. 32.-

Volume 8
Some aromatic azo compounds (1975)
357 pp.; US$14.40; Sw.fr. 36.-

Volume 9
Some aziridines, N-, S- *and* O-*mustards and selenium* (1975)
268 pp.; US$10.80; Sw.fr. 27.-

Volume 10
Some naturally occurring substances (1976)
353 pp.; US$15.00; Sw.fr. 38.-

Volume 11
Cadmium, nickel, some epoxides, miscellaneous industrial chemicals and general considerations on volatile anaesthetics (1976)
306 pp.; US$14.00; Sw.fr. 34.-

Volume 12
Some carbamates, thiocarbamates and carbazides (1976)
282 pp.; US$14.00; Sw.fr. 34.-

Volume 13
Some miscellaneous pharmaceutical substances (1977)
255 pp.; US$12.00; Sw.fr. 30.-

Volume 14
Asbestos (1977)
106 pp.; US$6.00; Sw.fr. 14.-

Volume 15
Some fumigants, the herbicides 2,4-D chlorinated dibenzodioxins and miscellaneous industrial chemicals (1977)
354 pp.; US$20.00; Sw.fr. 50.-

Volume 16
Some aromatic amines and related nitro compounds - hair dyes, colouring agents and miscellaneous industrial chemicals (1978)
400 pp.; US$20.00; Sw.fr. 50.-

Volume 17
Some N-*nitroso compounds* (1978)
365 pp.; US$25.00; Sw.fr. 50.-

Volume 18
Polychlorinated biphenyls and polybrominated biphenyls (1978)
140 pp.; US$13.00; Sw.fr. 20.-

IARC MONOGRAPHS SERIES

Volume 19
Some monomers, plastics and synthetic elastomers, and acrolein (1979)
513 pp.; US$35.00; Sw.fr. 60.-

Volume 20
Some halogenated hydrocarbons (1979)
609 pp.; US$35.00; Sw.fr. 60.-

Volume 21
Sex hormones (II) (1979)
583 pp.; US$35.00; Sw.fr. 60.-

Volume 22
Some non-nutritive sweetening agents (1980)
208 pp.; US$15.00; Sw.fr. 25.-

Volume 23
Some metals and metallic compounds (1980)
438 pp.; US$30.00; Sw.fr. 50.-

Volume 24
Some pharmaceutical drugs (1980)
337 pp.; US$25.00; Sw.fr. 40.-

Volume 25
Wood, leather and some associated industries (1981)
412 pp.; US$30.00; Sw.fr. 60.-

Volume 26
Some antineoplastic and immunosuppressive agents (1981)
411 pp.; US$30.00; Sw.fr. 62.-

Volume 27
Some aromatic amines, anthraquinones and nitroso compounds, and inorganic fluorides used in drinking-water and dental preparations (1982)
341 pp.; US$25.00; Sw.fr. 40.-

Volume 28
The rubber industry (1982)
486 pp.; US$35.00; Sw.fr. 70.-

Volume 29
Some industrial chemicals and dyestuffs (1982)
416 pp.; US$30.00; Sw.fr. 60.-

Volume 30
Miscellaneous pesticides (1983)
424 pp; US$30.00; Sw.fr. 60.-

Volume 31
Some food additives, feed additives and naturally occurring substances (1983)
314 pp.; US$30.00; Sw.fr. 60.-

Volume 32
Polynuclear aromatic compounds, Part 1, Environmental and experimental data (1984)
477 pp.; US$30.00; Sw.fr. 60.-

Volume 33
Polynuclear aromatic compounds, Part 2, Carbon blacks, mineral oils and some nitroarene compounds (1984)
245 pp.; US$25.00; Sw.fr. 50.-

Volume 34
Polynuclear aromatic compounds, Part 3, Some complex industrial exposures in aluminium production, coal gasification, coke production, and iron and steel founding (1984)
219 pages; US$20.00; Sw.fr. 48.-

Volume 35
Polynuclear aromatic compounds, Part 4, Bitumens, coal-tars and derived products, shale-oils and soots (1985)
271 pages; US$25.00; Sw.fr. 70.-

Supplement No. 1
Chemicals and industrial processes associated with cancer in humans (IARC Monographs, Volumes 1 to 20) (1979)
71 pp.; out of print

Supplement No. 2
Long-term and short-term screening assays for carcinogens: a critical appraisal (1980)
426 pp.; US$30.00; Sw.fr. 60.-

Supplement No. 3
Cross index of synonyms and trade names in Volumes 1 to 26 (1982)
199 pp.; US$30.00; Sw.fr. 60.-

Supplement No. 4
Chemicals, industrial processes and industries associated with cancer in humans (IARC Monographs, Volumes 1 to 29) (1982)
292 pp.; US$30.00; Sw.fr. 60.-

INFORMATION BULLETIN ON THE SURVEY OF CHEMICALS BEING TESTED FOR CARCINOGENICITY
No. 8 (1979)
Edited by M.-J. Ghess, H. Bartsch & L. Tomatis
604 pp.; US$20.00; Sw.fr. 40.-

INFORMATION BULLETIN ON THE SURVEY OF CHEMICALS BEING TESTED FOR CARCINOGENICITY
No. 9 (1981)
Edited by M.-J. Ghess, J.D. Wilbourn, H. Bartsch & L. Tomatis
294 pp.; US$20.00; Sw.fr. 41.-

INFORMATION BULLETIN ON THE SURVEY OF CHEMICALS BEING TESTED FOR CARCINOGENICITY
No. 10 (1982)
Edited by M.-J. Ghess, J.D. Wilbourn, H. Bartsch
326 pp.; US$20.00; Sw.fr. 42.-

INFORMATION BULLETIN ON THE SURVEY OF CHEMICALS BEING TESTED FOR CARCINOGENICITY
No. 11 (1984)
Edited by M.-J. Ghess, J.D. Wilbourn, H. Vainio & Bartsch
336 pp.; US$20.00; Sw.fr. 48.-